HENI

THE

PRINCIPLES, CONSTRUCTION, AND APPLICATION

OF

PUMPING MACHINERY

(STEAM AND WATER PRESSURE)

WITH PRACTICAL ILLUSTRATIONS OF ENGINES AND PUMPS APPLIED TO MINING, TOWN WATER SUPPLY, DRAINAGE OF LANDS, ETC.; ALSO ECONOMY AND EFFICIENCY TRIALS OF PUMPING MACHINERY

Elibron Classics
www.elibron.com

THE

PRINCIPLES, CONSTRUCTION, AND APPLICATION
OF PUMPING MACHINERY

WATT'S ENGINE.

Erected for the Birmingham Canal Navigations at Smethwick in 1777 ; removed in 1898, having been at work for nearly 120 years.
Cylinder, 32 inches diameter by 8 feet stroke.
A trial of this engine was made by Smeaton (see page 14).

THE

PRINCIPLES, CONSTRUCTION, AND APPLICATION

OF

PUMPING MACHINERY

(STEAM AND WATER PRESSURE)

With Practical Illustrations of Engines and Pumps applied to Mining, Town Water Supply, Drainage of Lands, etc.; also Economy and Efficiency Trials of Pumping Machinery.

BY

HENRY DAVEY

MEMBER OF THE INSTITUTION OF CIVIL ENGINEERS,
MEMBER OF THE INSTITUTION OF MECHANICAL ENGINEERS,
F.G.S., ETC.

WITH FRONTISPIECE, FIVE PLATES, AND OVER 250 ILLUSTRATIONS.

LONDON:

CHARLES GRIFFIN & COMPANY, LIMITED,

EXETER STREET, STRAND.

1900.

PREFACE.

THE purpose of this book is to present the information in such a form as will make it most useful to the practical engineer engaged in the application of pumping machinery in mines and elsewhere, or in circumstances under which large quantities of water have to be dealt with.

A large number of illustrations are therefore given, which will fully exemplify the best methods of procedure to be adopted in the general run of cases, and also the modifications required to overcome the special difficulties which are occasionally met with. In some of these special cases the machinery was designed by myself. A considerable proportion of the figures are reduced copies of the working drawings used in the execution of various important undertakings.

The information on 'pit work' is based on actual examples, and is put in the form best adapted to the needs of the resident engineer.

In the chapter on steam engines, the term 'efficiency ratio' has been adopted, as it allows of a comparison being made between what is actually done and what is theoretically possible, although the absolute thermal efficiency represents heat economy.

Types of machinery are described for the purpose of illustrating general principles only, no account being taken of the modifications by any particular individual.

I have to acknowledge the courtesy of the Institutions of *Civil* and of *Mechanical Engineers* for allowing me to make the extracts which I have taken from their Transactions, and my indebtedness to the authors who have supplied me with accounts of engine trials, and other information.

HENRY DAVEY.

TABLE OF CONTENTS.

CHAPTER I.

EARLY HISTORY OF PUMPING-ENGINES.

CHAPTER II.

STEAM ENGINES (PUMPING).

CHAPTER III.

PUMPS AND PUMP VALVES.

CHAPTER IV.

GENERAL PRINCIPLES OF NON-ROTATIVE PUMPING-ENGINES.

CHAPTER V.

THE CORNISH ENGINE: SIMPLE AND COMPOUND.

CHAPTER VI.

TYPES OF MINING ENGINES.

CHAPTER VII.

PIT-WORK.

CHAPTER VIII.

SHAFT SINKING THROUGH WATER-BEARING STRATA.

CHAPTER IX.

HYDRAULIC TRANSMISSION OF POWER IN MINES.

CHAPTER X.

Valve Gears of Pumping-Engines.

CHAPTER XI.

Water Pressure Pumping-Engines.

CHAPTER XII.

Waterworks Engines.

CHAPTER XIII.

Pumping-Engine Economy and Trials of Pumping Machinery.

CHAPTER XIV.

Centrifugal and other Low-Lift Pumps.

CHAPTER XV.

HYDRAULIC RAMS, PUMPING MAINS, ETC.

LIST OF FIGURES.

FRONTISPIECE.

PLATES.

PUMPING MACHINERY.

CHAPTER I.

History.—The elastic property of steam must have been observed in the earliest period of man's history, but we have no record of its application for power purposes earlier than 130 years before the Christian era. At that time Hero the Elder, a philosopher of Alexandria in the reign of Ptolemy Philadelphus, produced many mechanical contrivances, and among them a rotary steam engine, a mere toy, but an engine actuated by the reaction of a jet of steam.

The first person in modern times who applied the power of steam to moving mechanism was an Italian mathematician, Giovanni Branca, who lived in Rome in the beginning of the seventeenth century. His invention consisted in making a jet of steam from a boiler impinge on the vanes of a wheel constructed like a water-wheel. In both inventions we have rotary motion directly imparted to the mechanism by the reaction of a steam jet. These inventions were the forerunners of the steam turbine of to-day. It was not, however, by the extension of that principle that the steam engine came into existence as a commercially valuable machine. The first practical application of the power of steam, as far as we know, was to raise water. In very early times small water jets or fountains were produced by the elastic force of steam acting on the surface of water in a vessel from which a pipe conducted the water to a jet or nozzle. It would appear to be an easy step in the invention of the steam engine, from steam pressure on the surface of the water, to steam pressing on a piston transmitting the pressure to the water; but all the early attempts in employing steam to raise water were on the principle of the modern pulsometer, but were not self-acting.

In 1698 *Savery* obtained a patent for raising water by the elasticity of steam, and in 1699 he had erected some engines and made a trial of one before members of the Royal Society. We take the following illustration

A

and discussion on Savery's engine from Stuart's *History of the Steam Engine*, published in 1826 :—

"The first thing," says the ingenious inventor, "is to fix the engine

FIG. 1.—Savery's Engine.

(fig. 1) in a good double furnace, so contrived that the flame of your fire may circulate round and encompass your two boilers, as you do coppers for brewing. Before you make any fire unscrew G and N, being the two small gauge pipes and cocks belonging to the two boilers, and at the holes fill L, the great boiler, two-thirds full of water, and D, the small boiler, quite full. Then screw in the said pipes again as fast and as tight as possible. Then light the fire at b, and when the water in L boils, the handle of the regulator, marked Z, must be thrust from you, as far as it will go, which makes all the steam rising in the water in L pass with irresistible force through O into P, pushing out all the air before it, through the clack r, making a noise as it goes ; and when all is gone out, the bottom of the vessel P will be very hot. Then pull the handle of the regulator to you, by which means you stop O, and force your steam through O o into P p, until that vessel has discharged its air through the clack R up the force-pipe S. In the meantime, by the steam's condensing in the vessel P, vacuum or emptiness is created, so that the water must and will necessarily rise up through the sucking-pipe T, lifting up the clack M, and filling the vessel P.

"In the meantime, the vessel P p being emptied of its air, turn the handle of the regulator from you again, and the force is upon the surface of the water in P, which surface being only heated by the steam, it does not condense it, but the steam gravitates or presses with an elastic quality like air, still increasing its elasticity or spring till it counterpoises, or rather exceeds, the weight of the water ascending in S, the forcing pipe, out of which the water in it will be immediately discharged, when once gotten to the top, which takes up some time to recover that power ; which having once got, and being in work, it is easy for anyone that never saw the engine, after half an hour's experience, to keep a constant stream running out the full bore of the pipe. On the outside of the vessel you may see how the water goes as well as if the vessel were transparent ; for as far as the steam continues within the vessel, so far is the vessel dry without, and so very hot as scarce to endure the least touch of the hand. But as far as the water is, the said vessel will be cold and wet where any water has fallen on it, which cold and moisture vanishes as fast as the steam in its descent takes the place of the water ; but if you force all the water out, the steam, or a small part thereof, going through R will rattle the clack, so as to give sufficient notice to pull the handle of the regulator to you, which, at the same time, begins to force out the water from P p, without the least alteration of the stream ; only sometimes the stream of water will be somewhat stronger than before, if you pull the handle of the regulator before any considerable quantity of steam be gone up the clack R : but it is much better to let none of the steam go off (for that is but losing so much strength, and is easily prevented by pulling the regulator some little time before the vessel forcing is quite emptied). This being done, immediately turn the cock or pipe Y of the cistern X on P, so that the water proceeding from X through Y (which is never open but when turned on P, or P p, but when between them is tight and stanch), I say, the water falling on P causes, by its coolness, the steam (which had such great force just before, from its elastic power) to condense, and become a vacuum or empty space, so that the vessel P is, by the external air, or what is vulgarly called suction, completely refilled while P p is emptying. Which being done, you push the handle of the regulator from you, and throw the force on P, pulling the condensing pipe over P p, causing the steam in that vessel to condense, so that it fills while the other empties; the labour of turning these two parts of the engine—viz., the regulator and water-cock —and tending the fire being no more than what a boy's strength can perform for a day together, and is as easily learned as their driving of a horse in a tub-gin; yet, after all, I would have men, and those, too, the most apprehensive, employed in working the engine, supposing them more careful than boys."

"In case it should be objected that the boiler must in some certain time be emptied, so as the work of the engine must stop to replenish the boiler, or endanger the burning out or melting the bottom of the boiler: to obviate this, when it is thought fit by the person tending the engine to replenish the great boiler, which requires an hour and a half or two hours' time to the sinking of one foot of water, then, I say, by turning the cock E of the small boiler D, you cut off all communication between the great force-pipe S and the small boiler D; by which means D grows immediately hot, by throwing a little fire into B, and the water of which boils, and in a very little time it gains more strength than the great boiler, for the force of the great boiler being perpetually spending and going out, and the other winding up, or increasing, it is not long before the force in D exceeds that in L; so that the water in D, being depressed by its own steam or vapour, must necessarily rise through the pipe H, opening the clack I, and so go through the pipe K into L, running till the surface of the water in D is equal to the bottom of the pipe H. Then, steam and water going together, will, by a noise in the clack I, give sufficient assurance that D has discharged and emptied itself into L, to within eight inches of the bottom; and inasmuch as from the top of D to the bottom of its pipe H is contained about as much water as will replenish L one foot. Then you open the cock E, and refill D immediately, so that here is a constant motion, without fear or danger of disorder or decay. If you would at any time know if the great boiler be more than half exhausted, turn the small cock N, whose pipe will deliver water, if the water be above the level of its bottom, which is half way down the boiler; if not it will deliver steam. So, likewise, it will show you if you have more or less than eight inches of water in D, by which means nothing but a stupid and wilful neglect, or mischievous design, carried on for some hours, can any ways hurt the engine. And if a master is suspicious of the design of a servant to do mischief, it is easily discovered by these gauge pipes; for if he come when the engine is at work, and find the surface of the water in L below the bottom of the gauge-pipe N, or the water in D below the bottom of G, such a servant deserves correction; though, three hours after that, the working on would not damage or exhaust the boilers. So that, in a word, the clacks being, in all water-works, always found the better the longer they are used; and all the moving parts in our engine being of like nature, the furnace being made of Stourbridge, or Windsor brick, or fire stone. I do not see it possible for the engine to decay in many years; for the clacks, boxes and mitre pipes, regulator, and cocks are all of brass, and the vessels made of the best hammered copper, of sufficient thickness to sustain the force of the working of the engine. In short, the engine is so naturally adapted to perform what is required, that even those of the most ordinary and meanest capacity may work it for some years without its receiving any injury, if not hired or employed by some base person on purpose to destroy it."

This engine Savery applied for raising water for palaces, gentlemen's seats, draining fens, and supplying houses with water in general, and pumping water from ships; and he erected many of them in different parts of England. The power of his engine he limited only by the strength of the pipes and vessels; "for," he says, "I will raise you water 500 or 1000 feet high, could you find us a way to procure strength enough for such an immense weight as a pillar of water that height; but my engine at 60, 70, or 80 feet raises a full bore of water with much ease." And comparing this performance of his machine with that by manual labour, he continues: "I have known, in Cornwall, a work with three lifts of about 18 feet each, lift and carry a $3\frac{1}{2}$-inch bore, that costs forty-two shillings a day (reckoning twenty-four hours a day) for labour, besides the wear and tear of engines, each pump having four men working eight hours, at fourteenpence a man, and the men obliged to rest at least a third part of that time. I dare undertake that the engine shall raise you as much water for eightpence as will cost you a shilling to raise the like with your old engines, in coal pits, which is thirty-three pounds six shillings and eight-pence saved out of every hundred pounds; a brave estate gained in one year out of such great works! where £3000, £6000, or it may be £8000 per annum is expended for clearing their mines of water only, besides the charge and repair of engines, gins, horses, etc."

In 1705 we find Savery associated with Newcomen and Cawley in a patent for condensing the steam introduced under a piston and producing a reciprocating motion by attaching it to a lever. Here was a combination of Savery's method of forming a vacuum with Newcomen's piston and attachment for producing a reciprocating motion for the purpose of working pumps.

The first *Newcomen Engine* (fig. 2) consisted of an open-topped cylinder, the piston of which was attached by means of a chain to a quadrant end of a reciprocating beam, the opposite end of the beam being attached in a similar manner to the rod of the pump; the bottom of the steam cylinder had a pipe leading to the boiler, in which was placed a stop-cock; there was also a drain pipe from the bottom of the cylinder to drain away the water formed by the condensed steam; the cylinder itself was wholly enclosed in another cylinder of larger diameter, the annular space being filled with water for the purpose of condensing the steam. The engine was actuated in the following manner:—Steam was first admitted into the cylinder until it was filled with steam, then the communication with the boiler was shut off, until the cold water surrounding the cylinder had

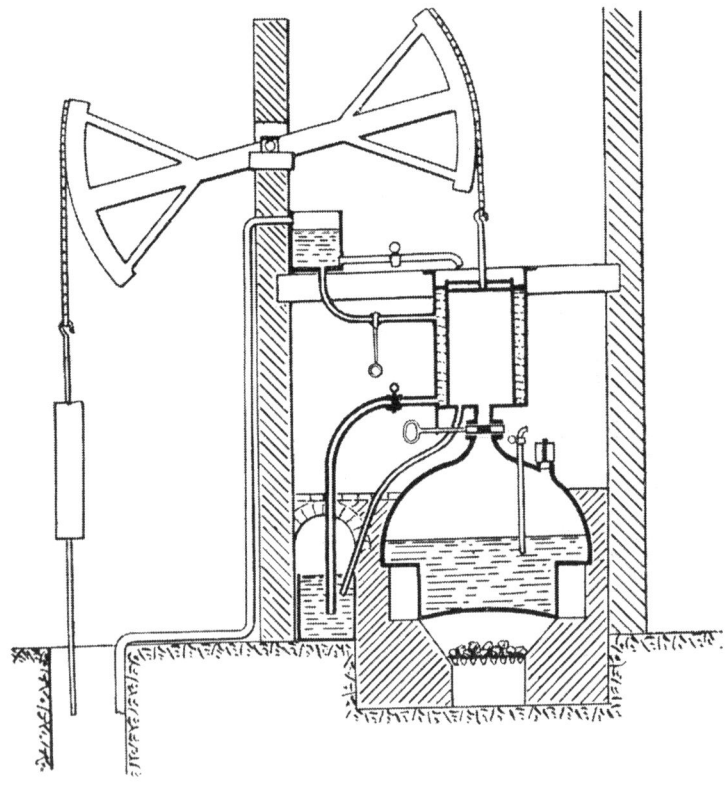

Fig. 2.—Newcomen's Engine.

condensed the steam, thus forming a vacuum; the atmospheric pressure on the top of the piston then caused its descent, and the operation was repeated. A layer of water was kept on the top of the piston for the purpose of making it air-tight. The motion of the engine was a very slow one, especially as regards the indoor or pumping stroke, because of the time required to condense the steam in the cylinder by the transmission of heat through the metal of the cylinder.

It accidentally happened, however, in the working of one of Newcomen's engines, that a defect in the piston allowed the water above it to pass in the form of a jet into the cylinder. This caused the condensation to become much more rapid, and the engine made a quicker stroke in consequence. This accident led to the abandonment of condensing by means of cooling the outside of the cylinder and to injection of water into the cylinder being adopted (fig 3); a further improvement was effected in working the injection cock automatically by means of a buoy enclosed in a pipe attached to the cylinder. Even with this imperfect mechanism the engines were able to make from six, to eight strokes per minute, and it is said that a boy named Humphrey Potter, who attended an engine, added a catch or scoggan, which increased the speed of the engine to fifteen or sixteen strokes per minute. We have no record what the contrivance was. The buoy which was employed to make the opening and closing of the injection valve self-acting might also have been made to lift a weight held by a catch and released by the motion of the beam, for the purpose of working the steam cock, and it is very possible that the word 'buoy' has been confounded with the word 'boy.' It is hardly likely that boys were entrusted with the working of engines requiring so much attention and adjustment.

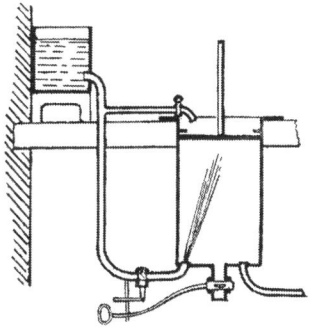

FIG. 3.—Newcomen's Engine with injection in the Cylinder.

The buoy employed to make the engine-cock self-acting was the beginning of the complete automatic action of the engine itself, and involved a principle which is embodied in all the automatic valve gears of non-rotative engines which have since been produced.

The first complete automatic engine of which we have any record is that by M. Francois, Professor of Philosophy at Lausanne, who proposed to work a pumping-engine on Savery's plan, giving motion to the stopcocks by means of a tumbling water bucket turning on a pivot. The bucket was filled by a constant stream from a pipe, and when full toppled over, giving motion to the stop-cock, and becoming empty, a counter weight brought it back to its original position.

We do not know that Savery's engines were ever made automatic in this or any other way, but the Newcomen engine was made completely automatic by the valve gear designed by Mr Henry Beighton, an engineer of Newcastle-on-Tyne, in 1718. The leading feature in this gear was that of a tumbling weight which was raised during the stroke of the engine, so as to tumble over when the engine had completed its stroke, thereby opening the cocks or valves for the commencement of the return stroke; the weight tumbling in the one direction opened and closed the valves for one stroke, and tumbling in the other direction opened and closed the valves for the return stroke.

The Newcomen engine had now been brought to a state of perfection, in which it remained, except as regards improvements in workmanship, up to the time when Watt made his great invention of the separate condenser. Watt's attention was first directed to the subject of the steam engine by Dr Robinson, then a student in the University of Glasgow, in 1759. In 1761 or 1762 he made some experiments with a small model of a steam cylinder, but the attention necessary to his business—that of a mathematical instrument maker—prevented his pursuing the subject any farther at that time. In the winter of 1763 and 1764, having occasion to repair a model of Newcomen's engine belonging to the Natural Philosophy Class of the Glasgow University, his mind was again directed to the subject. When the model was set to work, he was surprised to find the great amount of steam condensed with a moderate quantity of injection water. By subsequent experiments Watt ascertained that steam was 1800 times rarer than water. In another experiment, being astonished at the quantity of water required for injection and the great heat that it acquired from the small quantity of water in the form of steam, which had been used in filling the cylinder, and not understanding the reason of it, "I mentioned it," he says, "to my friend Dr Black, who then explained to me his doctrine of latent heat, which he had taught some time before this period, but having myself been occupied with pursuits of business, if I had heard of it, I had not attended to it, when I thus stumbled upon one of the material facts upon which this beautiful theory is founded." Dr Ure gives the conversation which he had with Watt on this subject, in which Watt describes the simple but decisive experiments by which he discovered the latent heat of steam. His means and leisure not then permitting an expensive or complex apparatus, he used apothecaries' phials ; with these he ascertained two more facts—first, that a cubic inch of water would form about 1 cubic foot of ordinary steam, or about 1728 cubic inches ; and that the condensation of that quantity of steam would heat 6 cubic inches of water from the atmospheric temperature to the boiling point. He also saw that six times the difference in temperature, or fully 800 degrees of heat, had been employed in giving elasticity to steam, and which must be all subtracted before a complete vacuum could be obtained under the piston of a steam engine. On reflecting further, it appeared to him that a cylinder should always be kept as hot as the steam that entered it.

It occurred to Watt in 1765 that for the purpose of avoiding the cooling of the steam cylinder, the condensation should be effected in a separate vessel, and in 1769 we find that he took out a patent for his invention of the condenser, and the use of oil and tallow for lubricating the piston.

It next occurred to Watt that the mouth of the cylinder being open, the air which entered above the piston would cool the cylinder. He therefore provided the top of the cylinder with a cover, and on the completion of the steam stroke, communication was opened between the space above the piston to that below it, thus placing the piston in equilibrium. The piston was then raised to the top of the cylinder by the weight of the pump rods. Communication between the top and the bottom of the cylinder was then closed, and whilst steam was being admitted to the top of the piston for the next stroke, the steam below it was admitted into the separate condensing vessel. Watt made a small model of this engine, in which the condenser was simply a tin tube with water on the outside forming a surface condenser; an air pump was added to keep the condenser evacuated.

Watt, however, found that the condensation arising from the cooling of the outside of the condenser would not be rapid enough unless the surface of the condenser was made very large; he therefore resorted to the method of injecting water into the condenser, as water had been injected into the Newcomen engine. The Watt engine was now complete.

In the year 1775 Watt entered into partnership with Matthew Boulton, of Birmingham, and soon after Boulton & Watt commenced the manufacture of Watt's engines, and some were erected in the coal mines of Staffordshire and Warwickshire and other places, but it was in Cornwall that Watt found the great field for the development of his invention. Coal was there very expensive, having to be brought from South Wales, and the mines were becoming deep and exceedingly costly to drain. The proprietors of the mines were, however, unwilling to be at the expense of taking out the old atmospheric engines, and of replacing them with Watt's improved engine; to meet that objection Messrs Boulton & Watt erected many engines at their own expense, taking as payment one-third the saving effected in coal, and it is said that in one mine at Chacewater three of Watt's engines were erected, and the proprietors of the mine engaged to pay £800 per annum for each engine as a compromise for the third part of the saving made in coals.

The expansive power of steam was understood by Watt in 1769, and was afterwards particularly described in his patent specification of 1782. Watt had a very clear idea of the economy to be effected by expansive working; in his patent specification he gives the following statement:—

When the cylinder is quite full, its performance will be as 1·0
When ½ full, its performance is increased as . . 1·7
  ,, ⅓ ,,   ,,   ,,   . . 2·1
  ,, ¼ ,,   ,,   ,,   . . 2·4
  ,, ⅕ ,,   ,,   ,,   . . 2·6
  ,, ⅙ ,,   ,,   ,,   . . 2·8
  ,, ⅐ ,,   ,,   ,,   . . 3·0
  ,, ⅛ ,,   ,,   ,,   . . 3·2

upon the supposition that steam contracts and expands by variation of pressure in the same ratio that air would do.

At that time the Watt engines were all employed in working bucket or lift pumps, and it was not until a much later date that the plunger pump was introduced. As long as bucket pumps only were used the extent to which expansive working could be carried was very limited, for reasons we have fully gone into in a subsequent chapter dealing with the principles of Cornish and other direct acting engines.

Cornwall formed the great nursery of the pumping-engine. In the early part of the last century many of the mines which were drained by means of water-wheels had reached such a depth that the power available was not sufficient, and the mines were in consequence on the point of being abandoned.

Savery failed in his attempts to introduce his engines into Cornwall, but it is said that a Savery engine was for many years employed at Kier's Manufactory, St Pancras, London. It was employed in raising water to turn a water-wheel 18 ft. in diameter, which wheel was applied to the purpose of giving motion to lathes, etc. It was stated to consume six bushels of good coals during twelve hours' work, and to make ten strokes per minute, lifting 7 cubic ft. of water per stroke through a vertical height of 20 ft. This in round numbers is at the rate of 70 cubic feet per minute, or $\dfrac{70 \times 63 \times 20}{33,000} = 2 \cdot 6$ H.P.

The consumption of coal per horse power per hour was then $\dfrac{42}{2 \cdot 6} = 16 \cdot 1$ lbs., and the dynamic duty on 84 lbs. of coal about 10,296,000. This must, however, be regarded as an incorrect statement of the actual duty. In 1774 Mr Rigby put up two of Savery's engines at Manchester to work water-wheels, which engines operated by means of suction. The duty was found to be five and a quarter millions with one, and five and a half millions with the other, for a consumption of one bushel of coal or 84 lbs.

About the year 1767 John *Smeaton* devoted himself to improving the condition of the atmospheric engine, and with very marked success. Smeaton had in his nature so admirably combined the virtues of untiring skill and enterprise, together with a comprehensive scientific knowledge, and a keen perception of the wants of the age, that any subject which he took in hand had thereby its guarantee of success. Smeaton cannot be remembered for any one invention, but his improvements in the various details of the engine completely metamorphosed it, and he succeeded in increasing the duty thereby nearly 50 per cent.

In 1775 Smeaton erected at Chacewater Mine, in Cornwall, an atmospheric engine with all his improvements ; this was the largest atmospheric engine probably ever put up. The cylinder was 72 in. diameter, and the stroke of the piston $9\frac{1}{2}$ ft. ; the water load was equal to $7\frac{2}{3}$ lbs. per square inch of the piston, and the lift was 306 ft. "It had, when originally erected, one boiler, 15 ft. diameter, placed immediately under the cylinder, and an extra one, constructed to collect and make use of the waste heat from the furnaces upon the works. This latter, however, being found a failure, it was removed after a very short use, and two new boilers of the same construction and dimensions as the centre one were added, being fixed in low buildings on each side of the engine-house. These were found successful in furnishing the engine with steam."[1]

This engine was erected to supply the place of two atmospheric engines which had previously been worked on the same mine—one with a 64-in. cylinder, and the other with a 62-in. cylinder, each having a stroke of 6 ft. Mr Smeaton reported the duty of the new engine to be greater than that of the old ones in the proportion of 7 to 4.

In 1769 Smeaton tabulated the duty of fifteen atmospheric engines at work in Newcastle, giving the mean duty of 5·59 millions per one bushel of coal. Soon after, he obtained from a new engine a duty of 9·45 millions. He ultimately raised the duty of the atmospheric engine from 7 to 10 millions. He arranged his pit-work much in the same way as it is done now as regards the disposition of the various lifts. The pumps were of the bucket type, with butterfly and clack pump valves. In his direction for the Chacewater engine, he says : "The durability of the leather of the bucket depends greatly on the proper proportions and construction of the bucket hoops. The bucket hoops on the outside should be a cylinder, that is, the same diameter above and below ; their taper inside must, therefore, be formed by the different thickness of the metal ; their external diameter should not be more than one quarter of an inch less than their respective working barrels ; the hoops to be made as broad as can be allowed."

Butterfly valves, with two semicircular flaps opening back to back on a joint or hinge, fixed across the middle of the bucket or clack, were used for all the common sizes. For pumps from 18 in. to 20 in. in diameter it was usual to divide the valves into four flaps.

The first engine erected by *Watt* in Cornwall was one with a cylinder 30 in. in diameter, fixed at Creegbraws, near Chacewater, very soon after Smeaton's engine at Chacewater, probably about 1776. It worked there a few months, and was then removed to Wheal Busy, where it remained in action for some time afterwards.

In 1777 the patentees erected three more engines—namely, at Ting-tang, Owan-vean, and Tregurtha Downs. These were of larger dimensions ; two of them had cylinders 63 in. in diameter, and were capable of working with a load of 11 lbs. or 12 lbs. on the square inch of the piston. The first-named of these three, that at Ting-tang, was put up with the concurrence of one of the Hornblowers, who was engineer to the mine at the time. In Watt's first engines—and in this among the number—the air-pump and condenser were exhausted previously to start-ing, by means of a small pump worked by hand, and on the first trial the engine suddenly started and killed the man who was working the

[1] Pole on *The Cornish Engine*.

pump. Watt, who was present at the time, afterwards added the arrangement of blowing through, in order to produce the vacuum at starting by condensation.

The success of these trials induced the rapid adoption of the improved engine, and rendered Cornwall the richest field of profit to the patentees; for in about twelve years the whole of the atmospheric engines in the district had been replaced by patent ones, whose dimensions varied from 24 in. to 66 in. cylinders, and many of them were double-acting. The extension of the mining operations kept up a constant demand upon the manufacturers of the engines, either for new machines, or for the enlargement of old ones, which lasted until the year 1800, when the monopoly expired, and the connection of Messrs Boulton & Watt with the county entirely ceased.

From the date of Watt's patent for expansive working—namely 1782—to the present time, the development of the pumping-engine has been, in the advances made in the application of the principle of expansive working under mechanical conditions, best suited to the nature of the work to be done. Watt proposed to work with increased expansion by using mechanical devices, by means of which the resistance of the pump was gradually lessened from the beginning to the end of the stroke of the engine, but he appears not to have carried his idea into practice. He, however, employed expansive working, as far as it was practicable with the system of bucket pumps then in use.

About the year 1738 Newcomen introduced into Cornwall an engineer and manufacturer of steam engines, one Jonathan *Hornblower*, who soon became one of the principal manufacturers of steam engines in that county. Young Jonathan, a son of the above-named Hornblower, after serving an apprenticeship with a pewterer and plumber at Penryn, turned his attention to engineering, and soon became one of the principal engineers of the Cornish mines, which occupation he held till his death in 1812. He was much esteemed by all who knew him, and displayed great talent as an engineer. In 1776 Hornblower turned his attention to the carrying out of the principle of expansion which he had previously conceived, and he made a large working model with cylinders 11 in. and 14 in. in diameter. He patented the invention in 1781, or about one year before the date of Watt's patent.

The mode of carrying out the expansive principle invented by Hornblower is identical with that now commonly known, incorrectly, as Woolf's invention, and consists in what is called the double cylinder expansive engine. The steam is first admitted into a small cylinder, and after doing its work there, without expansion, is allowed to escape into a much larger cylinder, where it becomes expanded. It is difficult to imagine that Hornblower could have worked out the details of the double-cylinder engine without assuming that the theoretical value of expansion would be the same whether one or two cylinders be used; and it is probable that he saw the mode of working expansively in a single cylinder, but imagined that practical difficulties were in the way of its being carried to any great extent without the use of the second cylinder. Watt's patent was subsequent to Hornblower's, so that Hornblower could not have introduced the second cylinder to evade Watt's claim.

Hornblower's engine is illustrated in fig. 4.

In 1800 Boulton & Watt's patent expired, and their connection with

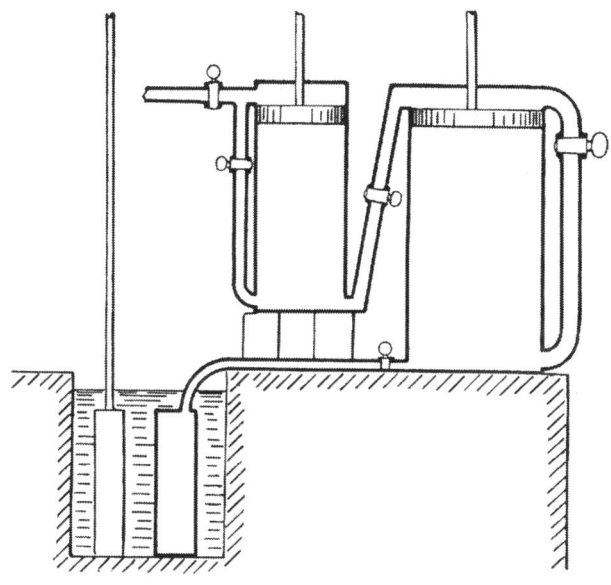

FIG. 4.—Hornblower's Compound Cornish Engine.

the mines in Cornwall ceased. The progress made up to that time is given in the following summary :—

1769.—Smeaton computed the effect of fifteen atmospheric engines working at Newcastle, and according to the data he furnished, the average duty was 5,590,000.

    Note.—The best was 7,440,000 : the worst was 3,220,000.

1772.—Smeaton began his alterations in the steam engines, and succeeded in performing 9,450,000.

1776.—Watt stated in a letter to Smeaton that his engine at Soho raised between 20,000 and 30,000 cubic feet of water 90 feet high with 120 lbs. of coal, which would be equal to 21,600,000. This was more, however, than Boulton & Watt would engage to perform, as, in a letter written by Boulton to the Carron Company in this year, which contained proposals for erecting an engine, he stated the performance as equal to about 19,000,000. Smeaton, about this time, after many experiments, laid it down as a rule that Watt's engines would do double as much as his own, which, as we see above, was 9·45 millions, consequently 18,900,000.

1778–1779.—Watt having stated that his engines should do 23,400,000, Smeaton made trial of two—one on the Birmingham Canal and one at Hull Waterworks, and found the duty of one equal to 18,000,000, and of the other 18,500,000.

1778.—Boulton & Watt erected an engine at Hawkesbury Colliery, near Coventry—cylinder, 58 ins. in diameter; stroke, 8 feet; load, 26,064 lbs.—which was found to do nearly 19,000,000.

1779–1788.—Watt introduced the improvement of working steam expansively, and he calculated that engines which would previously do 19,000,000 to 20,000,000 would thus perform 26,600,000 ; but we do not find any record of this duty having been attained in practice.

1785.—Boulton & Watt had engines in Cornwall working expansively, as at Wheal Gous and at Wheal Chance, in Camborne, but in these the steam was not raised higher than before, and the piston made a considerable part of the stroke, therefore, before the steam valve was closed.

1798.—As a consequence of a suit respecting their patent, which was carried on by Boulton & Watt and others, an account of the duty of all the engines in Cornwall was taken by Davies Gilbert and Captain Jenkins of Treworgie, and they found the average to be about 17,000,000.

An engine at Herland was found to be the best in the county, and was doing 27,000,000 duty, but, being so much above the average, some error was apprehended. This engine was probably the best then ever erected, and attracted, therefore, the particular attention of Boulton & Watt, who, on a visit to Cornwall, went to see it, and had by many experiments tried to ascertain its duty. It was under the care of Murdock. Captain John Davey, the manager of the mine, used to state that it usually did a duty of 20,000,000, and that Watt, at the time he inspected it, pronounced it perfect, and no further improvements could be expected.

1800.—About this time Boulton & Watt's patent expired, other persons began to construct engines, and Murdock left Cornwall, where he had been superintending most of the engines at the mines for sixteen years. The duty of the best engines was then stated to be 20,000,000.

Having thus brought the history of the Cornish pumping-engine up to the date 1800, we will briefly review the events of the subsequent period, which, although coming last, are not the least important. Watt left the engines of Cornwall doing a duty of not over twenty millions, but within forty years after the Cornish engineers raised the duty more than threefold.

From Lean's *Historical Statement* we have the following particulars of the period between 1801 and 1811 :—

"About the year 1801 Captain Joel *Lean* (who, besides being an experienced miner, was a good practical engineer) was appointed principal manager of Crenver and Oatfield Mine in the parish of Crowan, one of the deepest and most extensive mines then at work in the county. He found the engines and pit-work in a very bad condition. The mines were about 170 fathoms deep from the surface and 140 fathoms below the adit; and the water was drained by three steam engines—viz., a 63 in. cylinder double, on Bull's mode of construction, having the cylinder over the shaft and the piston rod working through the bottom, and 60 in. and 36 in. cylinder single engines on Boulton & Watt's plan. The consumption of coals by these engines was enormous, and the average duty was ten millions; Bull's engine, 63 in. double, consumed fourteen chaldrons of coal in one day. The pit-work, too, which consisted of leathern buckets with two or three pistons, such as were at that time in general use for plungers, was in a very bad state; and it may be safely asserted that the engines were idle at least one-third of the time, for the purpose of repairing the pit-work and changing the buckets.

"After he had assumed the management, Captain Lean's attention was immediately directed to the pit-work, and here he first introduced (what is now so generally used, and with so great advantage) the plunger pole, instead of the common box and piston, wherever he found it practicable. Equally bold and successful was the change which he made in the engines. The two smaller (which were erected in the same house and connected with the same rods) he threw aside, and put in their stead a 70 in. cylinder —the first of the size ever erected in the country—in which he adopted the expansive mode of working, at that time so little thought of and very partially practised. These improvements saved the mine. At that time they were burthened with a debt of many thousand pounds, which was continually augmenting; but as they consumed less than half the previous quantity of coal, and at the same time kept the mine effectually drained, so that the miners could work without hindrance, they not only discharged the debt, but obtained considerable profit."

We have been thus particular respecting what Captain Lean did at Crenver and Oatfield, because of the important consequences which resulted from the improvements introduced by him into the engines and pit-work of those mines. For being sensible that the defects which he had removed with so much advantage were not confined to the engines under his care, and convinced that it would be attended with much good, if the public generally, and more especially those who were adventurers in mines, had the means of comparing the different engines with each other, he endeavoured to bring some others of the principal managers into his views, and to awaken them to the necessity of registering and publishing the duty performed. It was, however, not until after many years that his wishes were accomplished.

The two governing facts which led to the great advance in duty was—first, the introduction of the plunger pump in place of the bucket, an alteration which gave quite a new character to the pit-work and thereby made increased expansion possible; second, the publishing of *Lean's Monthly Duty Records*, which held out an inducement to engineers to rival each other in the duty which they obtained from their engines, a matter which was of vital importance to the mine owners. The result of the rivalry was that the best engines were brought to such a high state of perfection that the efficiency ratio was as great as that of the best engines of to-day—that is to say, the Cornish engine made as good use of its opportunities as the triple-expansion engine does; but the opportunities of the Cornish engine were limited by steam pressure of 45 lbs., whereas the triple engine is now worked with pressures as high as 150 lbs. It was fully recognised that the economy of the engine depended largely on not wasting heat in radiation, or, in other ways, having dry steam, expanding the steam down to an absolute pressure of about 5 to 6 lbs. per square inch, and securing a good vacuum. The pressure of steam available was only from 20 to 30 lbs. per square inch, and although some engines were working with a boiler pressure of 40 lbs., that was the limit of pressure at which the full expansive power of the steam could be utilised in the single cylinder Cornish engine.

The idea of compounding the Cornish engine, originated by Hornblower, was taken up, and many compound engines were erected by Trevithick, Gribble, Sims, Woolf, and others, and although some of the engines did a good duty, no better results were obtained than with the best single cylinder engines; some of the compound engines were not properly designed for the best effect, but as long as the boiler pressure remained low, the compound engine had no *raison d'être*.

In 1824 Woolf erected a compound engine at Wheal Alfred Mine, intended to be worked with cast-iron boilers at 100 lbs. pressure, but owing to the failure of boilers the engine was removed.

It must be observed that the method of 'duty reporting' adopted in Cornwall applied to the engine and boiler as a whole; there was no indication of how far the results were influenced by a good or a bad boiler, or good or bad coal, nor were the observations which formed the basis of the reports taken with sufficient accuracy for strictly scientific purposes. There was the error of uncertainty of length of stroke, and the load of the engine was calculated from the size and length of the lifts. As long as the lifts were vertical no error could arise in that respect, but frequently the lifts were inclined, so that the length of the lift did not represent the height of water. It was, however, considered by the Cornish engineers that the extra friction of the inclined rods made up for the deficiency in height of water. As regards the coal used, it was probably fairly uniform in quality, as it was all South Wales coal. The efficiency of the boilers was also fairly uniform after the introduction of the Trevithick Cornish boiler, which took place about the year 1813.

Notwithstanding the defects in the system of 'Reporting,' the reports were of commercial value to the proprietors of the mines, and were useful to others as indicators of engineering progress, but if we had no better record of the thermal efficiency of the Cornish engine than that given in the reports, we could form no proper judgment in the matter; the Cornish engine has, however, had an extended application beyond the Cornish mines, and has been subject to rational and complete tests in the same manner as other engines.

The progress of events after Watt left Cornwall up to 1834 is given in the following summary :—

1810.—Woolf returned to Cornwall and introduced his engine working high-pressure steam in a small cylinder, and expanding into a larger one ; Captain Richard Trevithick also invented the simple high-pressure engine working without condensation. No immediate improvement in the duty seems, however, to have followed.

1811.—The first monthly report appeared of the duty of three engines at work at Wheal Alfred Mine, where Captain J. Davey was engineer.

The great improvements indicated by the rapid increase of duty in the period to which we shall now refer have resulted mainly from the use of high-pressure steam worked expansively, and the introduction of Trevithick's Cornish boiler.

1813.—The first year in which the duty papers appeared in their present form ; the number of engines reported was 24, of which the average duty was 19,456,000. In the early part of the year the best duty was about 26 millions, by Captain Trevithick at Wheal Prosper, Captain John Davey at Wheal Alfred, and Messrs Jeffery & Gribble at Stray Park. Towards the close of the year Davey first attained 27 millions, and Jeffery & Gribble reached 28 millions.*

1814.—Number of engines reported, 29 ; average duty, 20,534,232. During this year Jeffery & Gribble's engine at Stray Park performed the best duty, having reached 35 millions ; for twelve months the average was 32 millions. Woolf's engine at Wheal Abraham, first reported in October of this year, performed 34 millions.

1815.—Number of engines reported, 35 ; average duty, 20,526,110. Woolf's engine at Wheal Abraham attained a duty of 52·3 millions.

1816.—Number of engines reported, 32 ; average duty, 22,907,110. In

---

* The *duty* was calculated per bushel of 94 lbs. of coal up to July 1856 ; after that date the cwt. was substituted for the bushel.

this year an engine at Dolcoath, by Messrs Jeffery & Gribble, did a duty of 40 millions, which is the first record of such a duty done by a single-cylinder engine. In May of this year Woolf's engine at Wheal Abraham did a duty of nearly 57 million .

1817.—Number of engines reported, 31; average duty, 26,500,259. The general improvement was now beginning to be apparent, as may be observed from the average of this and the preceding year. Jeffery & Gribble continued to take the lead, and their engine at Dolcoath in some months reached 44 millions. Woolf's engine occasionally surpassed this, reaching to 51 and 52 millions, but after this year it did not exceed the average of the best single engines.

1818.—Number of engines reported, 32; average duty, 25,433,783. No improvement in this year, and the general rate of duty rather fell off.

1819.—Number of engines reported, 37; average duty, 26,252,620.

1820.—Number of engines reported, 37; average duty, 28,736,398. Among these were the engines lately erected at the Consolidated Mines, by Woolf, having cylinders of 90 in. diameter, and a stroke of 10 ft., the most powerful that had been constructed.

1821.—Number of engines reported, 39; average duty, 28,223,382.

1822.—   ,,   ,,   ,,   45;   ,,   28,887,216.

1823.—   ,,   ,,   ,,   45;   ,,   28,156,162.

1824.—   ,,   ,,   ,,   45;   ,,   28,326,140.

The best duty seems to have been done by Sims' engine.

1825.—Number of engines reported, 50; average duty, 32,000,741. General improvement was now observable. Sims' engine at Polgooth reached nearly 54 millions, and some new engines appeared in the first rank of merit, as those by Woolf, at Wheal Alfred and Wheal Sparnon, by Webb at Hirland, and by Grose at Wheal Hope. This latter engine deserves the more notice as it was first erected by this engineer, and its construction led to the great improvements in duty which he afterwards exhibited in his engines at Wheal Towan.

1826.—Number of engines reported, 48; average duty, 30,486,630.

1827.—   ,,   ,,   ,,   48;   ,,   32,100,000.

Although the average duty of all the engines in the county was not so much improved in this year as might be expected from the results exhibited by the performances in particular instances, yet this must be deemed an important epoch in the history of the steam-engine. One which had been erected by Grose, at Wheal Towan, of 80 in. cylinder, in which he had perfected all he had tried in his engine at Wheal Hope, was found to surpass all others for the first nine months of the year, and from April to September maintained a duty of more than 60 millions, reaching in the month of July to more than 62 millions.

1828.—Number of engines reported, 54; average duty, 37,000,000. Public attention having now been directed in Cornwall to the improvements which Grose had introduced, and the principles being applied with various modifications in many instances, the advantage was rendered obvious by the advance in the average duty of the whole. And in the meantime the engine at Wheal Towan was brought to a degree of perfection which had not been anticipated by the most sanguine, advancing in March to more than 80 millions, a rate of duty which it maintained, with some fluctuations, to the end of the year.

1829.—Number of engines reported, 53 ; average duty, 41,700,000.

1830.—	,,	,,	,,	56 ;	,,	43,300,000.
1831.—	,,	,	,,	58 ;	,,	43,400,000.
1832.—	,,	,,	,,	59 ;	,,	45,500,000.
1833.—	,,	,,	,,	56 ;	,,	46,600,000.
1834.—	,,	,,	,,	52 ;	,,	47,800,000.

During the year 1834 several engines were reported to do remarkable duty, notably West's engine at Fowey Consols and Taylor's engine at the United mines ; the former was said to do a duty of 125 millions on 94 lbs. of coal, and the latter 107 millions.

The author has indicator diagrams from both engines in his possession, and they show initial pressures of 50 and 36 lbs. absolute respectively, the expansion in both cases being a little over 4 to 1. If the above reported duty were correct, the Fowey Consols engine would have given at least an indicated H.P. per hour on 12 lbs. of feed water and Taylor's engine on 15·2 lbs.,* which is in the former case impossible.

Since 1834 little progress has been made in Cornwall ; the best engines have done about 60 millions duty, and, according to the records, in exceptional cases much higher, but of late years the recorded duty has fallen off, and there are but few engines at work in the county.

We have been induced to devote some considerable space to the progress of the Cornish engine, for it was with that engine that the principles governing steam-engine economy were first practically grappled with. The progress which has been made since has been great, but the possibility of advance has been afforded by the improvements in manufacturing processes and materials, by means of which we can safely use high-pressure steam.

In the early days of the development of the pumping-engine it became recognised that economy was to be secured by expansion, and it soon became known from the teaching of experience that considerable expansion could be secured in the single-cylinder Cornish engine.

As a matter of fact, the single-cylinder Cornish engine was quite

* Assuming an evaporation of 10 lbs. of water per 1 lb. of coal, and a mechanical efficiency of the engine and pumps combined, of 80 per cent.

capable of getting the best duty from steam of the pressure used. Attempts were, however, made by Hornblower, Trevithick, Woolf, Sims, and others to secure greater economy by the use of compound engines. Many compound engines were made, but the duty obtained was no better than that obtained with the single cylinder. The method of expanding in the compound engine was simply that of allowing the steam to expand from the small cylinder into the large cylinder. The total range of expansion was not more than in the single engine.

In figs. 5 and 6, sections of Hornblower and Trevithick's and also of Sims' engines are given. Neither the distribution of steam nor the boiler

FIG. 5.   FIG. 6.

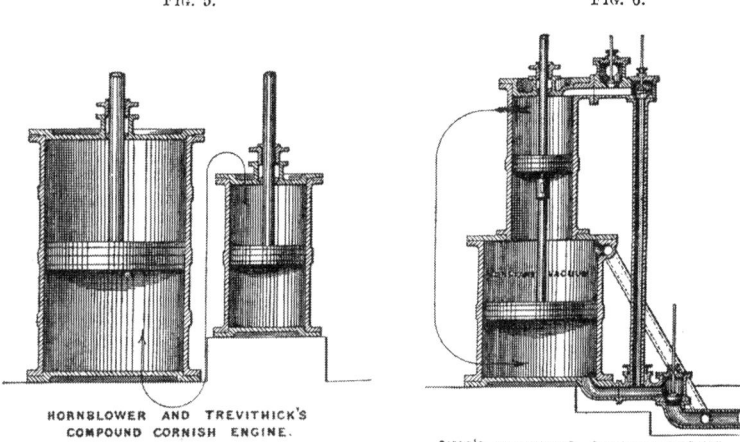

HORNBLOWER AND TREVITHICK'S
COMPOUND CORNISH ENGINE.

SIMS'S COMPOUND CORNISH ENGINE.

pressure available were conducive to economy in the compound engine. It is only quite recently that the development of the Cornish engine has advanced by the adoption of a second cylinder. From Watt's time—1800 to 1834—the boiler pressure was raised from 10 to 45 lbs., and the duty trebled; since 1834 the safe boiler pressure has been raised from 45 lbs. to 200 lbs., and the duty of our best pumping-engines nearly doubled. We have not space to trace further the history of the development of the pumping-engine. The practicability of making boilers to be worked with high-pressure steam and the new and extended requirements of modern times have led to the production of new forms of engines and pumps, many of which are noticed in the succeeding chapters of this book.

Note.—The duty of pumping-engines is now generally calculated on 112 lbs. of coal. Previous to 1856 it was expressed per bushel of 94 lbs. In Savery's and Smeaton's time the bushel was evidently taken to be 84 lbs.

PLATE I.] [*To face page* 21.

Aston Pumping Station, City of Birmingham Waterworks.

CHAPTER II.

As the steam engine plays so important a part in connection with pumping machinery, it will be well to devote this chapter to a consideration of the principles which govern and limit its application.

It will have been observed from the brief and incomplete history contained in the first chapter that it was as a pumping-engine that the steam engine as a practicable machine received its first application, and that the general principles which govern its economy were first grappled with in those engines. The greatest economy has always been obtained with pumping-engines, and it is so at the present day.

We need not consider the subject of quick reciprocating engines, because such engines have a limited application to pumping purposes. They are useful and convenient for driving centrifugal pumps direct, but for general pumping they are little used.* It may, however, be observed, that although certain economical conditions are secured by quick reciprocation, no quick reciprocating engine has been found to be so economical in steam as the slow-running pumping-engine.

It is the common experience that the quicker an engine runs within the limits of its application, the better should be its economy in steam, but that is only true when comparing the conditions of running of the particular engine or engines of its class. It is not true as a generalisation that quick-running engines are more economical in steam than slow-moving ones.

There are many factors which enter into the question of steam economy.

Fig. 7 is a diagram in which the curves are not true curves; they are used for illustrative purposes only.

A perfect steam engine, according to Rankine and others, was defined to be an engine working without loss, receiving the steam at the higher temperature and expanding it adiabatically to the lower temperature.

Let A (fig. 7) represent an indicator diagram, the upper line above it

---

* Professor Riedler has produced a quick reciprocating pump for quick-running engines ; it is described in the chapter on pumps.

representing the adiabatic curve for the weight of steam admitted to the
cylinder. Then the losses arise—

        a. From condensation in the cylinder.

        b. Clearance.

        c. Back pressure.

        d. Incomplete expansion.

The loss *a* is reduced by steam jacketing and superheating. By
steam jacketing, cylinder condensation is reduced, but experience has
shown that in jacketing the cylinders alone a less percentage of economy
is obtained in a modern economical triple compound engine than that
secured in ordinary compound or single-cylinder engines.

The efficiency of the steam jacket is also influenced by the quality of
the steam from the boiler. If water is admitted with the steam either
by priming or from condensation in the steam pipes, then the steam jacket

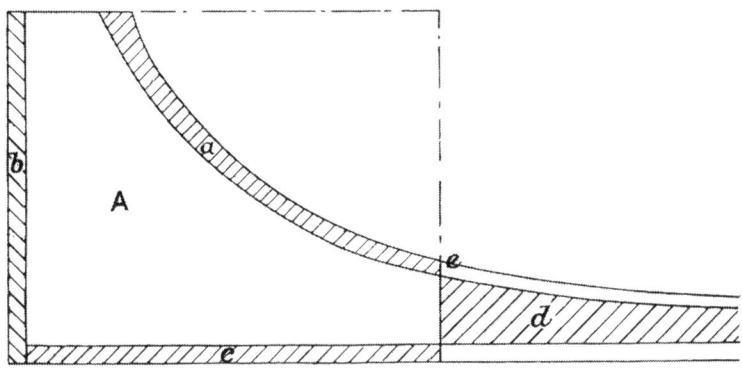

Fig. 7.—Diagram of Steam Engine Losses.

has additional work to do in evaporating the water deposited on the
surfaces by what is badly named 'wet steam.'

Superheating may be effected by utilising the waste heat from the
furnace gases, or by means of superheaters placed in the boiler flues.
In the latter case the superheat may be obtained at the expense of boiler
heat.

Superheat is readily dissipated by radiation from long lengths of steam
pipe.

Superheating, or adding heat, is also done by introducing super-heaters,
consisting of coils of pipe filled with steam at boiler pressure, into the
receivers between the cylinders of compound engines.

The loss from cylinder condensation is greater than that indicated by
the shaded part *a* (fig 7), because without condensation the expansion
might be carried further, and the loss from incomplete expansion, *d*,
thereby reduced—that is to say, the indicator diagram, if coinciding with
the upper line of the diagram might be extended till the terminal pressure

was only equal to the pressure required to overcome the frictional resistance of the engine, which may be assumed, for illustration, to be the terminal pressure of the diagram A. Re-evaporation takes place in the cylinder, because, during admission, the cylinder walls are receiving heat, and during expansion they are giving it out. If we assume, for the sake of illustration, that the re-evaporation was complete at the end of the stroke, and that the terminal pressure of the indicator diagram A coincided with the point *e* of the adiabatic curve, then the loss from cylinder condensation as compared with adiabatic expansion would be represented by the shaded portion *a*.

Clearance (*b*).—The loss from clearance spaces is sometimes considerable, especially in non-rotative engines, but when the non-rotative engine has a long stroke, the percentage loss may not be much greater than that of a short stroke rotative engine. Where considerable clearance is necessary, as in non-rotative engines, the loss is reduced by cushioning, especially in the Cornish engine, in which the *vis viva* of the falling mass is taken up by the cushioning.

Cushioning is also used in a large degree in the author's varying lever engine described in Chap. XII., and shown in detail in fig. 169, Chap. X.

In rotative engines, experience is in favour of reducing the clearance spaces, for, whatever may be the merits of cushioning in its influence on economy, it is undesirable to employ it in a large degree in pumping-engines controlled by a crank. Clearance spaces

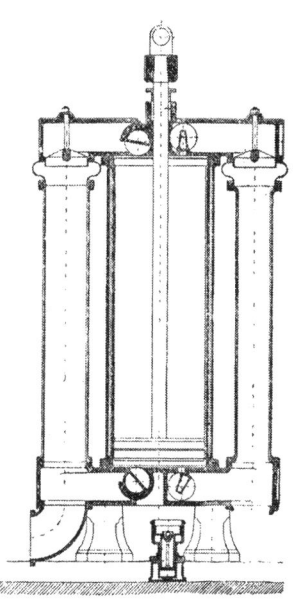

FIG. 8.—Section of Steam Cylinder with Corliss valves in covers for reducing clearance spaces.

may be reduced by putting the steam valves in the covers of the cylinders, as shown in fig 8.

Here the valves are of the Corliss type, and are clearly indicated in section.

Back Pressure (*c*).—This loss is an important one, and demands attention, not only in the designing of the engine, but also in its daily working. Let the average pressure on the low-pressure cylinder be 20 lbs. per square inch. A back pressure of 1 lb. will represent 5 per cent. A condensing engine working with a vacuum of 20 inches of mercury instead of 26 inches may be losing 15 to 20 per cent. in economy.

Condensers should be of ample capacity, and be provided with every facility for cleaning and repairs. We have heard it argued that a condenser is not part of a steam engine, but it is a good practical rule to look upon a steam engine with a bad condenser as a bad condensing engine.

Incomplete Expansion (d).—The loss from incomplete expansion may be looked at in two ways. Practically, there is no use in continuing the expansion beyond the point at which the terminal pressure of the indicator diagram is no more than that required to overcome the useless resistances of the engine. Theoretically, the loss is represented by the space bounded by the adiabatic curve beyond the point e (fig. 7), and the bottom line or line of no pressure. Practically, there must be back pressure, and the most perfect condensing engines in use have a back pressure of about 1 lb. absolute, corresponding to about 100° F. In calculating the theoretical work for a given weight of steam, it is usual to calculate from the limits of temperature, and as 100° F. is the lower practical limit for the most perfect condensing engines, that temperature will be adopted by the author. The loss from incomplete expansion is, then, that represented by the area bounded by the adiabatic curve beyond e, and the bottom line representing about 1 lb. back pressure or 100° F.

We shall see, further on, that the most economical condensing engines of to-day, compared with a theoretically perfect steam engine using saturated steam, give an efficiency ratio of over 60 per cent., so that the total of the losses which we have enumerated is under 40 per cent. It must be observed that by carrying expansion further the loss from incomplete expansion may be reduced, but the mechanical efficiency of the engine may, at the same time, be reduced; for not only may the friction be increased by the larger engine, but, as the mean pressure is reduced, the friction becomes a larger percentage of the total indicated power.

The indicator diagram, with all its defects, is the measure of the I.H.P. which forms the basis for scientific purposes of investigation or of comparison of different types of engines. It is therefore important that the diagram should be analysed as to the losses indicated in fig. 7, otherwise false conclusions may be arrived at.

For practical purposes the thermal units expended for the work done is the important result, but the losses as shown by an analysis of the indicator diagrams should not be ignored, because the result, good or bad, may have been greatly influenced, not by defects in the engine itself, but by the circumstances under which it was working.

Priming and Steam Pipe Radiation.—Engines are very seldom supplied with dry steam. Proper precautions should be taken to drain off the water from the steam pipe. This should be done quite close to the stop valve by means of automatic steam separators. Insufficient attention is given to this matter, which is of great importance as affecting the economy of the engine. Steam pipes should not be unnecessarily large or long, and should be thoroughly well covered to prevent radiation.

Frictional Losses.—It is usual to express this loss in percentages of the I.H.P. of the engine. The mechanical efficiency of an engine is said to be 70, 80, or 90 per cent., as found in the engine trial, but it is useful to know how many lbs. pressure per sq. inch of the piston that represents, for it is obvious that, within limits, the heavier the load on the engine, or the greater the mean pressure, the less may be the *percentage* of loss from friction. To obtain the greatest economy per I.H.P., it may be necessary in a particular case to work the engine with a low mean pressure, the frictional resistance forming a large percentage of the total I.H.P. The true economy is not expressed in terms of I.H.P., but in pump H.P., or, including the pump H.P., in the quantity of water pumped or what may be termed water H.P. Non-rotative engines have generally the advantage in mechanical efficiency, but rotative engines of the most approved designs, suitably applied, are superior in economy per I.H.P. Geared engines have a low mechanical efficiency and generally an inferior economy per I.H.P.

Leakage.—The question of leakage through valves and pistons is one of great importance, as an engine of superior design may fail in economy from that cause alone.

Good workmanship in the working parts is of more vital importance than polished columns, although we may here remark that a well-finished engine receives more care from its attendant than does a less sightly machine.

Standard of Comparison.—One steam engine may be shown to be better than another by its using less saturated steam per I.H.P. per hour or by its producing a greater number of work units per heat units expended. The thermal efficiency of the engine is thus expressed :—

$$\text{Thermal efficiency} \quad = \quad \frac{\text{heat utilised}}{\text{heat supplied}},$$

$$\text{Do.} \quad = \quad \frac{\text{foot lbs. of work done}}{\text{units of heat supplied} \times 778},$$

778 being the recognised equivalent in foot lbs. of one British thermal unit. The efficiency of engines thus expressed enables one to compare their relative economy, but as higher pressures possess greater possibilities of economy, and engines may be condensing or non-condensing, we require a standard of comparison which shall be applicable to the conditions under which the engine works, whereby we may be able to judge how far it has been advanced towards perfection. The true standard is the ideal perfect steam engine—an engine taking steam at the higher pressure and expanding it adiabatically to the lower or exhaust pressure without loss.

The formula for such an engine was determined by Rankine.

The B.T.U. per H.P. for the standard engine of comparison can be calculated by means of the formulæ which follow.

The formula for the thermal efficiency of the Rankine cycle for saturated steam is

$$\frac{(\mathrm{T}a - \mathrm{T}e)\left(1 + \dfrac{\mathrm{L}a}{\mathrm{T}a}\right) - \mathrm{T}e \text{ hyp log } \dfrac{\mathrm{T}a}{\mathrm{T}e}}{\mathrm{L}a + \mathrm{T}a - \mathrm{T}e} \quad *$$

in which formula the increase in the specific heat of water at higher temperatures affects the numerator and denominator nearly equally.

The B.T.U. per minute per H.P. for the standard engine of comparison is 42·4, divided by the thermal efficiency of the Rankine cycle thus:—

For saturated steam, the B.T.U. per minute per H.P. for the standard engine of comparison is:—

$$\frac{42 \cdot 4(\mathrm{L}a + \mathrm{T}a - \mathrm{T}e)}{(\mathrm{T}a - \mathrm{T}e)\left(1 + \dfrac{\mathrm{L}a}{\mathrm{T}a}\right) - \mathrm{T}e \text{ hyp log } \dfrac{\mathrm{T}a}{\mathrm{T}e}} \quad *$$

and similarly for superheated steam it is:—

$$\frac{42 \cdot 4\left\{\mathrm{L}a + \mathrm{T}a - \mathrm{T}e + 0 \cdot 48(\mathrm{T}as - \mathrm{T}a)\right\}}{(\mathrm{T}a - \mathrm{T}e)\left(1 + \dfrac{\mathrm{L}a}{\mathrm{T}a}\right) + 0 \cdot 48(\mathrm{T}as - \mathrm{T}a) - \mathrm{T}e\left(\text{hyp log } \dfrac{\mathrm{T}a}{\mathrm{T}e} + 0 \cdot 48 \text{ hyp log } \dfrac{\mathrm{T}as}{\mathrm{T}e}\right)} \cdot$$

The meanings of the letters used in the above formulæ are:—

$\mathrm{T}a$ = absolute temperature of saturated steam at stop-valve pressure.
$\mathrm{T}as$ = absolute temperature of superheated steam at stop-valve.
$\mathrm{T}e$ = absolute temperature in exhaust.
$\mathrm{L}a$ = latent heat of steam at temperature $\mathrm{T}a$.

As these formulæ are somewhat difficult and tedious to work out, use is made of the $\theta \phi$ chart, and for practical purposes when great accuracy is not required the author uses a simple approximation.

It will be observed that the leading factors are the higher temperature $\mathrm{T}a$ and the lower temperature $\mathrm{T}e$; the former that of the steam entering the engine, and the latter that of the steam leaving it.

$\mathrm{T}a$ = temperature of steam at inlet.
$\mathrm{T}e$ = temperature of steam in exhaust at outlet.

The engine may be condensing or non-condensing. In pumping-engine practice the value of $\mathrm{T}e$ may be taken.

$\mathrm{T}e$ = 212° non-condensing.
$\mathrm{T}e$ = 100° condensing.

The rational minimum lower temperature $\mathrm{T}e$ of the condenser for the perfect engine would appear to be the mean temperature of the atmosphere, 60°, but condensers must have a back pressure, and that corresponding to 100° Fahr., or say 1 lb. pressure, is a good working standard. If we take as a standard 100° as the lower temperature for condensing engines and 212° Fahr. (or that corresponding with the atmospheric pressure) for non-condensing engines, we may construct a very simple and useful approximate formula which will enable us to compare the performance of an actual engine with that theoretically possible, without the use of the Rankine formula.

* "Thermal Efficiency Report," *Proceedings Inst. C.E.*, 1898.

The following empirical formula is only applicable to saturated steam :—

Let L = lbs. of steam per I.H.P. per hour.

T = the higher and t the lower temperatures.

For condensing engines let $t = 100°$, and for non-condensing engines let it be $212°$.

Then approximately in the theoretically perfect engine the lbs. of steam required per I.H.P. per hour would be for condensing engines $\dfrac{2540}{T-40}$, for non-condensing engines $\dfrac{2540}{T-192}$.

This formula should only be used for pressures between 30 and 200 lbs. absolute ; but within these limits the errors are very small, as will be seen from the table below.

The following table gives the theoretical values according to the Rankine formula and the author's approximation :—

Condensing Engine's Exhaust Temperature 100° Fahr.				Non-condensing Engine's Exhaust Temperature 212° Fahr.			
T.	Boiler Press. Abs.	lbs. of Steam per I.H.P. per hour, Rankine Formula.	$\dfrac{2540}{T-40}$	T.	Boiler Press. Abs.	lbs. of Steam per I.H.P. per hour, Rankine Formula.	$\dfrac{2540}{T-192}$
228	20	13·34	13·51	292	60	25·10	25·40
267	40	10·82	11·19	312	80	20·69	21·16
292	60	9·80	10·08	328	100	18·25	18·68
312	80	9·13	9·34	341	120	16·72	17·05
328	100	8·68	8·82	353	140	15·55	15·78
341	120	8·36	8·43	363	160	14·72	14·85
353	140	8·09	8·11	373	180	13·99	14·03
363	160	7·88	7·86	382	200	13·40	13·37
373	180	7·69	7·63				
382	200	7·53	7·43				

The results given by the Rankine formula may, of course, be taken from the $\theta\ \phi$ chart.

The thermal equivalent of the work represented by 1 H.P. per hour is 2540 B.T.U.

Let S = lbs. of steam used by the engine per I.H.P. per hour.

T = Initial temperature of saturated steam (Fahr.).

H = Total heat of 1 lb. of the steam, saturated or superheated, supplied to the engine in B.T.U., minus feed heat.

Then, the approximate *efficiency ratio* for condensing engines $= \dfrac{2540}{T-40 \times S}$

and the *thermal efficiency* $= \dfrac{2540}{H \times S}$.

Example.—Let $T = 380°$ F, $H = 1300$ B.T.U., and $S = 12$ lbs.,

then $\dfrac{2540}{(380-40) \times 12} = 62·2\%$ = *efficiency ratio*

and $\dfrac{2540}{1300 \times 12} = 16·2\%$ = *thermal efficiency.*

The use of a theoretical standard is in practice that of comparison.

The nearer the theoretical standard approaches to the actual conditions of the steam engine, the more useful will it be in showing the defects or perfections of one engine as compared with another.

For practical use we have constructed diagrams (figs. 9 and 10) showing at a glance the theoretical weights of feed water per I.H.P. per hour

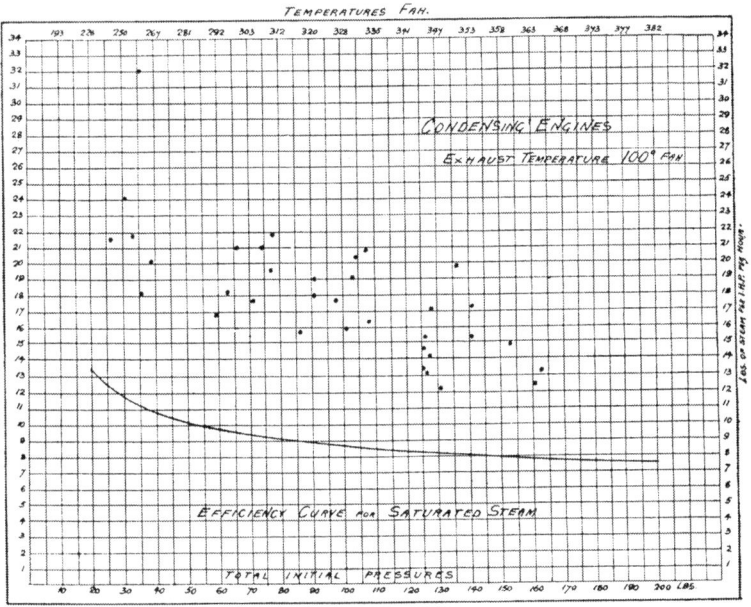

FIG. 9.—The curve in this diagram represents by the vertical ordinates from the base line the number of lbs. of steam required by a perfect engine, in which the steam is admitted at the temperatures and pressures indicated by the horizontal lines.

for different initial pressures in the engine, and above the curve are plotted the actual performances of engines from trials.

The basis of an engine trial is the number of lbs. of steam used per I.H.P. per hour: that must be known whatever is omitted. If superheated steam is employed, then the added heat must be taken into account.

It must be observed that the formula and charts above given are constructed on the supposition that the feed to the boiler is taken at the exhaust temperature, heated by the exhaust of the engine. That is to say, with condensing engines the feed is at 100°, and with non-condensing engines 212° Fahr. That is an ideal condition, and can very nearly be approached in practice. It will be observed that the ideal engine expands the steam adiabatically down to the back pressure of the condenser, but actual engines do not usually expand the steam to a lower pressure than 10 lbs. absolute.

With engines working in the Cornish cycle of steam distribution, it is

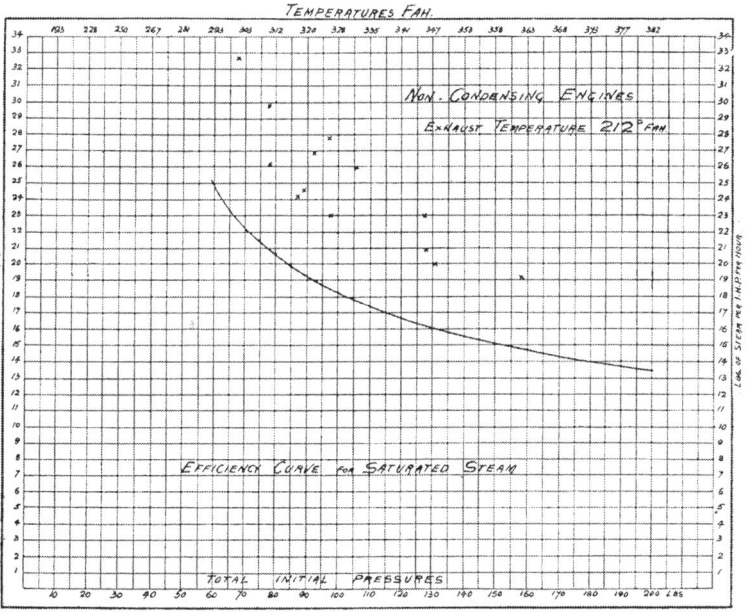

FIG. 10.—The curve in this diagram represents by the vertical ordinates from the base line the number of lbs. of steam required by a perfect engine in which the steam is admitted at the temperatures and pressures indicated by the horizontal lines.

possible to heat the feed water to a higher temperature than that of the condenser (fig. 14, Chap. II., and fig. 83, Chap. V.).

Efficiency Ratio.—The relation between the actual performance of an engine and that of a perfect steam engine working between the same limits of temperature is called the *efficiency ratio*.

$$\text{Efficiency Ratio} = \frac{\text{lbs. of steam per I.H.P. per hour for the perfect engine.}}{\text{lbs. of steam per I.H.P. per hour for the actual engine.}}$$

On reference to the efficiency curve (fig. 9, p. 28), it will be seen that a perfect steam engine, taking saturated steam at 382° Fahr., and exhausting it at 100° Fahr., requires 7·5 lbs. of steam per I.H.P. per hour. If the actual engine has been found to use 12 lbs. of steam under the same conditions as to initial and exhaust temperatures, then the efficiency ratio $= \dfrac{7·5}{12} = 62·5$ per cent.

In other words, the actual engine performance is 62·5 per cent. of that of a perfect steam engine using saturated steam without loss.

The number of thermal units per I.H.P. per hour expresses the economy of the engine, irrespective of steam pressure employed, but the efficiency ratio expresses the economy, taking the steam pressure into consideration. Having a given steam pressure, and knowing the weight of steam per I.H.P. per hour used by the engine, the question arises—Is the engine a good or a bad one ? The efficiency ratio gives the answer at once ; and if we know the best efficiency ratio which has been obtained with engines of that type, we at once know how far the engine in question departs from the best practice.

In this respect the efficiency ratio will be found to be of great practical use. To take an example. A triple expansion pumping-engine may have been found to give in working one I.H.P. per minute for 230 heat units. That may be considered a bad result as compared with another engine of the same type, but the two engines compared may have worked with different initial pressures, and the efficiency ratios may have been the same in both cases. The potential of the steam is taken into consideration in the efficiency ratio, just as the evaporative efficiency of the boiler and coal must enter into the calculation if we wish to estimate the relative economy of the engines from the coal consumption.

Practical Use of the Ratio.—If we refer to the table on page 31 it will be seen that the best efficiency ratios of the practical examples given for the different types of engines are approximately—

Condensing Engines.

Cornish, single acting,	60 per cent.
Single cylinder, double acting,	45 ,,
Compound double acting,	55 ,,
Triple compound double acting,	60 to 65 ,,

In the Table we have not been able to distinguish in all cases between the trials which give results in lbs. of feed water, and those which give lbs. of dry steam. The Cornish and marine examples of many of the others are expressed in lbs. of feed water, whilst the triple pumping-engine examples give lbs. of steam.

If all the trials were on the same footing in lbs. of steam, the efficiency ratios of the best examples would probably be represented approximately thus :—

Cornish, single acting,	65 per cent.
Single cylinder, double acting,	45 ,,
Compound double acting,	55 ,,
Triple compound double acting,	65 ,,

These figures must not be taken as standards of practical perfection, but if the best results are thus tabulated, a standard of comparison for each type of engine may be found; and any engine giving a lower efficiency may be said to fall short of the best modern practice.

The following tables contain the results of selected engine trials, and a column giving the efficiency ratio in each case.

STEAM ENGINE TRIALS.

TYPE OF ENGINE.	Reference Number of Trial.*	Initial Pressure.	Ratio of Expansion.	lbs. of Steam per I.H.P. per hour. †	Efficiency Ratio.
					%
Cornish, ...	2	31 J	2·83	24·00	49
,, ...	52	27 J	3·12	21·38	57
,, ...	55	34·25 J	3·87	20·72	55
,, ...	53	40 J	4·17	20·08	54
,, ...	54	36 J	4·28	18·82	60
Single Cylinder Rotative, .	1	34 J	2·63	26·69	42
,, ,, .	1a	36 N	2·34	32·14	35
,, ,, .	60	105·79 N	4·76	20·37	42
,, ,, .	58	104·77 N	4·93	19·15	45
,, ,, .	59	102·79 N	5·13	19·22	45
Marine Compound, . .	3	77·5	4·96	21·73	43
,, ,, . .	4	66·5	5·33	21·17	45
,, ,, . .	5	108·0	5·71	20·77	41
Three Cylinder Compound,	6	91 J	6·06	18·11	50
,, ,,	6a	91 N	5·98	19·15	46
Rotative ,,	7	78 J	6·36	19·52	47
,, ,,	7a	75 N	5·62	21·06	44
Non-rotative ,,	57	72·2 J	9·17	17·70	53
Rotative ,,	8	102 J	9·48	15·90	55
,, ,,	8a	99 N	8·61	17·67	50
,, Beam ,,	10	60 J	10·08	16·64	59
,, ,, ,,	10a	64 N	9·5	18·20	53
,, ,, ,,	56	109·2 J	10·16	16·24	52
Rotative Triple Expansion,	9	127 J	10·07	15·370	54
,, ,,	9a	129·5 N	9·54	17·170	50
Marine ,, ,,	12	137·5	11·60	19·830	41
,, ,, ,,	13	154·0	11·60	14·970	53
Pumping ,, ,,	15	128·5 J	15·30	14·240	60
,, ,, ,,	15a	126 N	14·40	14·690	56
,, ,, ,,	16	132·7	16·14	12·155	67
,, ,, ,,	17	127·5 J	16·30	13·160	63
,, ,, ,,	17a	126·5 N	15·60	13·470	61
Marine ,, ,,	19	164·5	18·90	13·350	58
Pumping ,, ,,	18	142 J	18·53	15·450	53
,, ,, ,,	18a	142 N	16·50	17·220	50

Note.—J and N in col. 2 indicate jacketed and non-jacketed respectively.

* See Table following.

† This column to be quite accurate should give lbs. of dry steam, but we have not been able to distinguish between the figures which give steam and those which give feed water.

Boiler pressure 200 pounds, piston speed 250 feet per minute. The size of steam cylinders are 19½, 29, 49½ and 57½ inches. The plungers are double acting, 14¾ inches in diameter, the plunger rod 4·5 inches diameter, the common stroke 42 inches.

"The valve gears are of the Corliss type, excepting the exhaust valves on No. III cylinder, which are single beat poppet valves, and all the valves on No. IV cylinder, which are also single beat poppet valves. The clearances of the engine as stated by the makers are as follows: In No. I, or high pressure, cylinder, 128 cubic inches equal 1·25 per cent. ; in No. II, first intermediate, cylinder, 360 cubic inches equal 1·3 per cent. ; No. III, cylinder, second intermediate, 444 cubic inches equal 0·55 per cent. ; No. IV, cylinder, low pressure, 390 cubic inches, 0·36 per cent. The dimensions as checked by measurement are given in the table of data and dimensions following the report.

"The principal peculiarity of the pumping-engine consists in the arrangement of a series of heaters through which the condensed steam is successively passed on its journey from the hot well to the boiler, and in which it is warmed by steam drawn from the low-pressure cylinder, the receivers and the jackets, the detailed arrangement of this system for heating the feed water being as follows:—A surface condenser is employed which is kept at low temperature by large quantities of injection water drawn from and discharged into the suction main. The condensed steam is delivered by the air pump into a hot well, from which it is pumped into an oil purifying tank located on the outside of the building and standing at a comparatively high level. From this tank the feed water passes to a heater in which is arranged a series of shelving over which it falls in a series of drops through the ascending current of exhaust steam from the low-pressure cylinder; this heater is under the same vacuum as the condenser, and is termed the exhaust or pre-heater. The water discharged from the pre-heater is pumped into heater No. I, which is of similar construction to the pre-heater, heat being taken from the low-pressure cylinder, however, for warming the feed water. The steam is drawn into the heater through a pair of auxiliary valves in the low-pressure cylinder which are opened at about the ⅞ part of the stroke and after cut-off. From heater I the feed water is discharged into heater II by gravity. In heater II the feed water is further warmed by discharge of steam from receiver III. The water is then pumped from heater II to heater III, where it is warmed by steam from receiver II and by the jacket discharge from cylinders III and IV. It is thence pumped to heater IV, where it is further warmed by steam from receiver I and by the discharge of water from the jackets of cylinders I and II, and the re-heaters of the receivers.

"The temperatures which were actually obtained by the admission of steam as described in the various heaters were as follows:—Starting with the water in the outside tank at 88° it is raised in the exhaust heater to

C

105°; in heater I to 136°; in heater II to 193°; in heater III to 260°, and in heater IV to 311°. The temperature rise actually obtained was about 15° less than expected by Mr Nordberg. The various pumps required for forcing the feed water through the various heaters are compactly arranged, especially constructed for pumping hot water, and are all mechanically driven by attachment to the main pumping-engine. The work of operating the pumps is included in the test as a portion of the friction of the main engine. The actual work of lifting the feed water from the point of discharge to an elevation equivalent to boiler pressure is about equivalent to lifting 150 lbs. per minute against a head of 500 feet, or about 2·3 horse power. The additional work caused by the friction of the various pumps cannot be separated from the friction of the pumping-engine.

"The steam for the engine was supplied by a battery of Hogan boilers and kept during the test at a gauge pressure of about 200 lbs. per square inch. The steam for the main engines passed through a separator, the drip of which was removed by a tank pump and returned to the boilers. During the duty test the drip of the separator was cooled and weighed for information regarding its efficiency. The various cylinders of the engines were jacketed on the barrels, and the receivers between the cylinders were provided with reheating pipes. Steam for the jackets was drawn from the main steam pipe between the boiler and the separator. The jackets of cylinder I were supplied with steam of boiler pressure, it thence passed through a reducing valve and was reduced in pressure to 116 lbs., thence into the jacket of cylinder II; it thence passed into the reheating tubes of receiver III at a pressure of 105 lbs., thence into the receiver tubes of reheater II at a pressure of 103 lbs., thence into reheater tubes of receiver I at a pressure of 102 lbs., thence the discharge led through a trap into heater IV. Jacket steam for the barrels of cylinder III was taken from the main steam line at the same point as the line previously described, and passed through a reducing valve which lowered the pressure to 40 lbs. per square inch, it thence passed to jacket of cylinder IV, in which it had a pressure of 39 lbs., and from thence was discharged through a trap into heater No. III.

WORK OF STEAM CYLINDERS.

Number of steam cylinders,				I	III	II	IV
M.E.P., top,				69·35	12·85	35·16	8·85
M.E.P., bottom,				70·68	14·02	33·98	9·33
I.H.P., top,				77·56	96·58	88·64	89·14
I.H.P., bottom,				77·55	104·09	84·99	93·63
I.H.P., top and bottom,			155·11	200·67	173·63	182·77
Boiler pressure, .							200 lbs.
I.H.P., total, .						712·18	,,
Feed water per hour, total,						8850·4	,,
Dry steam per hour, total,						8732·4	,,
Dry steam per I.H.P., .						12·263	,,
B.T.U. per I.H.P. per minute (above feed water temp.),			185·96	
Total boiler pressure, .						215 lbs.	

It will be seen that the total steam was 12·26 lbs. per I.H.P. per hour, and that the total boiler pressure was about 215 lbs. If we compare the result with the trial of the Milwaukee engine,* also by Professor Carpenter, we shall see that the efficiency ratios taken on feed water are approximately—

Nordberg engine, . . . 60 per cent.
Milwaukee engine, . . . 66 ,,

The absolute boiler pressure in the former case was 215 lbs., and in the latter 140 lbs., whilst the total steam per I.H.P. per hour was :—

Nordberg, ... 12·26 lbs.
Milwaukee, ... 12·15 ,,

From the description of the Nordberg engine we conclude that it was a very expensive engine compared with the ordinary triple compound, and although the boiler pressure was higher than that of the Milwaukee engine, there was no gain in the actual consumption of steam, the efficiency ratio becoming less. Had it been the same, the Nordberg engine would have only required 11 lbs. of steam as against the actual consumption of 12·26 lbs.

Superheating, as we know, increases the economy, but the ordinary methods of superheating are not yet brought to a state of practical perfection for every-day use, under the varying conditions of pumping-engine installation, and as the percentage gain in economy to be expected from higher initial pressures alone is smaller for a given rise of pressure, as the pressure increases, the author thinks that probably it will be found that the best practical result will be obtained by not exceeding 215 lbs. as the initial pressure in the engine, and working the boiler to say 260 lbs., throttling the steam between the boiler and engine, and maintaining the full boiler pressure in re-heaters between the cylinders, and in the jackets also where practicable. The superheating is then done in the engine itself, and is not wasted in radiation from the steam pipe.

It must be observed that the only gain that can be expected from superheating is that secured by reducing condensation in the cylinder.

When the boiler gives a high thermal efficiency, the waste heat is not so great that a high degree of superheat can be got from it. If additional fuel is to be used in superheating, then the heat so applied must be debited to the superheater. As an engine question alone the value of the super-heat may be gauged by the increased efficiency ratio, although that ratio is calculated on a standard for saturated steam.

If *thermal efficiency* is required, then the total heat of the superheated steam must be found.

Example.—Let the superheated steam contain 1300 B.T.U. above the feed heat per lb.; if the engine is found to use 12 lbs. of steam per I.H.P. per hour ;

* Triple expansion engine, Milwaukee Waterworks. Cylinders 28 + 48 + 74 in. 5 feet oke.

Then—thermal efficiency $= \dfrac{2540}{1300 \times 12} = 16\,\%$

Table showing the number of lbs. of saturated steam required per I.H.P. per hour, with various steam pressures and efficiency ratios :—

CONDENSING ENGINES, Lower Temperature 100° Fahr.				
Steam Pressure Absolute in lbs. per sq. in.	lbs. of Steam per I.H.P. per hour with an Efficiency Ratio of 40 %.	lbs. of Steam per I.H.P. per hour with an Efficiency Ratio of 50 %.	lbs. of Steam per I.H.P. per hour with an Efficiency Ratio of 60 %.	lbs. of Steam per I.H.P. per hour with an Efficiency Ratio of 70 %.
40	27·05	21·64	18·03	15·45
60	24·50	19·60	16·33	14·00
80	22·82	18·26	15·22	13·04
100	21·70	17·36	14·46	12·40
120	20·90	16·72	13·93	11·94
140	20·22	16·18	13·48	11·56
160	19·70	15·76	13·13	11·26
180	19·22	15·38	12·82	10·98
200	18·82	15·06	12·55	10·76

Table showing the number of lbs. of saturated steam required per I.H.P. per hour, with various steam pressures and efficiency ratios :—

NON-CONDENSING ENGINES, Lower Temperature 212° Fahr.				
Steam Pressure Absolute in lbs. per sq. in.	lbs. of Steam per I.H.P. per hour with an Efficiency Ratio of 40 %.	lbs. of Steam per I.H.P. per hour with an Efficiency Ratio of 50 %.	lbs. of Steam per I.H.P. per hour with an Efficiency Ratio of 60 %.	lbs. of Steam per I.H.P. per hour with an Efficiency Ratio of 70 %.
40	84·67	67·74	56·45	48·39
60	62·75	50·20	41·83	35·86
80	51·72	41·38	34·48	29·56
100	45·62	36·50	30·42	26·07
120	41·80	33·44	27·86	23·89
140	38·87	31·10	25·92	22·21
160	36·80	29·44	24·53	21·03
180	34·97	27·98	23·32	19·99
200	33·50	26·80	22·33	19·14

Referring to the table on the opposite page, it will be observed that the Cornish engine efficiency ratio in the last column of the table is very low. That is accounted for by the fact that the initial pressure in the cylinder was low : the expansion was small, and the diagrams show a great loss from throttling during the equilibrium stroke, and considerable back pressure in the condenser. The engine, however, is not a true Cornish engine, but a Boulton & Watt engine, lifting the water on the steam stroke. This is a type of engine which could not possibly give a high efficiency ratio. The true Cornish engine gives a ratio as high as 60 %

Table compiled from Mr Mair's paper on the independent testing of steam engines in the *Proceedings of the Inst. of Civil Engineers*, vol. lxx. To which is added the efficiency ratios of the examples given, the lower temperature Te being taken as 100° Fahr.

CONDENSING ENGINES.

	A.	B.	C.	D.	E.	F.	G.	H.	J.	K.
Date of trial	Feb. 1881	Aug. 1881	Aug. 1881	Dec. 1881	Dec. 1881	Dec. 1881	Feb. 1882	Feb. 1882	Oct. 1881	Nov. 1881
Type of engine	Single Cylinder Beam	Compound Woolf Beam	Compound Woolf Beam	Compound Woolf	Compound Woolf	Beam	Compound Woolf Tandem	Horizontal Tandem	Compound Receiver Beam	Cornish
Time running	6 h. 40 m.	6 h. 0 m.	10 h. 0 m.	10 h. 0 m.	11 h. 0 m.	10 h. 0 m.	3 h. 30 m.		6 h. 0 m.	10 h. 30 m.
Revolutions per minute	14·63	17·84	19·62	33·73	34·22	34·52	80·45	81·51	23·98	strokes 12·84
Indicated H.P.	101·70	75·9	75·2	267·86	267·94	267·91	120·30	149·75	127·40	146·02
Temperature of feed water	66°	63°	63·2°	47·3°	48·45°	48°	54°	54·04°	59·05°	51·05°
Temperature of injection water	..	54·57°	59·00°	51·50°	54°	53°	53·1°	53·3°	50°	50·45°
Temperature of air pump discharge	70°	98·67°	86·40°	..	..	63°	08·66°	107·54°	73·4°	86·2°
Percentage of priming water in steam	4	8½	2½	2½	2¼	2½	4	4	4	3
Height of barometer during trial, in inches	29·60	29·82	29·73	30·25	29·8	29·75	30·67	30·67	30·20	30·04
Absolute boiler pressure in lbs. per square inch	48	58	62	84·8	88·14	88·95	86·05	84·77	76	46·4
Speed of piston per minute in feet	160·82	100·24	215·82	371·03	376·42	379·72	683·82	692·83	263·78	201·33
Total number of expansions	6·85	9·3	15·70	9·64	9·50	7·77	11·48	11·64	13·61	..
Percentage of supply from boiler passed through jacket	4·0	Jacket not in use	9·9	7·1	8·3	Jacket not in use	No jackets on cylinders		13·2	4·5
Percentage of water present in cylinder at cut-off	43·8	51·1	40·8	37·6	37·7	34·0	42·6	43·0	34·1	
Percentage of water present in cylinder at end of high-pressure stroke	28·4	35·1	18·1	24·3	21·6	28·0	25·4	25·6	22·3	
Percentage of water present in cylinder at end of low-pressure stroke	..	38·8	8·7	17·5	14·3	35·2	26·3	26·3	19·0	
Thermal units per I.H.P. per minute	429·11	519·5	338·8	348·1	341·8	378·1	425·2	410·2	291·0	460·5
Pounds of dry saturated steam per I.H.P. per hour	22·00	26·02	17·34	17·73	17·30	10·24	21·05	20·88	14·84	24·15
Efficiency ratio, Lower Temperature T taken as 100° Fahr.	46·8%	36·8%	56·07%	50·8%	51·1%	46·3%	41·6%	43·3%	62·8%	43·3%

The efficiency ratio is here calculated from the boiler pressure; but to be quite accurate it should be calculated from the initial pressure in the engine.

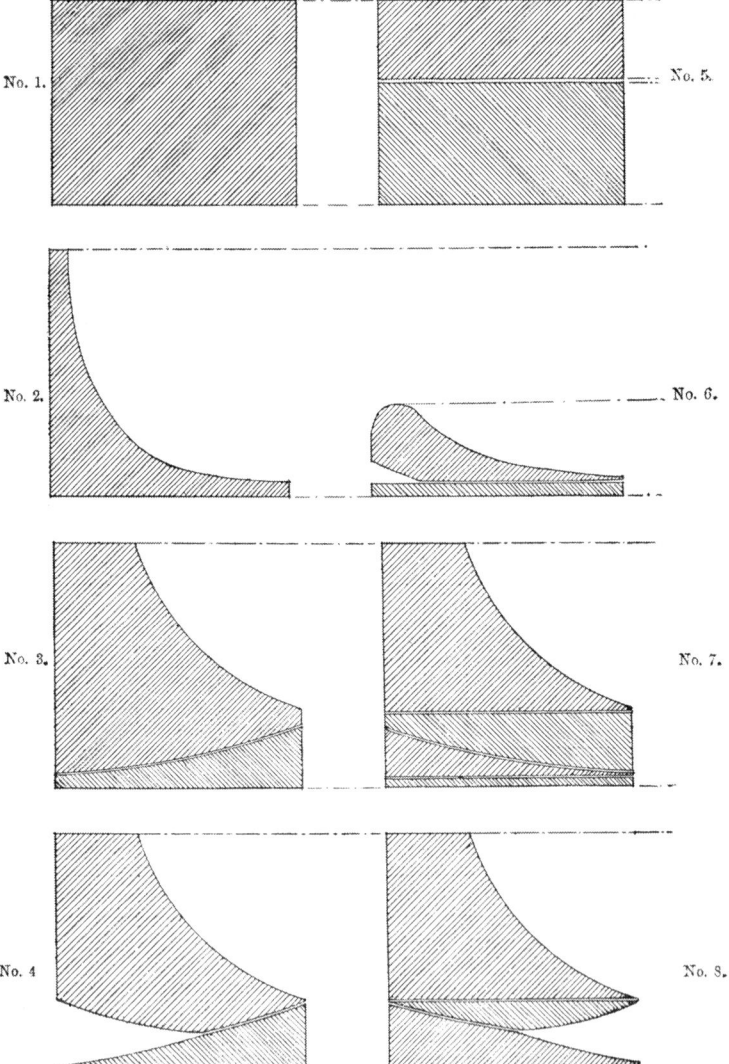

Fig. 11.—Diagrams illustrating the distribution of steam in various types of Pumping-Engines.

-or more. Here, then, is the use of the ratio in discovering how imperfect an engine is of its type.

Steam Distribution.—On the opposite page we illustrate by diagrams the different systems of steam distribution used in pumping-engines (fig. 11).

No. 1 is that of the *ordinary non-expansive single cylinder steam pump*. It is difficult to say what is the usual consumption of steam per H.P. When non-condensing, it is probably 100 lbs. or more per pump H.P. per hour, but notwithstanding this wastefulness, such pumps have a commercial value in the cheap and ready way in which they are applied to all sorts of temporary and, in fewer cases, permanent work. Considerable improvement is obtained by compounding and condensing.

The author has applied the single non-condensing steam pump in special cases in a way which has resulted in considerable economy of steam. The system is explained by the following diagram (fig. 12).

At the Whitacre pumping station of the Birmingham Water Works there are two sets of steam boilers, one supplying steam at 70 lbs. pressure to the compound engines, and the other at 30 lbs. pressure supplying steam to the Cornish engines.

The low lift pumps for lifting the river water to the filter beds are of the simple duplex steam pump type, and are worked from the high-pressure boilers, exhausting into the low-pressure ones.

At the Aston station of the Birmingham Water Works, where there are six large low-pressure engines working with 20 lbs. boiler pressure, it became necessary to increase the pumping power with a small capital -outlay, and the author adopted the system illustrated in fig. 13.

A large non-condensing compound steam pump (illustrated in fig. 214, Chap. XII.), capable of raising 3,000,000 gallons per day 250 ft. high, was erected, together with a battery of boilers having a working pressure of 130 lbs.

The low-pressure engines were all connected to one low-pressure steam main, and the new steam pump was made to exhaust into the low-pressure system; in that way the low-pressure engines were supplied with steam largely from the exhaust of the compound steam pump.

No. 2, fig. 11, illustrates the distribution of steam in a single cylinder engine.

No. 3, fig. 11. *Woolf compound engine*, in which there is a cut-off valve on the high-pressure cylinder only. In such engines there is generally a considerable drop in pressure between the cylinders, especially if the pistons move in the same direction, because then the steam port capacities become large.

No. 4, fig. 11. *Receiver engine*. That is a compound engine having a cut-off valve on both cylinders. A receiver between the cylinders becomes necessary, and the cut-off should be so arranged that the initial pressure in the low-pressure cylinder is about equal to the terminal pressure in the high-pressure cylinder.

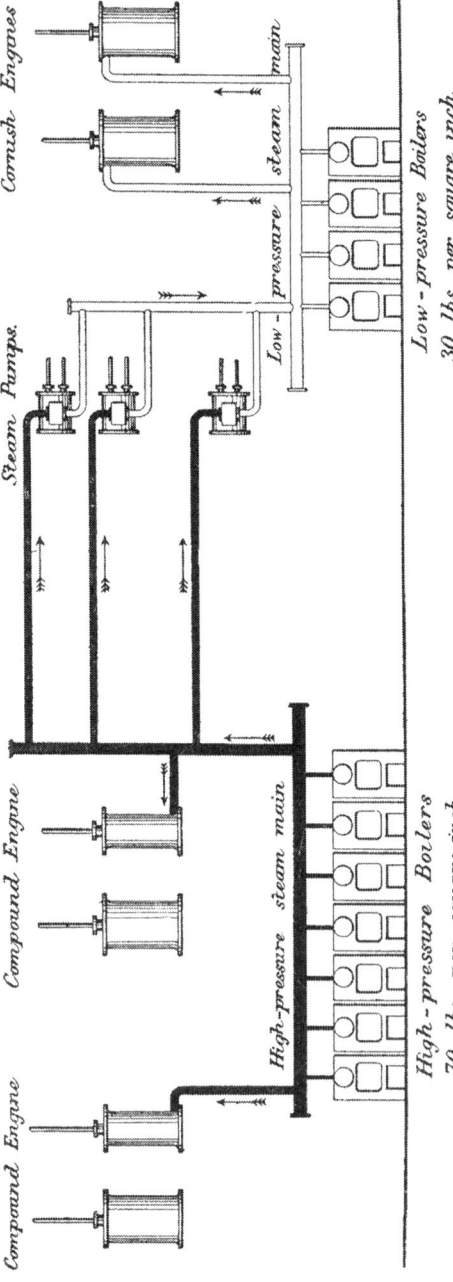

Fig. 12.—Distribution of Steam—Whitacre Pumping Station.

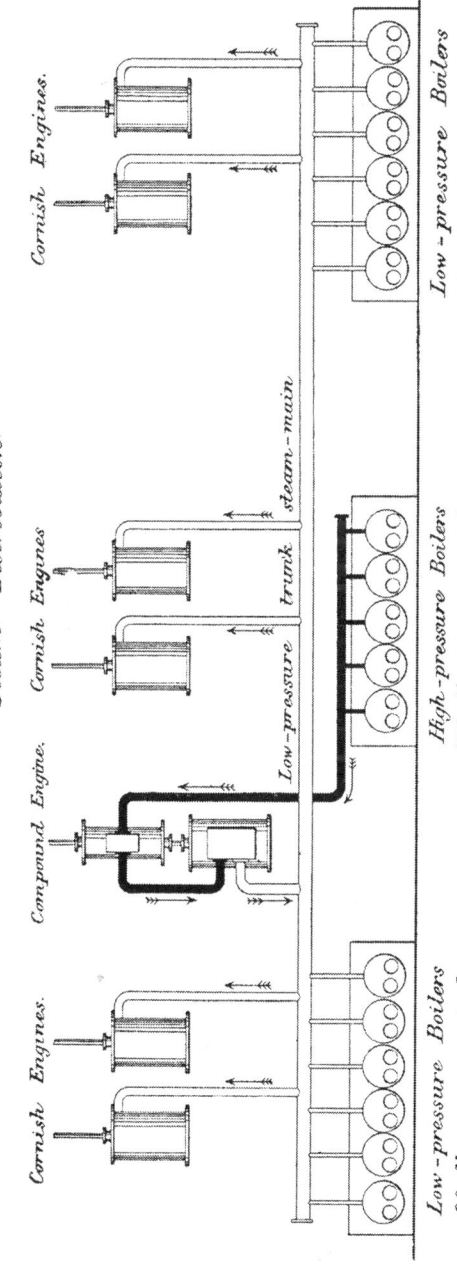

FIG. 13.—Distribution of Steam—Aston Pumping Station.

Cornish Cycle Engines.

All the four systems of steam distribution admit of the Cornish cycle. In No. 5, fig. 11, where there is no expansion, the steam is admitted to one side of the piston, and, after doing its work, is passed to the other

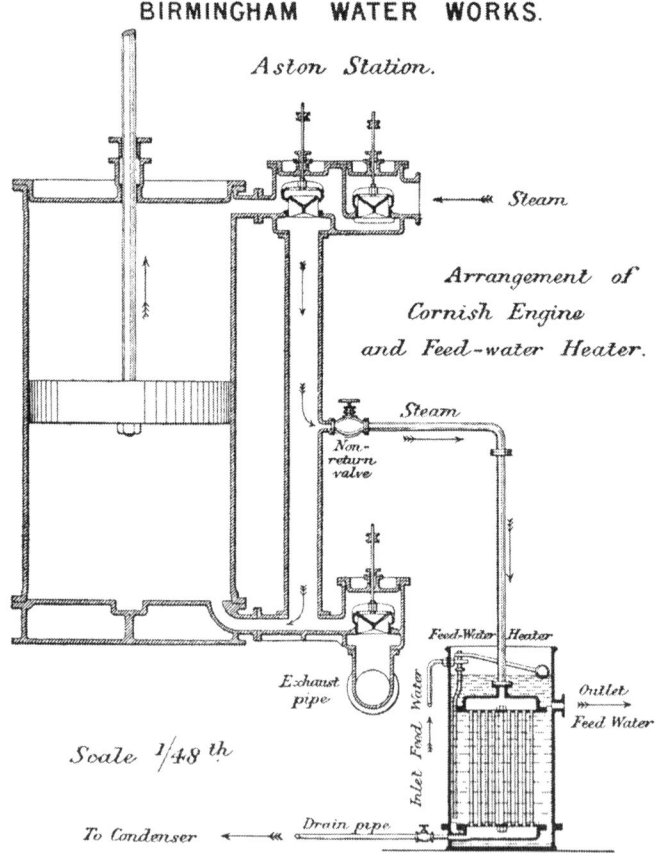

FIG. 14.—Arrangement of Feed-water Heater ; Aston Pumping Station.

side of the piston to be discharged to the atmosphere or to a condenser on the next steam stroke.

No. 6, fig. 11, is a diagram of the ordinary Cornish engine distribution. The limit of expansion in the ordinary Cornish engines is from three to four times. The engine is therefore worked with low-pressure steam.

No. 7, fig. 11, illustrates the distribution in a compound Cornish cycle engine—that is, an engine in which both the high- and low-pressure cylinders are single acting, both taking steam together, and both taking the equilibrium stroke together. There is a cut-off in the high-pressure cylinder only.

No. 8, fig. 11, is from a similar engine having a cut-off in the low-pressure as well as in the high-pressure cylinder.

The Cornish cycle admits of a system of heating feed water advantageously to a temperature above that of the exhaust steam. During the equilibrium stroke of the engine the steam is but little below the pressure of release, generally about 10 lbs. absolute or 194°. Now if a pipe having a non-return valve be taken from the equilibrium pipe of the engine to a feed-water heater, as shown in fig. 14, the feed water will be heated advantageously to a temperature far above that of the exhaust (100°) by steam which has done its work, but which is retained in the cylinder during the equilibrium stroke.

The feed heater shown on previous page was applied to a Boulton & Watt engine from which the diagrams fig. 15 were taken.

Indicator Diagrams.

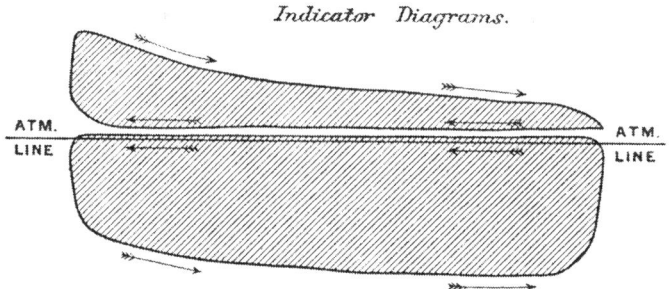

Fig. 15.—Indicator Diagrams.

It will be seen that the equilibrium steam happened to be at about atmospheric pressure.

Other applications of the system are shown in figs. 16 and 17.

Fig. 17 illustrates the construction of a condensing steam pump working without expansion, and heating the feed water to a very high temperature.

Let the steam pressure be 100 lbs. per square inch, equal to 328° Fahr. During the equilibrium stroke the pressure might fall to say 80 lbs., equal to 312° Fahr.

On the opening of the exhaust valve, the exhaust steam would fall to say 100° Fahr.; that steam would pass through the first section, A, of the feed heater. . During the equilibrium stroke some of the equilibrium steam would pass into B, the second section of the heater, raising the temperature of the water to say 300° Fahr.

Fig. 16 illustrates one pair of cylinders of a double-acting Cornish cycle rotative engine having cranks opposite each other. Steam is admitted

and expanded above the top piston, then it is passed to the bottom side,
where there might be placed a coil constituting a superheater.   The steam
from there passes during the next steam stroke to the top of the bottom
piston : from there it passes during the up stroke to the under side of the
piston to be discharged through the heater A to the condenser on the
down stroke.

The down stroke is the steam and exhaust stroke, and the up the

Fig. 16.   Fig. 17.

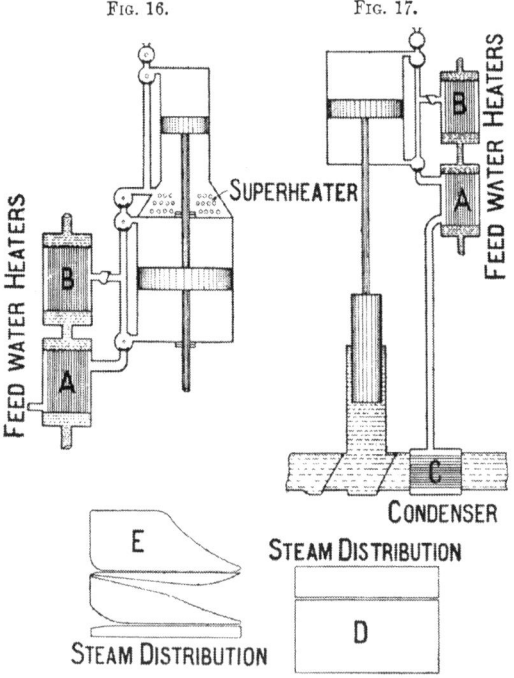

Distribution of Steam.   Heaters.

equilibrium stroke.   This engine may be worked on the receiver system
of steam distribution as indicated by the diagram E, and the feed-water
heaters applied as shown.   The engine is made double-acting by having
two pairs of cylinders like those shown in the illustration with the pistons
connected to cranks opposite to each other.

This form of engine has the advantage of having no lifting strain on
the cranks, so that it may be run with loose bearings, the wear being
always in one direction, and the bearings free from knock.

CHAPTER III.

PUMPS AND PUMP VALVES.

Types of Pumps.—The following illustrations of pumps are mere diagrams showing the peculiarities and characteristics of different types.

Bucket Pump, No. 1 (fig. 18).—This is the ordinary form of bucket pump. It consists of a hollow piston provided with a valve, the combination being called a 'bucket.' Below the bucket is a valve called a 'foot' or 'retaining valve,' commonly called a 'clack'; below the foot valve is the suction or supply pipe, and above the bucket is the rising main or delivery pipe. This pump may have an open top, the delivery of the water taking place over the top of the rising main, or the top may be closed by a cover having a stuffing box through which the rod passes. The water is then delivered through a branch in the rising main, from which a pipe may be taken in any direction.

The common house pump is of this type; when open topped it is called a 'lift pump,' but when the top is closed and the water is forced to a cistern above the pump, it is then called a 'force pump'; but the general and proper distinction between lift and force pumps is that bucket pumps of all kinds are lift pumps, and plunger pumps force pumps. When the plunger is combined with the bucket, it is called a 'lift and force pump.'

Fig. 18. Bucket Pump.

In the larger applications of the bucket pump it is called a 'bucket lift,' meaning that the water is lifted on the bucket in contradistinction to its being forced by a plunger or piston. In the bucket pump the total work in pumping the water is done on the up or lifting stroke, for that is both the suction and the delivery stroke; the return or downward stroke is 'idle,' no work being done, and the bucket simply passes through the water to take its position for the next active stroke. The work is nearly all done on the up stroke, the resistance to the downward stroke is simply the friction of the bucket and the resistance of the water in passing through the bucket. This was the earliest form of mine pump worked by the Newcomen and Watt engines. It suited the engines in being single acting, and was attached to the outer end of the engine beam whilst the steam piston was attached to the inner end.

The unbalanced weight of the pump rods overcame the friction of the engine and pump on the down stroke, but the down stroke was a source of accident from what is termed a 'riding column.'

Assume that the pump has been working properly and that the rising main or pump above the bucket is full of water, and that the bucket makes an up stroke without taking in any water below it, either from the failure of water supply or from an air leak in the pump below the bucket, then on the commencement of the down stroke the bucket will be forced down not only by the unbalanced weight of the pump rods, but by the addition of the full water load on the bucket, which in the case of large pumps may be 10 to 20 tons or more. Such a weight falling freely through only a few feet is a serious matter as regards the safety of the plant. This and other questions affecting the safety of pumps will be dealt with in other chapters. We have seen that this bucket pump is entirely single acting.

Bucket Pump, No. 2 (fig. 19).—It is obvious that if two bucket pumps are employed, one making the up stroke whilst the other makes the down stroke, then double action is secured, and such arrangement is admissible where sufficient space is available, but it often happens that pumps have to be put in very confined spaces, such as bore holes, and then it is important to get the greatest delivery from a given sized pump. The pump now under notice is made double acting by placing two buckets in the same pump barrel, one below the other. To insure the double action one bucket must be going up whilst the other is going down ; the buckets must therefore have separate rods ; this is usually accomplished by making the rod of the upper bucket a tube through which the rod of the lower bucket passes. A little reasoning on the action of the two buckets thus actuated will enable one to see that a foot valve is not required and that we really have a double-acting pump with only two valves, one on each bucket.

It will be seen that during the down stroke of the bottom bucket (which takes place simultaneously with the up stroke of the top bucket) water equal to the full displacement of the double stroke is taken through the bottom bucket valve, so that the speed of water through this valve is double that through the valves of an ordinary bucket pump, and the speed through the upper bucket is the same as through the lower one.

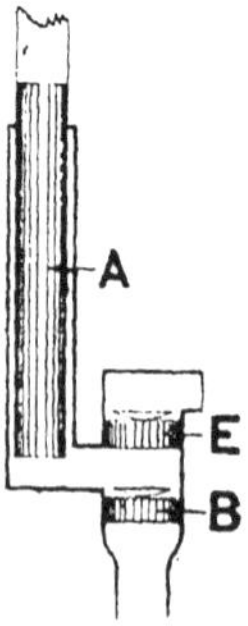

Plunger Pump.—The ordinary plunger pump is indicated in fig. 20.

In this pump the displacement of the water is by means of a rod or plunger A working in a barrel provided with suction and delivery valves,

FIG. 19.
Bucket
Pump.

FIG. 20.
Plunger Pump.

B and E. The work done on the suction side is accomplished during the up stroke, whilst that of the delivery is effected during the down stroke. One of the advantages of this type of pump over the bucket or piston type is that the packing of the plunger can be applied from the outside, and the superior form of packing makes the pump suitable for greater columns or pressures of water than are possible with buckets. As the work done on the delivery stroke is generally, with this pump, immensely greater than that done on the suction stroke, the pump plunger is chiefly under compression strain, and when applied to mining purposes, it is expedient to make the weight of the pump rods above the plunger sufficient to overcome the resistance to the plunger on the down stroke. On the up stroke the work done by the engine is that of lifting the weight of the plunger and rods, and overcoming the resistance of the column on the suction side of the pump. The height of the suction column is that from the level of the water to the top of the plunger barrel.

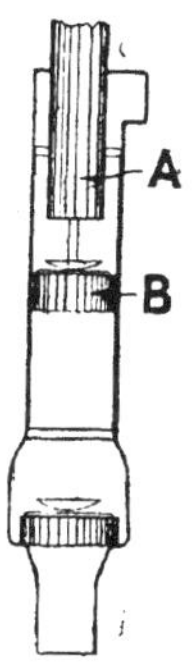

We have already noticed in the history of the pumping-engine that the introduction of this type of pump changed the character of the Boulton & Watt engine, and constituted it the Cornish engine for reasons explained elsewhere. This pump in its various modifications is applied to almost all purposes where very high pressures have to be dealt with.

FIG. 21.
Bucket and
Plunger Pump.

The packing of the plunger is simple enough for almost any pressure in practice. There are many kinds of packing made, but a hemp 'gasket,' plaited square, and steeped in boiling tallow, is almost as good as anything. The valves are of various kinds, governed by the speed of the pump, the pressure, the state or nature of the water, and other practical considerations. This remark applies to all other pumps.

Bucket and Plunger Pumps (fig. 21).—This is a combination of the two types of pumps already noticed, and its action will be readily understood from the figure. This pump is double acting on the delivery, and single acting on the suction side. The area of the plunger A is usually half that of the bucket B; in that case the work done on each stroke is not equal, because all the work of the suction side is done on the up stroke. It has been observed in the working of this pump, as also in the bucket pump, that when (in exceptional cases) the suction pipe is of considerable length, and the height to which the water is pumped is small, that the pump when driven fast may deliver more water than is accounted for by the displacement of the bucket. This is interesting and is worth notice. The reason may be thus given. The suction pipe, having no air vessel connected to it, contains a body of water which is put into motion by the action of the bucket, and if the weight and velocity of the water are sufficient, the momentum will keep the water in motion after the bucket has completed its suction stroke and is making the return stroke. A fuller explanation will make this clear.

Let the suction pipe be 100 ft. long with a capacity of 1 gallon or 10 lbs. per foot, and let the water in the pipe be put in motion with a velocity of, say, 6 ft. per second ; then the stored energy in the water

$$= \frac{Wv^2}{2 \times g}$$ where W = the weight of the water which in this case is 1000 lbs.

The stored energy is then $\frac{1000 \times 36}{64} = 560$ ft. lbs. about. So that before the water could be brought to rest, 560 foot lbs. of energy would have to be expended. Now let the top of the delivery pipe be 10 ft. above the source of supply, then 5·6 gallons flowing over the top of the pipe would represent 560 foot lbs. This is more than would actually flow over, because of the friction of the pipe, and the work done in moving the water entering the suction pipe, but the illustration suffices to show the reason for what has been observed in exceptional cases. It must be explained that the work done by the pump is represented by the water actually delivered, notwithstanding the action just described, because the work done in imparting velocity to the water is part of the total work in all cases, but with ordinary reciprocating pumps, it forms but a small part of the total power expended on the water.

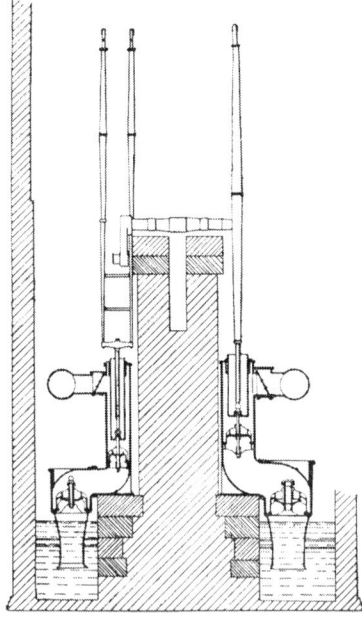

FIG. 22.—Bucket and Plunger Pumps—worked from Beam Engine.

Fig. 22 shows a form of bucket and plunger pump used with beam engines at the New River Water Works.

Piston and Plunger Pump (fig. 23).—This is a pump of the character of the bucket and plunger, the bucket having been replaced by a piston.

There are two valves only, as in the plunger pump. The internal packing of the piston is an objection, but sometimes the piston is replaced by a plunger and packing boxes introduced, as in the double-acting plunger pump (figs. 25 and 26).

Double-acting Piston Box Pump (fig. 24).—This is a form of pump much used for low lifts, and is generally provided with multiple valves.

Double-acting Plunger Pumps (figs. 25 and 26).—Both forms are much used, especially in mining work.

Three-throw Plunger Pumps (fig. 27) and *Three-throw Bucket Pumps*

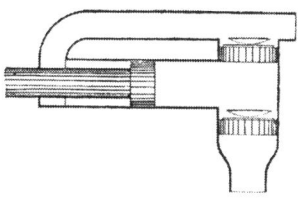

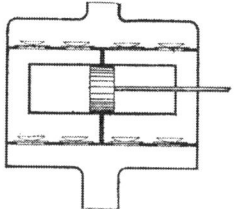

Fig. 23.—Piston and Plunger Pump.   Fig. 24.—Double-acting Piston Pump.

(fig. 28).—The general construction is indicated in the figures. The buckets or plungers are driven from a crank shaft with cranks placed 120° apart. These pumps are also driven direct from the piston rods

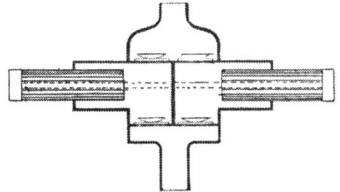

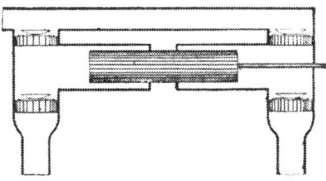

Fig. 25.—Double-acting Plunger Pump.   Fig. 26.—Double-acting Plunger Pump.

of triple-expansion engines. The general idea in favour of three-throw pumps is that of a nearly uniform flow of water. It is, however, advisable to use an air vessel of about the same capacity as would be used with

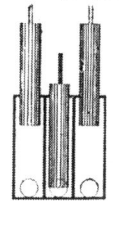
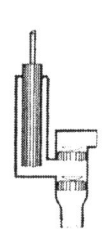
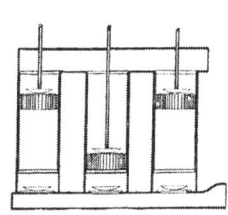
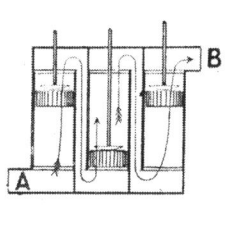

Fig. 27.—Three-throw Plunger Pump.   Fig. 28.—Three-throw Bucket Pump.   Fig. 29.—Three-throw Bucket Pump without foot valves.

ordinary double-acting pumps, because one of the three pumps may by accident for a few strokes, or for a longer time, become idle, and then the flow becomes very irregular, as will be seen from the pump displacement diagrams (fig. 55, page 65).

Three-throw Bucket Pump without foot valves (fig. 29).—In this pump

D

the water enters at A and is discharged at B, passing in an unbroken stream through the buckets, as indicated by the arrows. The displacement of this pump is peculiar, as each bucket is displacing water in a forward direction for one-third of a revolution, so that there are three deliveries in immediate succession in one revolution of the crank shaft, and the total displacement is the capacity of one of the barrels multiplied by 2·6.

Pumps and Pump Valves.—The practical difficulties in the working of pumps are chiefly connected with keeping working parts watertight, and securing efficient working of the valves. Plungers are easier kept in order than any form of piston or bucket, but piston and bucket pumps are cheaper, more compact, and occupy much less space; they are advantageously used for low lifts.

Almost the earliest form of valve in practical use was the flap valve, similar to that shown in connection with Cornish pit work (fig. 122, Chap. VII.), which is the valve now frequently adopted in sinking pumps where the water is charged with grit and other substances. The valve consists of a sheet of thick leather cut to form the valve : one part of the leather is fastened down to the seat to form the hinge, and the back and front of that portion of the leather which rises and falls is fortified by means of wrought-iron plates, generally secured by means of copper rivets, which are easily cut out when new leather is required. This valve is improved by so hinging it that it can rise above the seating at the hinge.

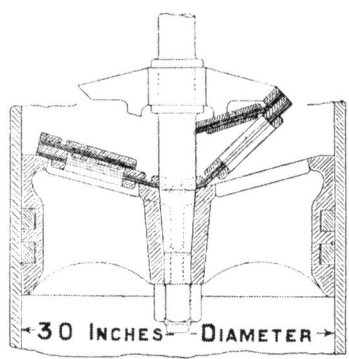

FIG. 30.—Pump Bucket with butterfly valves.

Sometimes the valve is hinged in the centre ; it is then called a 'butterfly valve,' because it resembles the wings of a butterfly.

Flap valves are sometimes provided with an additional flap on the back, as in fig. 30.

This valve is known as 'Teague's valve.' In this particular case it has been applied to a double-acting sewage pump (fig. 31).

The pump barrels (fig. 31) are 30 in. in diameter, have a 3 ft. stroke, and are so constructed as to require no foot valves. One bucket makes the up whilst the other makes the down stroke, and the displacement is equal to a pair of single-acting pumps with cranks opposite to each other. The pumps make fifteen double strokes per minute. As all the sewage pumped has to pass through both buckets, the velocity through each bucket is twice as great as that through the buckets of a pair of single-acting pumps, which is a drawback to this type of pump, but there is a practical advantage in having no foot or suction valves, so that this pump may be usefully employed if the speed of the buckets is not too great.

Figs. 32 and 33 are examples of this type of pump designed for deep and shallow wells, which are suitable for use with small motors. In

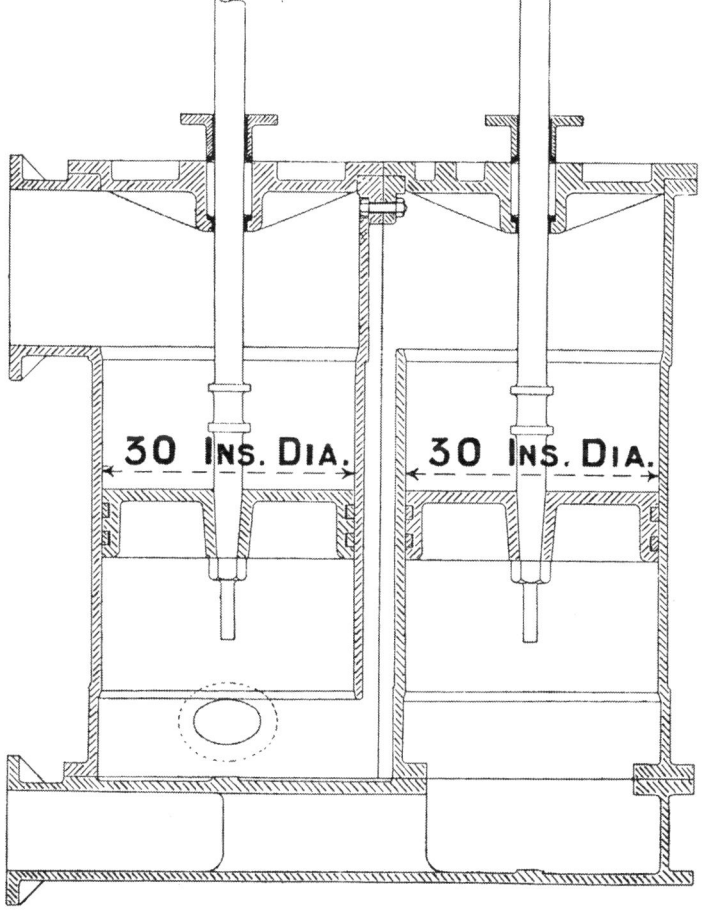

FIG. 31.—Double-acting Bucket Pumps without foot valves.

the deep well pump (fig. 32) a cover is placed on the rod of the bottom bucket. The cover falls into a conical seating, and the weight of the water column keeps it in place.

Ashley's Pump.—A peculiar kind of pump, in which the foot or suction valve is combined with the bucket, has been invented by Mr Ashley; it is shown in fig. 34.

The bucket has a pipe extending downwards, forming at the bottom a hollow plunger moving in a cylinder fixed in the pump.

In the pipe joining the bucket and the hollow plunger are placed small

FIG. 32.   FIG. 33.

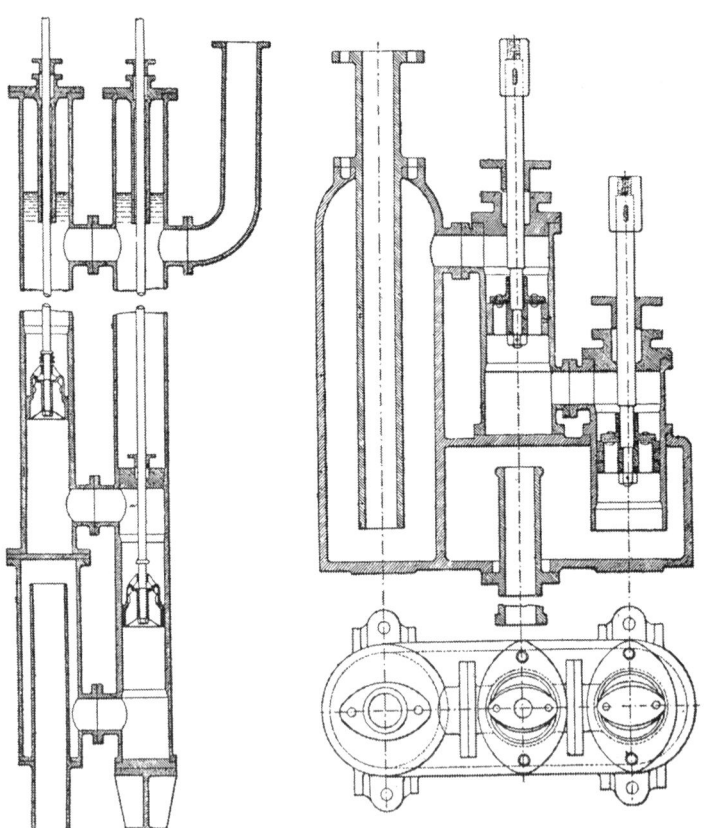

Double-acting Well Pumps without foot valves.

valves opening into the pipe. These valves perform the function of a foot valve. During the up stroke of the bucket water is taken through the small valves into the pipe and bottom cylinder, and during the down stroke that water is displaced through the bucket into the rising main, hence the action is that of an ordinary bucket pump. The object of the arrangement is to enable the suction valves to be drawn with the bucket.

Restler's Pump.—Pumps have been made with the valve attached to the pump rod and opening below the bucket; such a pump is illustrated in fig. 35.

This pump is of the type already described, a double-acting bucket pump with no foot valve. The valve, as will be seen from the figure, is opened by the downward and closed by the upward motion of the pump

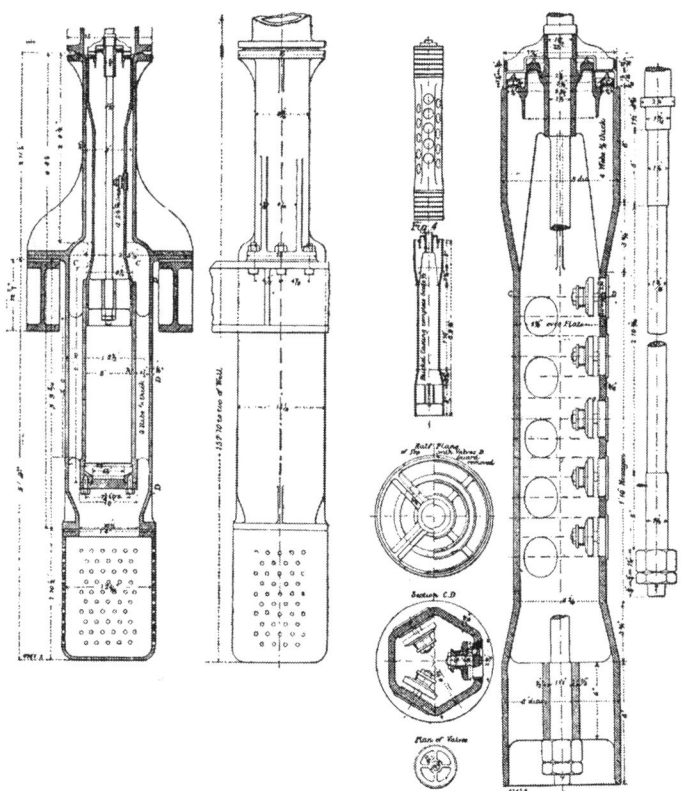

FIG. 34.—Ashley's Well Pump.

rods, but as one valve is closing whilst the other is opening, there would appear to be 'slip' or a backward flow of the water during that time unless the buckets are loose enough in the barrels to perform the function of a free-falling valve, and fall on the fixture on the rod during the turn of the stroke.

Mechanically-moved pump valves are employed in the *Riedler Pump* (fig. 36).

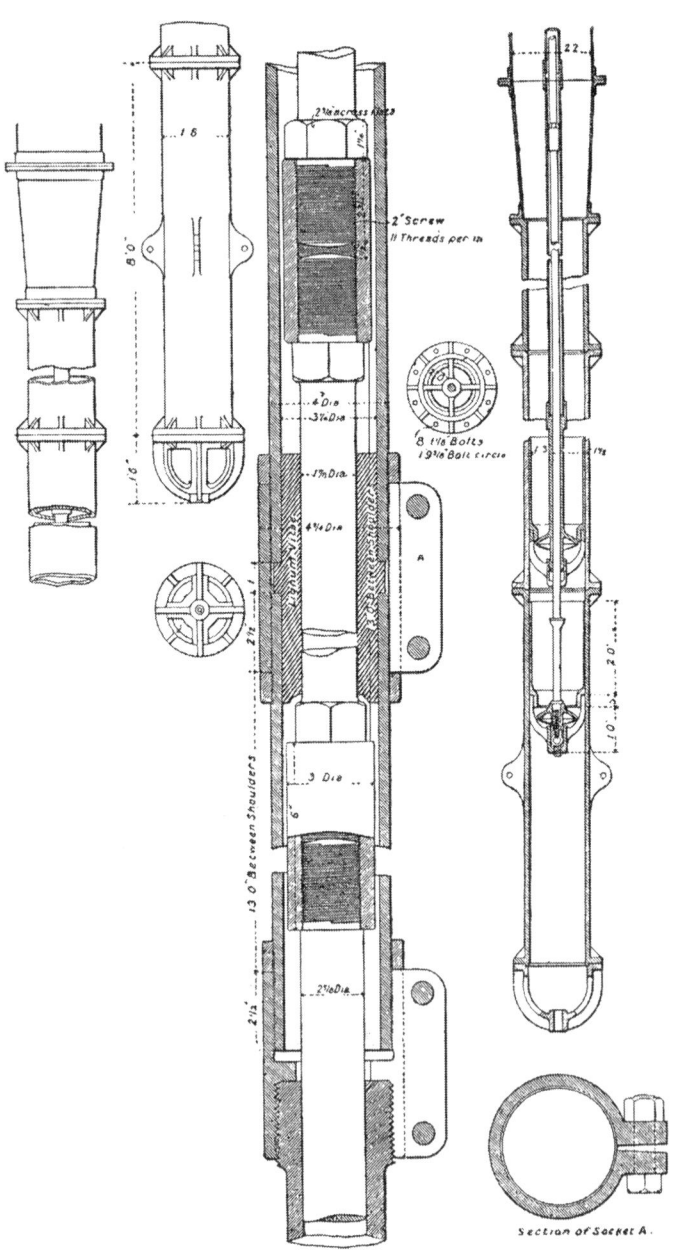

FIG. 35.—Restler's Double-acting Well Pump.

In this pump the valves are timed in opening and closing by a
mechanical connection to the engine by means of an eccentric on the
engine shaft, and the necessary attachments indicated in the illustration,
the object being to enable the pump to be worked at a greater speed than
if the valves had a free movement; but the speed at which a pump may be
worked with free falling valves is largely dependent on the proportion of
valve area to the pump displacement, and on the lift of the valve. Prof.
Riedler has lately designed a quick reciprocating pump, illustrated in

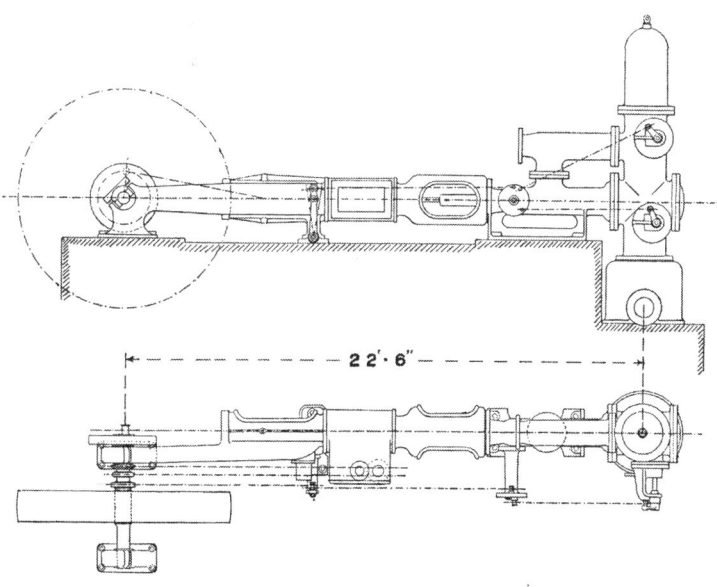

Fig. 36.—Riedler's Piston and Plunger Pump with mechanically-controlled valves.

figs. 37 and 38. In this pump the suction valve (clearly shown in the
sections, figs. 39, 40, and 41) is caused to close at the end of the stroke by
means of a hollow cylindrical fixture on the end of the pump plunger.
The pump is of the differential plunger variety, single-acting on the
suction side.

When the pump is completing its suction or outward stroke, the ring
carried by the plunger strikes the suction valve, carrying it to its seating.

Pumps with multiple valves giving a large area and small lift are worked at considerable speed without having the valves mechanically controlled.   *Non-rotative* pumps having a pause at the end of the stroke allow

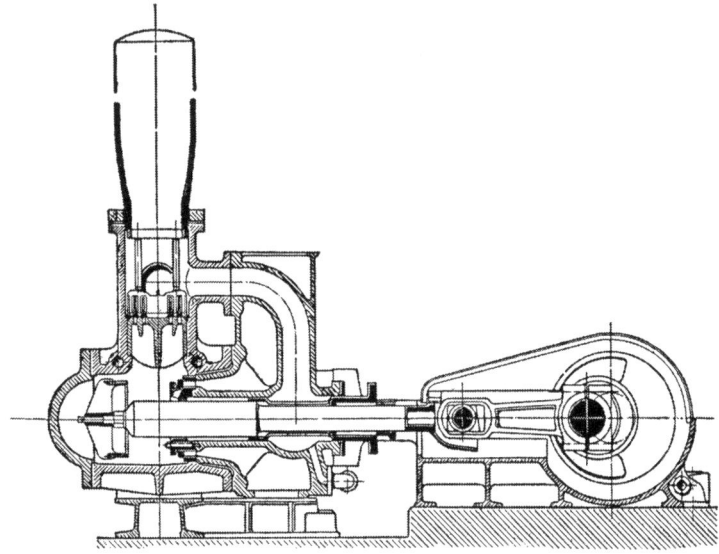

Fig. 37.—Riedler's Quick Reciprocating Pump—vertical section.

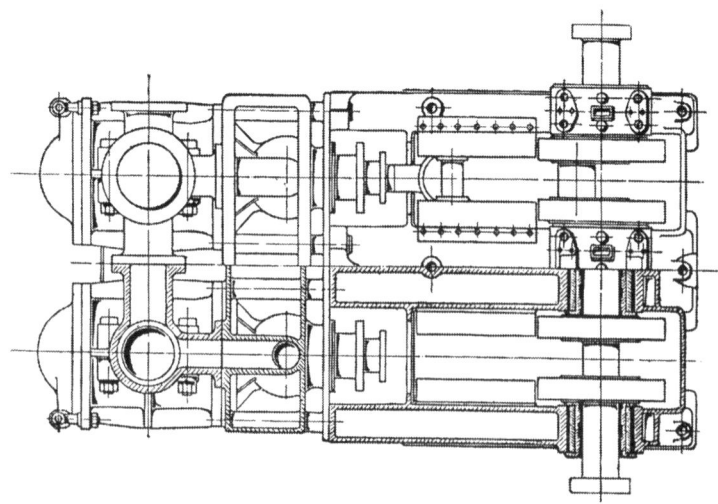

Fig. 38.—Riedler's Quick Reciprocating Pump—plan.

time for the valves to drop to their seats before the return stroke commences, and when such pumps have a long stroke, a considerable piston speed is obtained.

Valves.—The flap and butterfly valves already noticed must have a

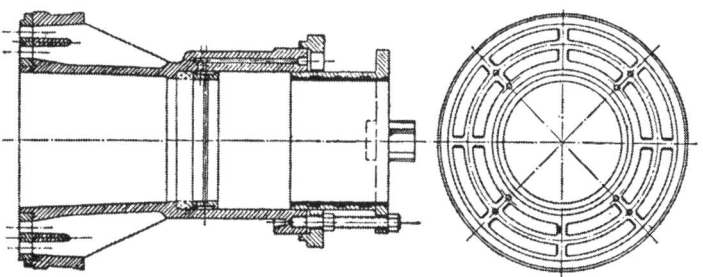

FIG. 39.—Suction Valve Seating of Riedler's Quick Reciprocating Pump.

considerable rise to give a large opening. They are only suitable for pumping against low lifts and at very moderate speeds.

Figs. 42 and 43 show various types of valves which do not need any further explanation.

In the early days of the application of the Cornish engine for the water

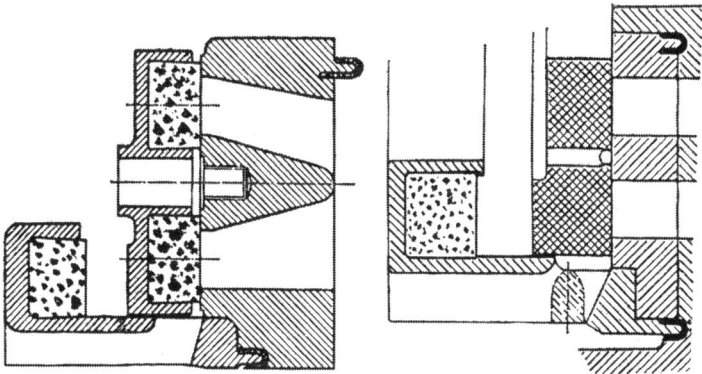

FIG. 40.—Suction Valve Riedler Pump—section. FIG. 41.—Suction Valve Riedler Pump—section.

supply of London, West, a Cornish engineer, invented what is known as the 'double beat valve,' and soon after Husband invented his 'four beat valve' (fig. 44).

The multiple valve may be made in various forms, such as those shown in figs. 45 and 46.

All valves which rise vertically on a spindle are improved in working by the application of a spring consisting of the india-rubber rings shown in several of the illustrations. The valve has a free lift before compressing

the rubber.   The spring forms a buffer to the valve in rising, and facilitates
the descent.   The 'beat' or contact of the valve with the seat is made

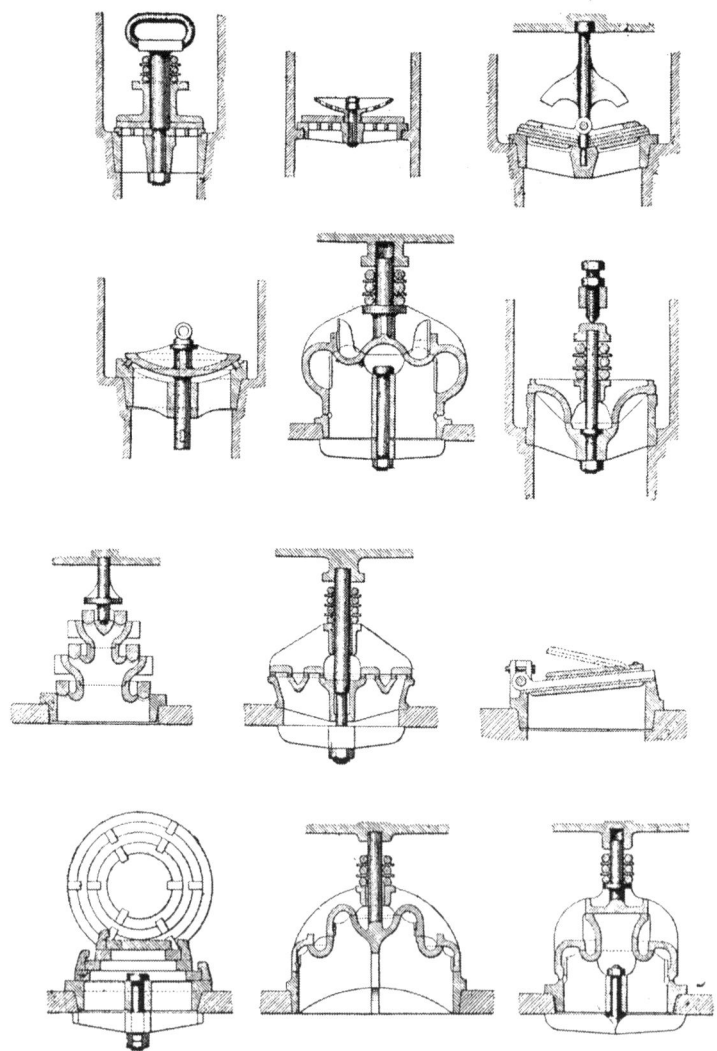

Fig. 42.—Types of Pump Valves.

watertight in a variety of ways; sometimes by fitting the surfaces of the
metal watertight, and sometimes by the introduction of an elastic material
either in the valve or seat.

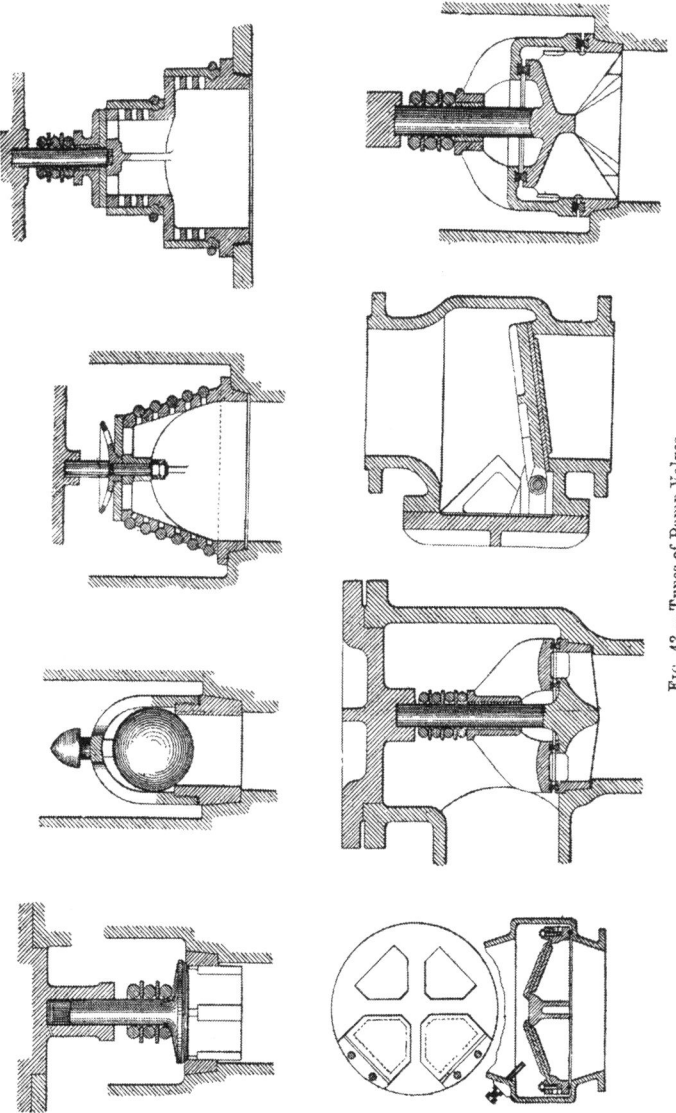

FIG. 43.—Types of Pump Valves.

Valves are made as in fig. 47, when the weight of the water on the

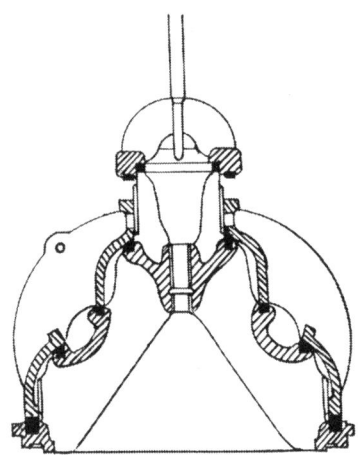

FIG. 44.—Husband's Four Beat Valve.

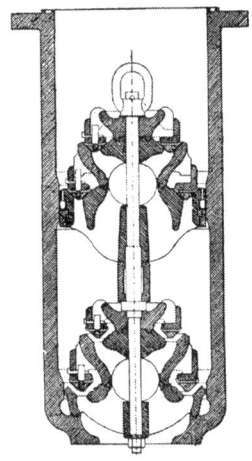

FIG. 45.—Suction and Delivery
Valves of multiple ring type.

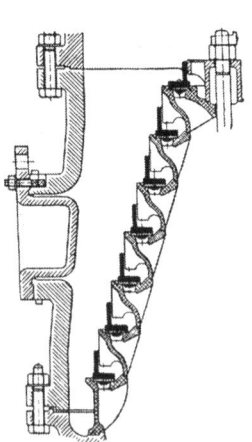

FIG. 46.—Half-section of Multiple Ring Valve.

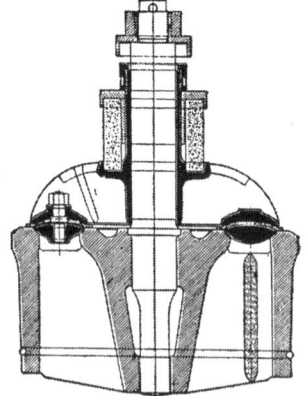

FIG. 47.—Double Beat Valve in which
the weight of the water is taken on
metallic surfaces, whilst the valve
is made watertight by means of
leather.

valve is sustained by metal beats, but the valve is made watertight by
means of a ring of leather above the beat.

India-rubber is largely used for valves, especially those in the air pumps of condensers (fig. 48), and the hat-band valve (figs. 49 and 50).

These latter are illustrations of valves and buckets of well or bore-hole pumps. In fig. 50 is shown the 'fishing' apparatus for drawing up the foot valve. It will be seen that the valve seating does not fit into a conical seating, such as that shown in valve seatings of many of the other valves, but seats itself and makes a watertight joint on a gutta-percha beat let

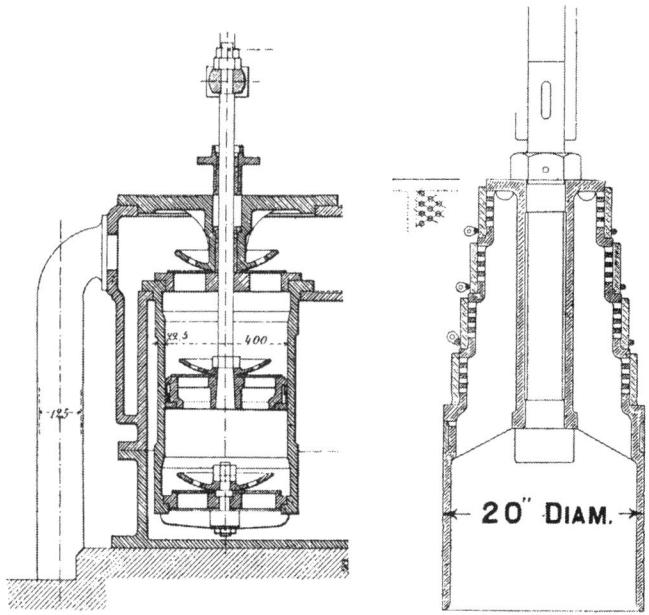

FIG. 48.—Section of Air Pump of Condenser, showing application of disc rubber valves.

FIG. 49.—Bucket fitted with hat-band rubber valves.

into the bottom of the seating. It is thus very readily drawn without putting much strain on the 'fishing tackle.'

Speed of Pumps.—The number of strokes per minute at which pumps may be driven depends on the type of pump, and the promptness with which the valves seat themselves before the commencement of the return stroke.

By means of multiple valves which give a large water-way with a small lift, quick reciprocation may be secured.

Bucket pumps do not admit of the application of multiple valves, and with such pumps a high bucket speed is best secured by means of a long stroke with a pause at the end to give time for the valves to seat themselves.

Rotative engines have necessarily a shorter stroke than non-rotative ones.

Pump Displacement Diagrams.—It is useful to know what is the variation in displacement per unit of time in pumps of different designs and combinations. The Cornish engine usually performs its 'indoor' or suction stroke in one-half the time occupied by the outdoor stroke; hence arises the desirability of giving more water-way through the suction than through the delivery valves.

Sometimes two suction valves have been provided on the suction and one only on the delivery side; see the engine shown in fig. 205, Chap. XII. The variations in the pause between the strokes have also to be taken into consideration, for it is by varying the pause that the rate of pumping is regulated.

Air vessels, when required on Cornish engine pumps, should therefore be very large. Non-rotative direct-acting engines are now usually made double-acting, as also are the pumps; so that air vessels on such pumps are not required to be so large as with the single-acting pump.

As an air vessel is not costly it may be made very large, especially as it is a very perfect means of regulating the flow and avoiding shocks, and more advantageous than any other method in use. Three-throw pumps have been employed with a view to avoiding having a large air vessel, but a large air vessel is important even with three-throw pumps, for it sometimes happens that one of the three pumps ceases to act because a valve has failed or for some other reason, and then the variation in displacement per unit of time becomes so great, that the shocks would be severe if a small air vessel only were present.

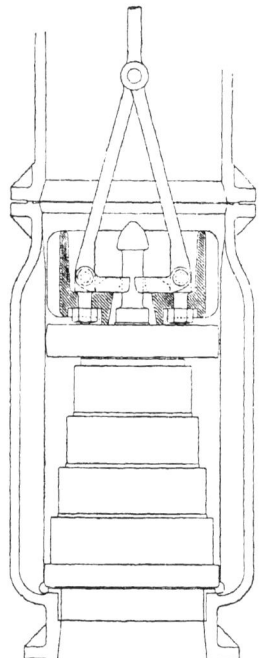

Fig. 50.—Foot Valve of Bucket Pump, with hat-band valves showing application of 'fishing tackle' for drawing the valve and seating.

When pumps are driven from rotative engines and cranks, the pump displacement per unit of time is easily determined. If we set out a curve the ordinates of which represent the displacement to a time base for a single-acting pump, then this curve will indicate the variation in displacement for a single pump delivery, and a combination of such curves may be made to give the variation for any combination of pumps. Fig. 51 is for a pair of double-acting pumps with cranks at right angles; fig. 52 is for two single-acting pumps with cranks at right angles;

fig. 53, single-acting three-throw pumps; fig. 54, double-acting three-throw pumps; and fig. 55, double-acting three-throw pumps with one of the

FIG. 51.—Displacement Diagram of a pair of Double-acting Pumps with cranks at right angles.

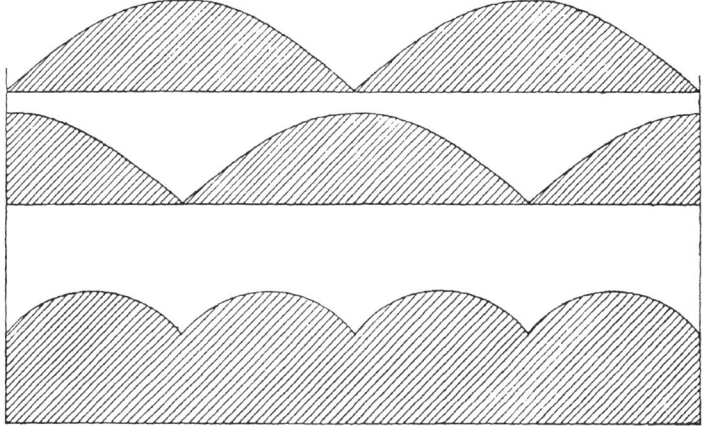

This lower diagram shows the variation in displacement per revolution.

pumps idle. Air vessels require to be proportioned with due regard to the water pressure and the variation in pump displacement. It is very

FIG. 52.—Displacement Diagram of Single-acting Pumps with cranks at right angles.

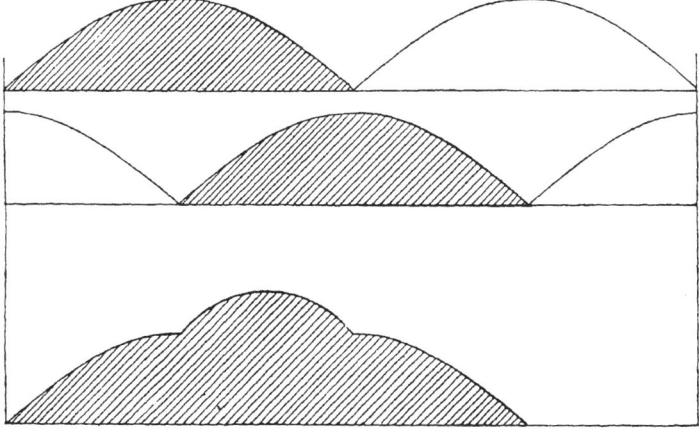

This lower diagram shows the variation in displacement per revolution.

usual to make the air vessel twelve to eighteen times the capacity of the pump. Where the air vessel is not readily charged with air, a large

Fig. 53.—Displacement Diagram of a Three-throw Single-acting Pump.

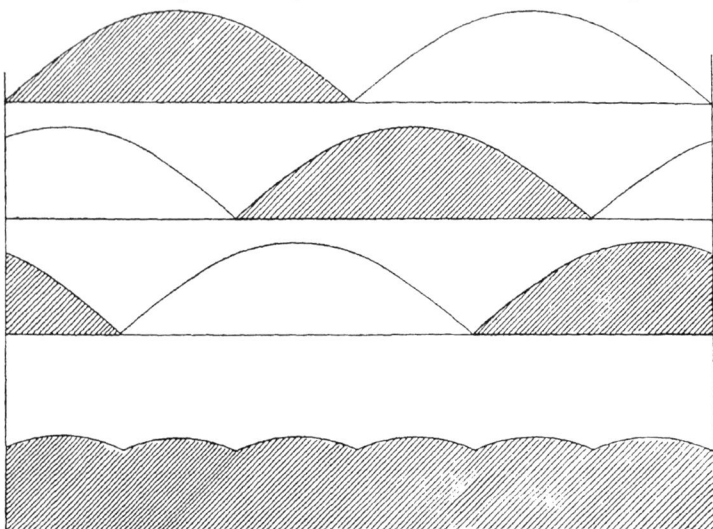

This lower diagram shows the variation in displacement per revolution.

Fig. 54.—Displacement Diagram of a Three-throw Double-acting Pump.

This lower diagram shows the variation in displacement per revolution.

capacity should be provided. There are several ways of charging air vessels. A small air valve may be provided on the pump with a screw-down adjustment for admitting air with the water, but it is not to be generally recommended.

The device now usually adopted is the Wippermann air charger (fig. 56). This is really a pistonless air pump actuated by water pressure from the pump.

The vessel A is attached to the pump barrel by means of the pipe B, which is kept open when the charger is at work.

During the suction stroke air is drawn into the vessel A, and during

FIG. 55.—Displacement Diagram of a Three-throw Double-acting Pump with one pump out of action.

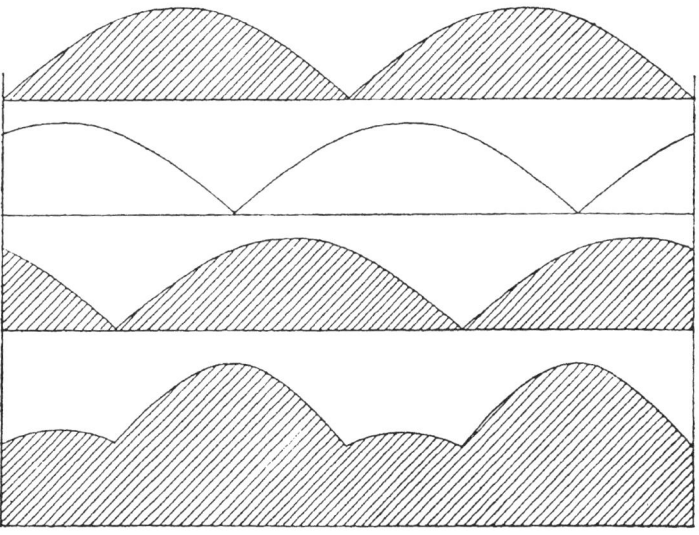

This lower diagram shows the variation in displacement per revolution.

the delivery stroke that air is forced into the air vessel C through the small delivery pipe, small suction and delivery valves being provided in the top of the air-charging vessel.

Another form of charger is that of a small air pump worked by a hydraulic engine with pressure from the pumping main.

Another method for charging by hand is to have a charging vessel provided with pipes and valves attached to the air vessel, the pipes and valves being so arranged that the vessel may be first filled with air by opening it to the atmosphere; then communication with the atmosphere is shut off, and communication opened to the air vessel. The operation may be repeated as often as necessary.

E

In large pumping stations where there are several air vessels, it is convenient to have an air-compressor worked by steam to enable the air vessels to be fully charged before or whilst the pumping-engines are being put to work.

Air vessels require to be charged continuously or very frequently, as the water absorbs air, and there may be leakages to be compensated for.

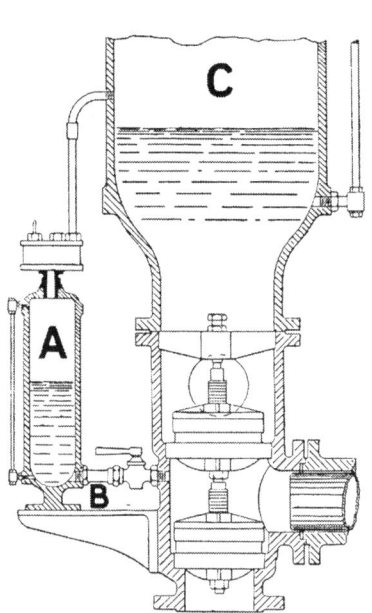

FIG. 56.—Wippermann's Air Charger for air vessels of pumping-engines.

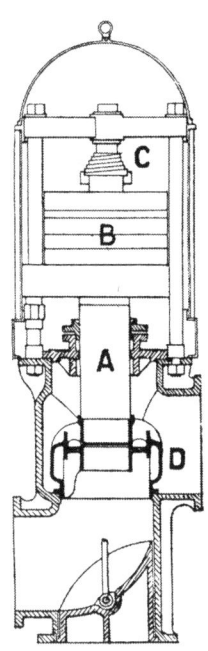

FIG. 57.—Balance Valve for maintaining a constant pressure in pumping mains.

Fig. 57 represents a valve used to preserve a constant pressure in the main against the pumping-engine, when a uniform pressure is not secured by use of a service reservoir.

Referring to the figure, D is an equilibrium valve attached to the plunger A. The plunger is loaded with weights B to the pressure it is required to maintain. C is a buffer spring to limit the rise of the weight. By this means the pressure in the main at D may vary, but on the other side of the valve it is kept constant.

PLATE II.]

Compensating-lever Well Engine, South Staffordshire Waterworks.—Davey System.

CHAPTER IV.

WE must now consider in detail the principles employed in the Cornish engine, because this will explain the leading characteristics in non-rotative pumping-engines generally.

Each stroke of the engine, it will be observed, is a distinct and separate operation; the engine starts from a state of rest and returns to it. The pump rod is put in motion and comes to rest. The pump rod is of sufficient weight to overcome the resistance of the pump, so that the steam has only to lift the pump rod, and thereby to store in it sufficient energy for doing the work on the return stroke. We have already pointed out in the history of the Cornish engine that the introduction of the plunger pump so altered the condition under which the Cornish engine worked that it at once became possible to work the engine with a far greater degree of expansion than had been possible as long as it remained in the stage Watt left it, which was only suited for working bucket pumps. With the bucket pump the water was lifted on the steam stroke; with the plunger pump the pump rod was lifted. There was, therefore, with the plunger pump a greater mass to be put in motion, which might have imparted to it a much higher velocity than would be safe for the bucket of a bucket pump.

The energy of motion is expressed by the formula $\dfrac{W V^2}{2 g}$, in which W is the weight of the rod and its attachments, V the velocity it attains in feet per second, and $2 g = 64 \cdot 4$.

For the sake of brevity the formula may be thus approximately written $\dfrac{W V^2}{64}$. The fact that the pump rod possessed weight, and was put in motion and brought to rest each stroke, made it possible to use steam expansively on the piston.

Figs. 58 and 59 represent steam indicator diagrams from Cornish engines. The line A B is that of the average pressure throughout the stroke; the shaded portion from A to C above the line represents the energy expended in putting the pump rods in motion, and the shaded portion below the line from C to B represents the energy recovered, whilst the pump rods are decreasing in velocity and coming to rest. If there had been no mass to put in motion, but merely a resistance to be overcome, the indicator diagram would have been a rectangle, and no expansion would have been possible.

The reciprocating pump rod performs the function of a fly-wheel, storing energy whilst its motion is being accelerated, and giving out the stored energy whilst the velocity is being retarded.

The point C, then, is the point of maximum velocity V. By increasing either V or W the degree of expansion may be increased, but as in the Cornish engine W is constant, any increase in the degree of expansion employed increases the value of V. As the energy varies with V^2, an increase in the velocity with which the pump rods are lifted enables an increase in the degree of expansion to be employed. The action of the pump valves limits the number of strokes which the engine can make in one minute, but the stroke may be long or short. In the old Cornish engine the stroke of

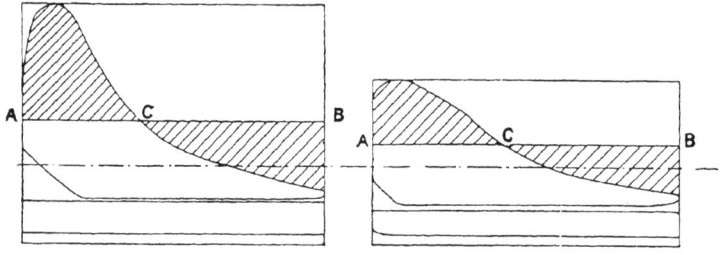

Indicator diagrams from Cornish Engines.

FIG. 58.—Mining Engine.   FIG. 59.—Water Works Engine.

the pump was usually made less than that of the engine. In the new type, or compound engine, the stroke of the pump is made much longer than that of the engine (see fig. 80, Chap. V.).

This lengthening of the stroke increases the value of V, and therefore enables, even in a single cylinder, a greater expansion to be safely employed, but in the double cylinder it is of still greater value.

For the purpose of expansive working, the Cornish engine is usually made to traverse its steam stroke in one-half the time occupied by the water stroke ; the maximum speed of the piston during the steam stroke is often as much as 600 ft. per minute ; for that reason the suction valves and pipes should be large ; sometimes two suction and one delivery valve are employed ; this gives double the water-way on the suction side of the pump.

The water stroke performed by the descent of the pump rod is made slowly, but with increasing velocity ; the energy does not, however, become very great because of the slow velocity, and the rod is brought to rest by the closing of the equilibrium valve. The steam above the piston is then compressed into the clearance space, thus reducing the loss which would otherwise occur.

By the use of a velocity indicator of special construction we have been able to take velocity diagrams from pumping-engines, showing the variation in velocity during the stroke, the length of the pause, etc.

Fig. 60 was taken from a Cornish engine, and at the same time the rise and fall of the pump valves were indicated as shown on the diagram, fig. 60.

When the Cornish engine is applied for town water supply from a

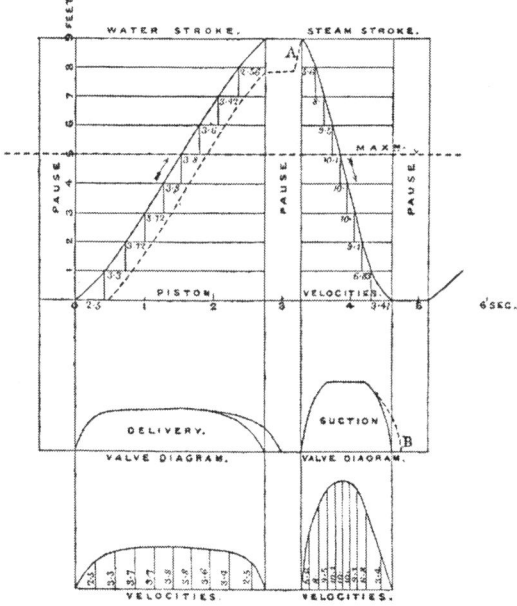

Cornish Engine, 60" Cylinder.

FIG. 60.—Velocity Diagram.

reservoir, the load on the plunger is simply that required to overcome the resistance of the pump, but in mines the load is often greater, owing to the great length of pump rods, part of the weight of which is taken up by balance bobs as illustrated in the Cornish pit-work, fig. 121, Chap. VII.

The value of W being thus greatly increased, the greater degree of expansion can be employed; in some cases we have known the pump rods to weigh over 200 tons, whilst the load on the piston was only 45 tons. This is the reason why the Cornish engine in the Cornish mines has been able to give a higher duty than Cornish engines generally; this is very easily seen on inspecting the Cornish engine diagrams (figs. 58 and 59).

The left-hand diagram (fig. 58) was taken from a mining engine in which W was very much greater than the water load on the engine; the right-hand diagram (fig. 59) was taken from a waterworks engine in which W was very little more than the water load on the piston.

Although increasing the weight enables greater expansion to be employed, greater expansion means greater strains on the rods and connections, and it must be observed that, without increasing the weight to an impracticable degree, the Cornish engine is not capable of using effectively the expansive force of steam of a greater pressure than about 40 lbs. per square inch.

To be able to secure greater economy by using higher pressure steam with the proper degree of expansion, it becomes necessary to work with more than one cylinder. In the early days of the Cornish engine com-

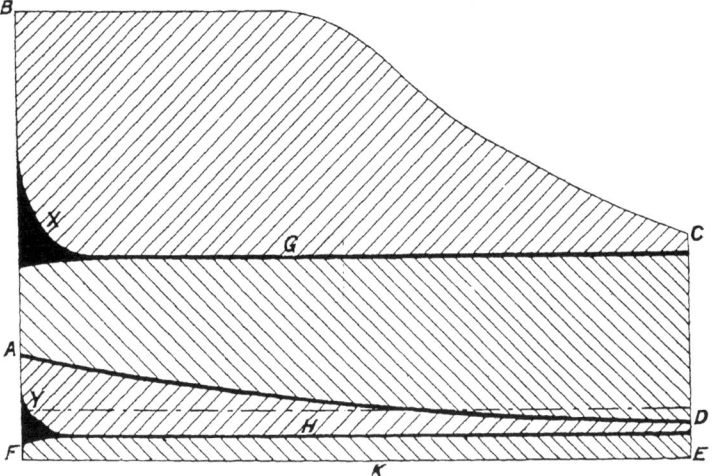

Fig. 61.—Diagram illustrating distribution of steam in a Compound Cornish Engine.

pound engines were used, but no better result was secured than that obtained with single cylinder engines, for the simple reason that the boiler pressure in use was limited to about 45 lbs. per square inch. Now that boilers can safely carry pressures exceeding 150 lbs. per square inch, the Cornish engine may be compounded with great advantage. By compounding, steam can be used at 150 lbs. initial pressure and more, expanded down to 8 or 10 lbs. terminal pressure with engines having pump rods of the usual weight; this is clearly seen in the diagrams (figs. 61 and 62), which represent the distribution of steam in the Basset engine illustrated in fig. 80, Chap. V.

Let steam be cut off at about half stroke in the small cylinder, as in fig. 61. During the equilibrium stroke, the pressure is indicated along the line G, and the cushioning takes place at X; at the same time the equilibrium steam pressure for the large cylinder is indicated along the line H, and the cushioning at Y, the condenser pressure being indicated at K. The work done in both cylinders may be taken to be the for area

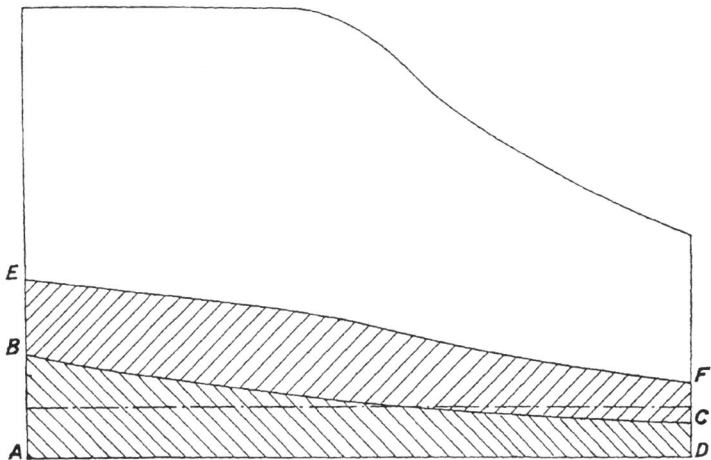

FIG. 62.—Diagram showing combined power diagrams of a Compound Cornish Engine.

Of the diagrams, *minus* the small losses caused by the passage of the steam during the equilibrium strokes, and between the two cylinders. For the sake of illustration the diagram may be taken to represent

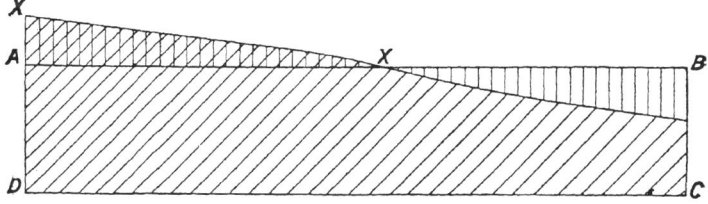

FIG. 63.—Diagram showing variation of force or strain during the stroke for a Compound Cornish Engine.

the total work; the area A, B, C, D, A that done in the small; and the area F, A, D, E, F that done in the large cylinder. In order to refer all to the low-pressure cylinder, the ordinates of the upper figure must be divided by the ratio of the capacities of the two cylinders. In this case the capacities are 4·6 to 1.

In fig. 62 the work done in the small cylinder is represented by B, E, F, C, B, and that in the large cylinder by A, B, C, D, A.

The total effect referred to the large cylinder is A, E, F, D, A.

In fig. 63 the line A B is drawn through the point of mean pressure, so that the rectangle A, B, C, D, A represents the total work done by the engine on the steam stroke and the area A, X, X, A, the work represented by the *vis viva* acquired by the load during the former part of the stroke and given out again during the latter part. On comparing this with the single Cornish engine diagram (fig. 58, page 68) it will be seen that the compound engine with more than three times the steam pressure, greatly increased expansion, and greater mean pressure does not require such a heavy mass or so great a velocity during the steam stroke for its effective working. It will also be seen that the strains on the engine are reduced. The initial strain is only 1·3 times the mean, whilst in the ordinary Cornish engine it is twice the mean. It will also be observed that the work done by the small piston is applied to the pump end of the beam, so that less than one-half of the work of the engine comes on the beam centre.*

The variation of strain in the engine itself is not so great in this engine as it is in the ordinary Cornish engine, that is to say, the maximum strain does not so greatly exceed the mean resistance. The following table contains a comparison of the initial, terminal, and average pressures, maximum piston speeds, and proportion of weight of mass to water load in several practical examples which were tested some years ago, and to it has been added an example from the compound Cornish engine.

| TYPE OF ENGINE. | STEAM PRESSURES. | | | | | FORCES. | PISTON VELOCITIES. | | |
	Initial, Absolute.	Terminal, Absolute.	Average.	Initial ÷ Terminal.	Average ÷ Terminal.	Initial ÷ Mean.	Maximum Velocity per Minute.	Mean Plunger Speed per Minute.	Mass ÷ Water Load.
	lbs.	lbs.	lbs.	lbs.	lbs.		ft.	ft.	
Cornish,	45	10	19	4·5	1·9	2·26	500	80	3 about
,,	31	10	16·1	3·1	1·6	1·8	600	...	1·7
,,	. 25	9	12·2	2·77	1·35	1·72	600	...	1·7
Compound Differential,	80	10	24·1	10·6	2·4	1·37	220	150	2·0
,, ,,	43·7	7	12·75	6·24	1·8	1·4	228	168	0·66
,, ,,	85	11·3	20·25	7·5	1·78	1·3	...	...	1·1
,, ,,	25	8·5	14	...	1·6	...	210	144	1·2
Cornish,	...	...	13 about	...	...	...	600	100	1·7
							570	112	1·7
,, Compound Cornish,	150	...	35	...	...	1·3	...	...	1·7 about

To be able to utilise higher pressures and greater ranges of expansion in the Cornish engine, it is absolutely necessary (as we have seen) to use more than one cylinder. Referring to the diagram (fig. 58) it will be noticed that if we put E for the energy of the mass, A for the area of the

* See illustration of the engine, fig. 80, Chapter V.

cylinder, L for the length from A to C, and P for the mean pressure of the top shaded portion of the diagram, then $E = A \times P \times L$,

$$W = \frac{E\ 64}{V^2}$$

$$V = \sqrt{\frac{E\ 64}{W}}$$

from which we can get the necessary quantities in further consideration of the subject of non-rotative pumping-engines.

If we increase W in proportion to the water load on the piston, then higher pressure and greater expansion can be employed without increasing V, and by compounding we can carry the advantage immensely further (as already pointed out). This leads to the consideration of the balance

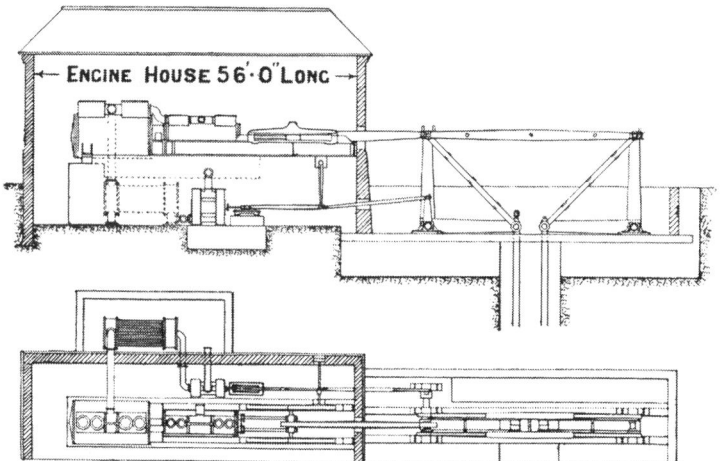

Fig. 64.—Differential Engine with double quadrants and double system of plunger pumps.

system of pump rods as compared with the Cornish. This system is illustrated by fig. 64, in which it will be seen that there are two quadrants or angle bobs for giving motion to the pump rods ; the two quadrants are coupled together in such a manner that one pump rod balances the other, one making the up while the other makes the down stroke. The weight to be put in motion is thus double that of the Cornish engine, with the same water load, whilst a double-acting engine is employed to work the pumps. With a compound or triple engine this system utilises the expansive force of very high-pressure steam. A very important practical advantage arises from the fact that one pump rod balances the other, so that at the end of the stroke and during the pause, when the engine is in equilibrium, no motion can take place till steam is admitted to the engine, whereas, with the Cornish engine, the weight of the rod is partly unsupported at the end of the stroke, if the water has failed to follow close to the plunger so as to completely fill the pump. In that case the plunger falls back on the water. With the double cylinder Cornish engine the unsupported

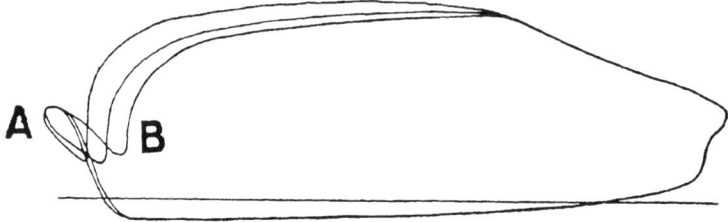

FIG. 65.—Differential Pumping-Engine—diagram taken during sinking operations with pumps "working on air."

weight becomes less than that with a single cylinder. With the double rod system, should the water not have followed up the plunger, a shock can only occur by the rods being put in motion by the admission of steam,

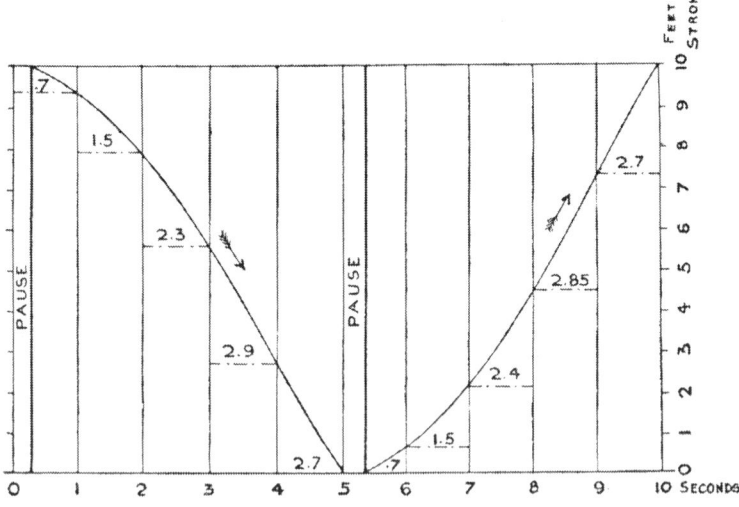

FIG. 66.—Velocity Diagram, South Durham Colliery—differential compound pumping-engine cylinders 45 in. and 72 in. diam. × 10 feet stroke.

and it is possible by the use of the differential gear to so control the admission of steam to the engine that it is throttled during the time that the engine has no pump resistance ; by this means only sufficient steam is admitted to the engine to overcome the inertia and frictional resistance up to the point where the plunger encounters the water, thereby minimising the shock. This action is illustrated by an actual indicator diagram (fig. 65), where it will be seen that the full pressure of steam does not come on the piston until it has reached the point B, where the plunger meets the water; in moving from A to B only sufficient steam has been

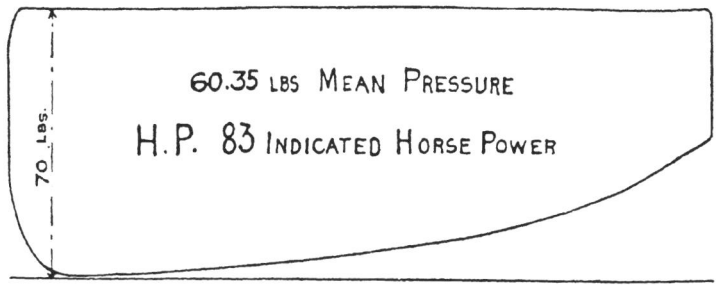

FIG. 67.—Indicator Diagram.
South Durham Colliery—Differential Compound Pumping-engine.   Ratio of Cylinders 6 to 1.

admitted to move the engine itself, and it has been found in practice that on this system it is quite possible to work heavy lifts on air without

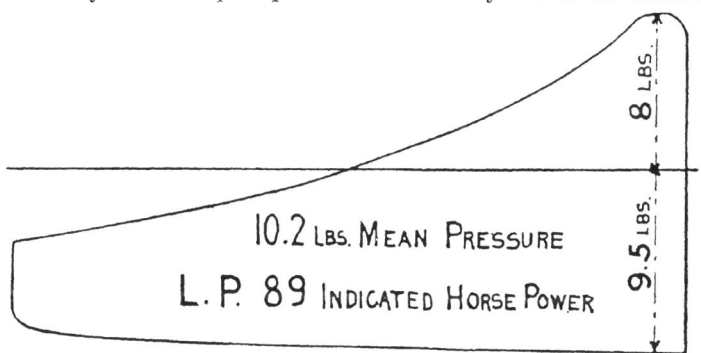

FIG. 68.—Indicator Diagram.
H.P. cylinder 22 in. diam. × 6 feet stroke.   L.P. cylinder 54 in. diam. × 6 feet stroke.   10½ strokes per minute.

producing any serious shock.   A steam piston velocity diagram from an engine of this type is given in fig. 66.

Pumping-engines such as we are considering are subject to a sudden

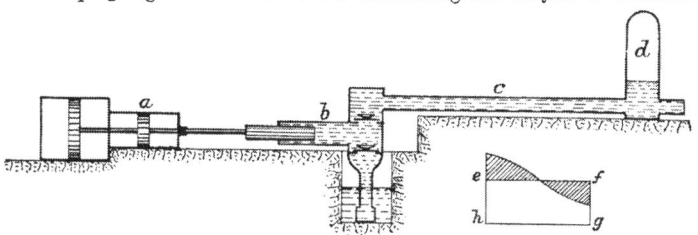

FIG. 69.—Diagram illustrating effect of inertia in water columns.

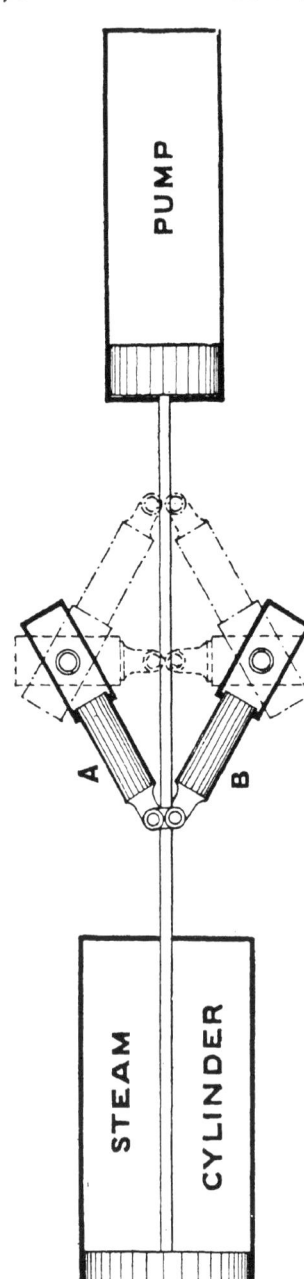

FIG. 70.—Diagram of Worthington Compensating Device.

loss of load either from the breakage of pumps or rods, failure of water supply to the pumps, or defective action of the valves; when any of these things occur, it is desirable that the engine should be brought to rest as quickly as possible; the continuance of working with a broken rod in the shaft, or anything wrong with the pumps, only incurs the risk of additional breakage. Non-rotative engines are more easily brought to rest in the case of accidents than rotative engines.

In the early days of the application of steam pumps to mining work underground, the author made use of the inertia of the water column to secure increased expansion. The indicator diagrams (figs. 67 and 68) were taken from a compound steam pump at the South Durham Colliery. The pump was placed at the bottom of the mine; it forced the water to the surface through a column of pipes which had no air vessel. The water in the column was thus put in motion at the beginning of the stroke, storing up energy, which energy was expended during the latter part of the stroke.

The resistance to the engine by the water column was therefore greater at the beginning and less towards the end of the stroke, thus enabling a considerable degree of expansion to be secured.

The function of inertia in a water column may be illustrated by fig. 69.

Let a be the steam engine, b a plunger pump, c the delivery pipe from the pump. In the case of an underground engine the pipe c would be the water column reaching to the surface, but it might be a horizontal pipe terminating at an air vessel d.

Let e, f, g, h, e represent the pump resistance where the air vessel d was part of the pump. Then when the air vessel d is removed to a distance and connected to the pump by the pipe c, the inertia of the water in the pipe c would cause the pump-resistance diagram to be altered in shape. The *plus* resistance at the beginning and the *minus* resistance at the end of the stroke are represented by the shaded portions of the diagram.

This function of inertia of water column is of importance in the application of non-rotative engines to well pumps, especially because it enables

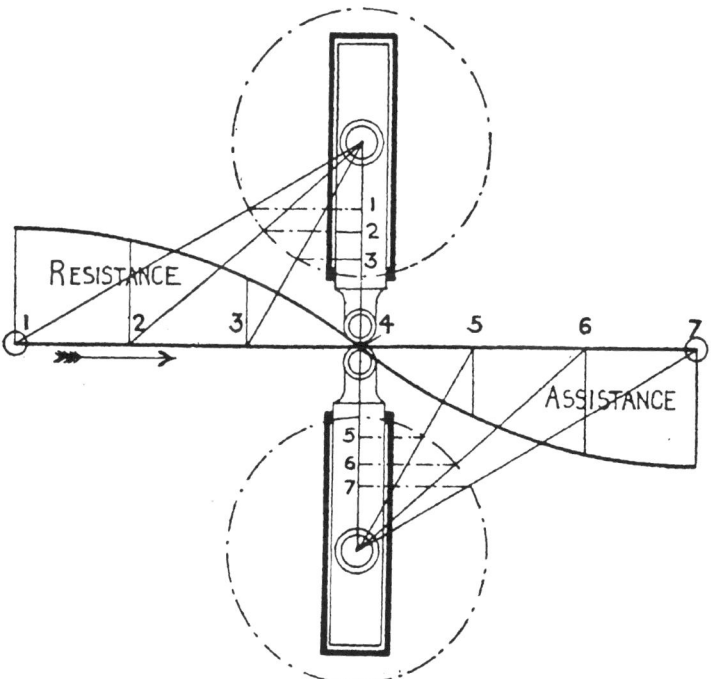

Fig. 71.—Worthington Compensating Device. Curve of Effect.

the engine-power and pump-resistance diagrams to be equated when the mechanical compensating device does not in itself do it, which is the case in the lever device hereafter described.

Waterworks Engines.—Cornish and beam rotative engines have been largely used, but of late years quite new types have been adopted. Self-contained compound and triple direct-acting engines, both rotative and non-rotative, are now more in favour.

To enable the non-rotative engine to work with the necessary expansion for economical results, compensating devices have been introduced, where the inertia of the moving parts are not sufficient for the purpose.

The Worthington device (fig. 70) consists of two single-acting hydraulic plungers, A and B, working in oscillating cylinders, the plungers being attached to the piston rod of the engine. The required pressure is main-

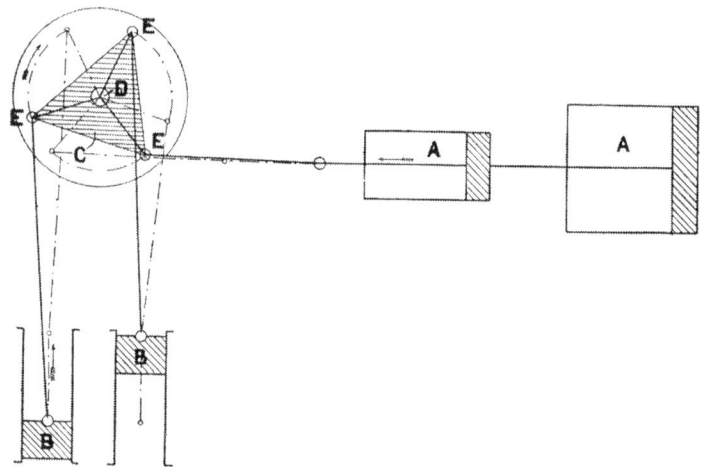

FIG. 72.—Davey Compensating Device applied to well pumps.

tained in the hydraulic cylinders by means of an accumulator kept charged by a small pump attached to the engine.

The effect of this arrangement is that an artificial resistance is put on the engine during the first half, and a corresponding assistance during the

FIG. 74.   FIG. 73.

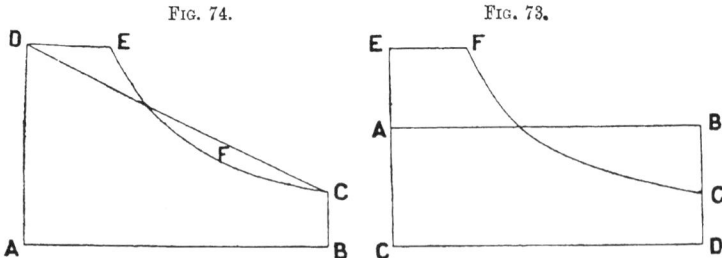

FIGS. 73 and 74.—Diagrams illustrating function of Davey Compensating Device.

second half of the stroke, the curve of effect being similar to that in fig. 71.

The Davey device is that of attaching the engine to the pump by means of a triangular lever in such a way that the pump resistance becomes less as the stroke proceeds. In fig. 72 A A represent the steam cylinders of the engine, D the triangular lever, E, E, E points of attachment of engine and pumps, and B B the pumps. In this illustration the work is done on the up stroke of the pumps. It will be seen that the pump B, which is finishing its up stroke, is acted on by a greater leverage effect than was exerted at the beginning of the stroke. In other words, the stroke is made with a decreasing resistance.   In fig. 73 let C, A, B, D, C be the pump resistance and C, E, F, G, D, C the steam diagram.   The leverage device makes the pump resistance more nearly correspond with A, D, C, B, A in fig. 74, the inertia of the mass of the moving parts of the engine and pumps being sufficient to equate the two diagrams, the area E representing the inertia of the mass at the beginning and the area F that towards the end of the stroke.   This engine has been applied largely to pumping from wells and bore-holes.

Examples of such engines will be found in Chap. XII., figs. 212, 219, 223, and Plate II.

CHAPTER V.

The Cornish Engine.—This engine is fully illustrated in fig. 75.

It consists of a single cylinder A having a piston communicating its motion through an overhead beam B to a single pump rod C. The engine beam B is carried in bearings on a thick wall, forming one of the walls of the engine house. The outer end of the beam is directly over the shaft and is connected direct to the pump rod. The cylinder is bolted down to a massive loading L of masonry, which must be made heavy enough to resist the total initial force of the steam on the piston.

Attached to the pump rod by means of 'set-offs' are the plungers of the pumps, the pumps and pump rods, etc., being illustrated in fig. 121, Chap. VII.

The rods are made heavy enough to overcome the resistance of the pumps.

The engine, which is single-acting, is employed to lift the rods, and the rods are allowed to fall by their own weight, lifting the water. To secure sufficient strength it is often necessary to make the rods much heavier than is necessary to overcome the resistance of the pumps. In such cases beams or balance bobs are attached to the rods either above or below ground, having counterweights to take up the surplus weight (see pit-work, fig. 121, Chap. VII.). The distribution of steam in the engine will be understood from the section of the valves and nozzles given in fig. 76, in which A is the steam valve, B the equilibrium valve, and C the eduction or exhaust valve. D is a pipe connecting the two ends of the cylinder, commonly known as the 'equilibrium pipe.' The working of the engine is thus described. Steam is admitted above the piston through the steam valve A (fig. 76), the equilibrium valve B remaining closed, and the exhaust valve C open. The valve A remains open during the time the piston makes about one-third of its stroke; it is then closed by the action of the valve gear, and the remainder of the stroke is completed partly by the decreasing pressure of the expanding steam and partly by the momentum of the pump rod and attachments. At the termination of the stroke the

eduction valve C is closed and also the injection valve shown at N in
fig. 75.

The pump-rod having been raised to the full extent of its stroke is

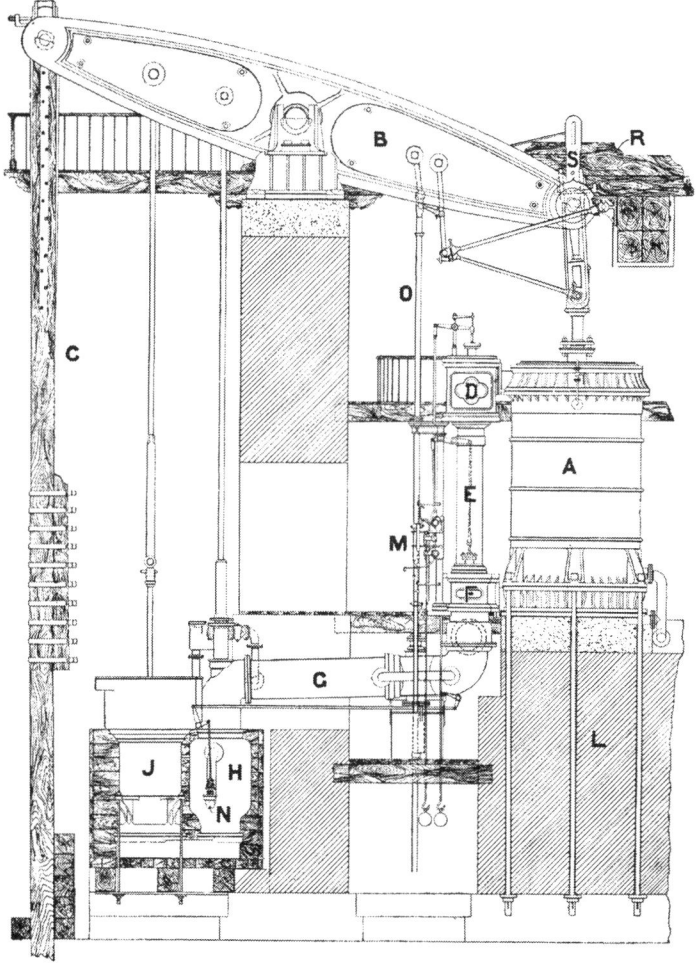

FIG. 75.—Cornish Engine, as applied to a mining shaft.

partly supported by the water in the pumps under the plungers, and
partly by the terminal pressure of the steam above the piston. The
equilibrium valve B is then opened, and the steam above the piston is
allowed to pass down the equilibrium pipe D to the under side of the

piston ; the pump-rods being free to fall by their own weight, descend, and in doing so lift the water in the pumps. The rods falling under the influence of gravity have an accelerating motion, but they are brought to rest at the completion of the stroke by the closing of the equilibrium valve B ; that valve having been closed a little before the end of the stroke, the momentum of the moving weight is then expended in cushioning the steam confined above the piston. The piston is now brought to rest, and the

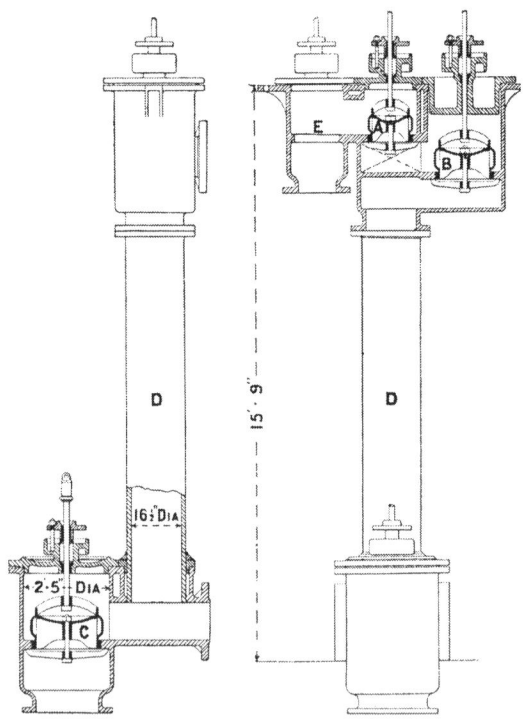

FIG. 76.—Cornish Engine.   Steam nozzles and valves.

next stroke is performed by the opening of the steam valve A and the eduction or exhaust valve C. The eduction valve C is made to open a little before the steam valve, and at the same time the injection valve (N in fig. 75) of the condenser is opened.

It will be seen that when steam is admitted above the piston, the steam from below is exhausted into the condenser. Referring to fig. 75, which is a general illustration of the engine, D is the steam and equilibrium valve box, E the equilibrium pipe, F the exhaust valve box, G the exhaust pipe to the condenser, H the condenser, J the air pump, and N the injection valve.

The air pump and condenser are placed in a tank of water from which the injection water is taken through the valve N. This valve is opened at the same time as the exhaust valve by the action of the valve gear. The valve gear itself is shown at M, and is fully described and illustrated in Chap. X., figs. 161 and 162.

The steam pressure on the piston at the termination of the steam stroke is not sufficient to support the weight of the pump rods.

If, therefore, the water in the pump has not closely followed the plunger, the rods will be unsupported at the completion of the stroke. When this happens, the rods fall back, causing the plunger to strike the water, and producing a shock. On the other hand, during the steam stroke the loss of resistance arising from the breaking of a rod or other cause may result in the piston making its stroke with undue velocity. To provide against contingencies of this kind, banging beams are provided in and over the shaft with catch-pieces on the rods to prevent the piston striking the cover on the outdoor

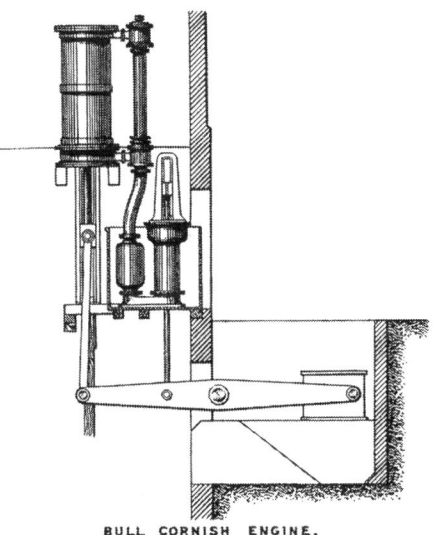

BULL CORNISH ENGINE.

Fig. 77.

stroke, and spring beams and catch-pieces are provided at the engine end of the beam to limit the indoor stroke.

The spring beams are shown at R (fig. 75), and the catch-pieces on the engine beam at S. Devices are employed such as the differential valve gear (see *Valve Gear*) to cause the valves of the engine to become closed, when, owing to loss of resistance, the engine has exceeded its normal speed.

Sometimes the Cornish engine is made without an overhead beam, the cylinder being placed over the pump with the piston rod attached directly to the pump rod. Such engines are known as Bull engines.

An illustration of such an engine is given in fig. 77.

Cornish Engine Details.—A section of the cylinder is given in fig. 78.

In the stuffing-box between the packing is placed what is termed a *lanthorn brass or lush* A, consisting of a perforated and movable cylinder of brass fitting around the piston rod with the openings through the side.

Opposite this lanthorn bush and in the side of the stuffing-box is fitted a small pipe connected to the main steam pipe.

The object of this arrangement is to prevent air leaking into the cylinder around the piston rod and vitiating the vacuum. With steam pressure in the middle of the packing, steam might leak into the cylinder, but air could not. This was of considerable importance in the Cornish engine, because the mean pressure on the piston was not more than 12 or 13 lbs. per square inch, and a small back pressure in the condenser caused by air leaks, represented a considerable percentage of loss. The lanthorn brass is very usefully and frequently employed in the stuffing-boxes of air-pumps of modern engines, the pipe being taken to the hot-well instead of to the steam pipe, and water being used instead of steam to prevent air leaking into the air-pump.

Air-pump and Condenser. — A section of the Cornish engine air-pump and condenser is given in fig. 79.

The air-pump A, and bucket B, are shown in section. The bucket is provided with a valve D, usually of india-rubber working on a brass grid C. A head valve or swimming cover F is placed in the hot-well closing on a wooden beat let into the top of the air-pump. The bucket rod E passes through a stuffing-box.

G is the condenser, H the exhaust pipe, and J the inlet for injection water.

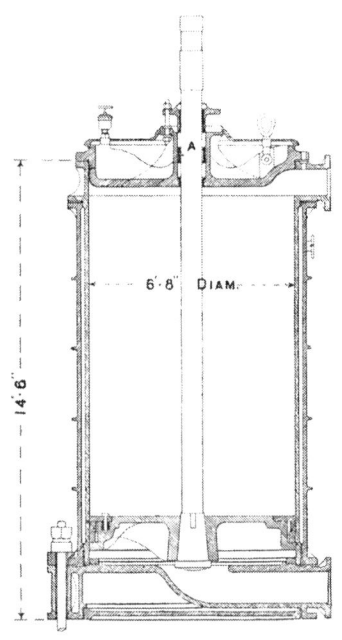

FIG. 78.—Cornish Engine Steam Cylinder.

Valve Gear.—The Cornish form of valve gear, and the application of the differential gear to the Cornish engine, are fully described in Chap. X.

Duty of Cornish Engines.—In Chap. I. we have noticed the progress of economy effected in the Cornish engine. The maximum economy was obtained by using the greatest boiler pressure and the highest rate of expansion which the principle of the engine would permit, the limits being about 40 lbs. boiler pressure and 4 to 5 expansions. The average duty was about 50 millions on 112 lbs. of coal, or in the best examples from 60 to 70 millions. In some cases 90 millions have been obtained on 112 lbs. of coal with a high evaporation, and results as high as 130 millions have been claimed on 94 lbs. of coal, but such a high result is impossible.

One H.P. per hour is equal to 1,980,000 foot lbs., or, say, in round numbers, 2 millions.

$$130 \text{ millions duty} = \frac{130}{2} = 65 \text{ H.P. hours.}$$

$$\frac{94}{65} = 1\cdot45 \text{ lbs. of coal per P.H.P. per hour.}$$

With an evaporation of 10 lbs. of water per lb. of coal, the feed water

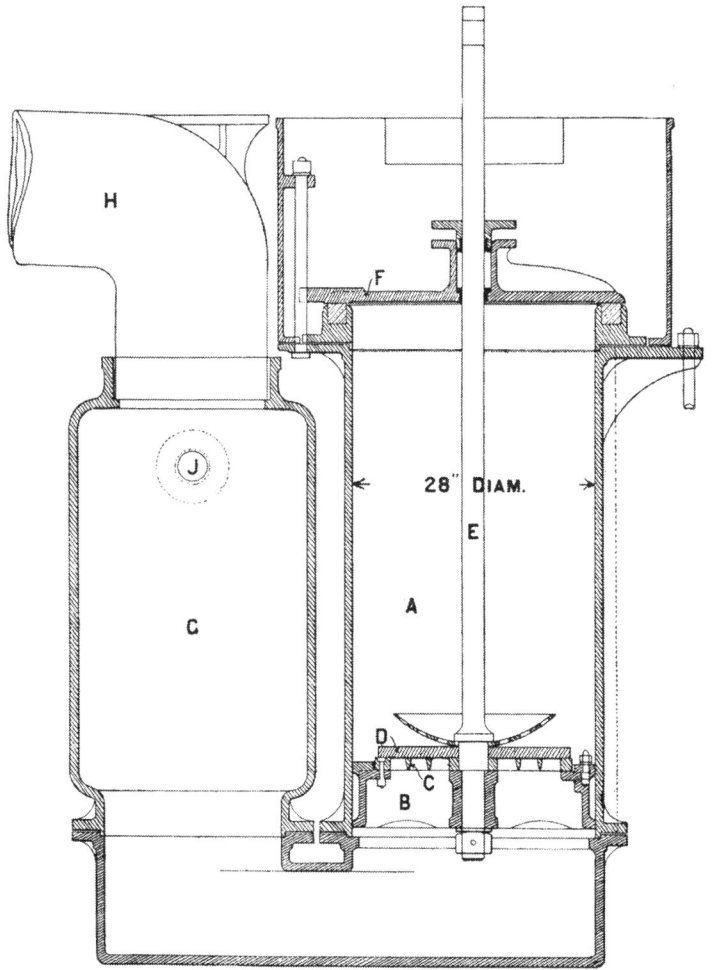

Fig. 79.—Cornish Engine Air-pump and Condenser.

per pump H.P. per hour would then be 14·2, and assuming a mechanical efficiency of 80 per cent., we have 11·36 as the consumption of steam per

I.H.P. per hour. Taking the initial pressure as 40 lbs., we have as the efficiency ratio $\dfrac{10\cdot7}{11\cdot36} = 90\,\%$. The best efficiency ratio obtained with Cornish engines is about 60 % to 65 %.

Expressed in weight of steam, the average engines used from 26 to 30 lbs. of steam per I.H.P. per hour. A better result has been obtained on short trials, and a consumption as low as 18 to 20 lbs. per I.H.P. per hour has been recorded, but no further economy could be obtained with a single cylinder, nor could better results be secured by compound engines with the same boiler pressure. It is only quite recently that high steam pressure has been applied to the Cornish engine.

The single cylinder Cornish engine is limited to 4 expansions, and even then the velocity of the piston during the indoor stroke is very great (see fig. 60, Chap. IV.).

Four expansions with 40 lbs. initial steam pressure gives 10 lbs. as the terminal pressure.

The only method of increasing the initial pressure and the ratio of expansion is by the use of a second cylinder.

By the use of a second cylinder the strains on the engine and pump rods are reduced, and a greater range of expansion is secured, thus adapting the engine to the use of high-pressure steam.

The author having recently to design a pumping-engine for the Basset Mines, Cornwall, adopted the Cornish system of steam distribution. By employing two cylinders he succeeded in using steam of 150 lbs. boiler pressure.

Fig. 80 represents an elevation of the engine. A is the high and B the low pressure cylinder, both cylinders having the Cornish cycle of steam distribution. The pistons of both cylinders are connected to a steel rocking beam C, one piston making the up whilst the other makes the down stroke.

The beam extends beyond the small cylinder over the shaft, and gives motion to the pump rod. The inner end of the beam is made of box section to receive balance weights. The cylinders are 40 and 80 in. in diameter, and have 9 and 10 ft. strokes respectively. The stroke of the pumps is 13 ft. ; the pump plungers are 18 in. in diameter, and the present pit-work extends to a total depth of 1000 ft. At 8 strokes per minute the actual H.P. is 360. The engine has only been at work a few months, and no trials have yet been made as to the consumption of steam, but the consumption of coal has been found from week to week to be one-half that consumed by a good Cornish engine on the same mine doing similar work. Practically the duty is over 90 millions on the coal used. The evaporation is probably 8 lbs. of water per lb. of coal. The duty then on an evaporation of 10 lbs. of water per lb. of coal would be 110 millions.

It is not likely that this engine will give as high an efficiency ratio as

60 per cent., but an efficiency ratio of 50 per cent. with 150 lbs. steam would make the feed water per I.H.P. per hour equal to 16 lbs. An efficiency ratio of 60 per cent. would reduce the feed water to 13·3 lbs. In this engine a disused cylinder and nozzles of an old Cornish engine were used as the low pressure cylinder. An entirely new engine may be expected to give a much better result.

The water load on the engine (all referred to the large cylinder) is

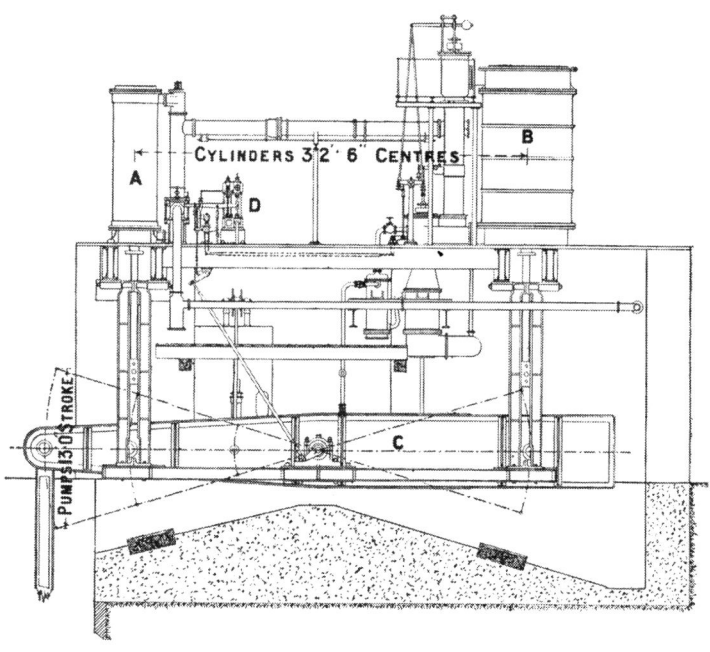

Fig. 80.—Davey's Compound Cornish Engine, Basset Mines.

H.P. cylinder, 40 in. diameter; length of stroke, 9 ft. L.P. cylinder, 80 in. diameter; length of stroke, 10 ft. Pumps, 18 in. diameter; length of stroke, 13 ft. Total height of lift 1000 ft.

equal to 30 lbs. per square inch. This is double the proper load of the old Cornish engine.

An 80-inch engine in this system may be made double the power of the ordinary Cornish engine, and as the first cost of the engine is not greatly in excess of that of the single cylinder engine, the cost of the new engine per H.P. is lower.

It will be seen that by placing the cylinder A on the pump side of the beam centre instead of by the side of the cylinder B, the strain on the beam and the weight on the centre are greatly reduced. The cylinders may, however, be brought together by adopting a modification of the parallel motion beams, designed by the author for the differential engines illustrated in fig. 97, Chap. VI., and fig. 211, Chap. XII. The modification necessary is shown in fig. 81.

The main beam Q is actuated through the secondary beams forming a parallel motion for the piston rods. The outer ends of the secondary beams are anchored to the foundation by means of links. One end of the

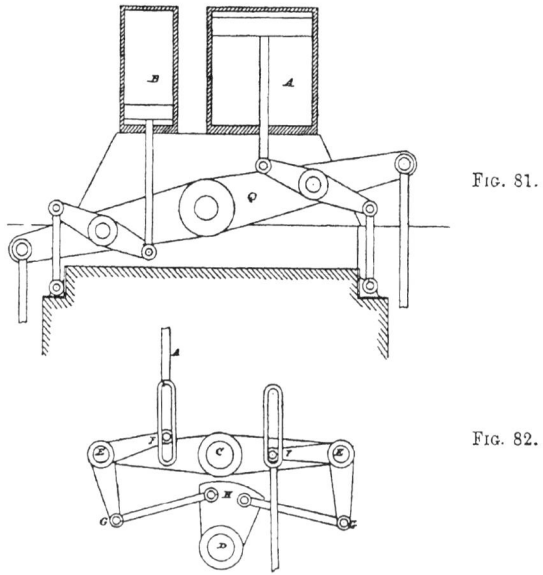

FIG. 81.

FIG. 82.

FIGS. 81 and 82.—Compound Cornish Engine, with parallel motion beams.

The lower figure represents the cut-off motion for the steam valves ; a description of a form of this motion is also given in Chap. X., on valve gears.

main beam would be over the shaft, and the other end made to carry a balance weight.

Fig. 82 represents the gear for giving motion to the steam valves.

C is a rocking shaft actuated by means of the differential gear, and D a rocking shaft partaking of the motion of the engine beam.

A A are valve rods to the admission valves to the engine cylinders receiving motion from the bell-crank levers E E by connections to G G and F F.

The motion of the shaft C operates to open the valves, whilst the

motion of D closes them. By suitable adjustments the cut-off may be
made to take place at any point of the stroke.

The Basset engine has a surface condenser with 900 square feet of tube
surface.

It is also provided with the author's compound system of heating the
feed water, by means of which the water is heated to a much higher
temperature than can be obtained from the exhaust steam.

That the construction of the engine may be understood, we have pro-

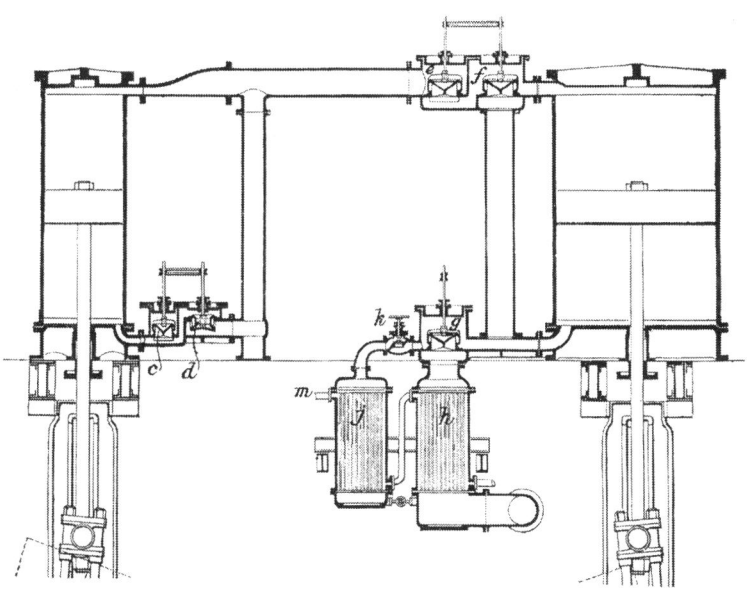

Fig. 83.—Section of Compound Cornish Engine—Waihi Mine.

duced in fig. 83 a section of an engine on the same principle, designed by
the author for the Waihi Mines.

In this example the valves, cylinders, and feed heater are all in a line,
so that one section shows them all. Steam from the boiler is admitted
through the valve c under the small piston ; at the same time the steam
above that piston is admitted to the large cylinder through the valve e ;
the steam below the large piston passes to the condenser through the
exhaust valve g and the feed-water heater h. At the completion of the

steam stroke the valves c, e, and g close, and the equilibrium valves d and f are opened. The steam below the small piston then passes to the other side, and the steam above the large piston passes into the space below it. Both pistons now being in equilibrium, the pump rods fall by their own weight, lifting the water in the pumps. Towards the completion of this stroke the equilibrium valves d and f are closed, and the steam above the large and below the small piston being cushioned, the pump rods are thereby brought to rest.

The feed-water heater is divided into two sections, j and h: the cold feed-water enters at n and passes around the tubes in h; then it passes in a pipe from the top of h to the bottom of the heater j, in which it passes around the tubes and then through the pipe m to the boilers. In h the water takes up heat from the exhaust steam from the large cylinder, whilst in j it is further heated by steam taken from the equilibrium stroke of the large cylinder. It will be seen that the steam passing the equilibrium valve f has done its work, and is only retained below the piston to keep the cylinder hot and to be discharged into the condenser during the next stroke.

In the pipe to the second section of the heater j is a non-return valve k, so that steam may pass to j, but cannot return.

In this way the steam pressure in the heater j is practically maintained at the pressure of the equilibrium steam. The effect of this arrangement is that the water in the heater h is subject to a temperature of 100° Fahr., whilst in j the temperature depends on the temperature of the steam at the end of the stroke of the large piston. In the case of the Basset engine that temperature is 190°. The steam condensed in the heaters passes through a small pipe at the bottom to the exhaust pipe. Practically all the heat thus put into the feed-water is taken from the waste heat of the engine.

The steam valves of the engine are actuated by the differential gear.

Duty Reports.—The economy of an engine, as ascertained by an accurate test of twelve or twenty-four hours' duration, enables us to compare its performance with that of another engine tested under similar conditions, but the actual consumption of steam or coal per unit of time as deduced from the consumption during long periods is much greater than that ascertained by a scientific trial. In such a trial it is always necessary to preserve a uniformity of conditions throughout. In actual work the general results are influenced by change of speed, stoppages, and irregularities of all sorts, but for an engine trial to have a scientific value, the irregularities of ordinary working must be excluded.

In our history of the pumping-engine we have already noticed the system of duty reporting instituted in Cornwall in the year 1811. At that time Welsh steam coal was exclusively used in the Cornish mines, so that the evaporative efficiency of the coal may be assumed to be fairly uniform. The type of boiler in use was the single flue or Cornish boiler,

invented by Trevithick.   The setting of the boiler was always the same,
the gases passing from the fire along the flue tube, then back through the
side flues, and finally through the bottom flue to the chimney.

The engines were fixed with the bottom of the cylinder a few feet above
the top of the boiler, and as near the boiler as possible to minimise the loss
of radiation from the steam pipe.   The steam supply pipe to the jacket
was large, and the drain pipe, from 3 to 4 in. in diameter, went direct
to the boiler.   Cylinder nozzles and steam pipes were well clothed, and the
engines were worked continuously at practically a uniform speed.   The
working conditions were therefore fairly uniform, and that being so, the
duty expressed in millions of foot lbs. per cwt. of coal had a commercial
value.   The duty reports, although defective in construction, served a
good purpose by stimulating an ambition for each engineer to do his best
to secure economy.

From a scientific point of view the construction of the duty reports was
bad and inaccurate.   The following is a copy of the duty report as at
first constructed, and even now used notwithstanding the fact that the
working conditions are much more varied than they were when the reports
were first instituted.   It is from *Lean's Engine Reporter and Advertiser*,
February 1899 :—

In the calculation of the comparative Horse Power, inserted in the third column, the Load is taken at 20 lbs. per square inch on the Piston—the Velocity at 220 feet per minute (equal to 110 feet effective velocity for single-acting Engines)—and the Horse Power as equal to 33,000 lbs. lifted 1 foot high per minute.

The length of the Lifts is given without any allowance for head water, or any addition for the stroke in the pump of the Forcing Lift—neither is any allowance made for Horizontal Rods, Dry Rods in the shaft, etc. When a Lift has been idle or ineffective during the period, a proportionate deduction is made in calculating the Duty.

The variation in the quality of coal, even from the same pit, is great, and the economy in attending to this will be found to vary from $\frac{1}{10}$th to $\frac{1}{5}$th of the consumption. This presents ample inducement for purchasing the best coal and thus securing more regular and greater duty from the Engines.

N.B.—In comparing the 'Duty' performed by the Engines (per 112 lbs. of coal), as now calculated, with that reported previous to July 1856 (per bushel of 94 lbs. of coal), nearly *one-fifth must be added* to the amount of the latter.

In the sixth column *p* denotes a Plunger Lift, *d* a Drawing Lift.

WORK PERFORMED BY THE STEAM ENGINES—PUMPING-ENGINES.

MINES (and Parish in which situate).	Time.	ENGINES, their Size and Horse-power.	Length of the Stroke in the Cylinder.	Length of the Stroke in the Pump.	Number and Description of the Lifts.	Length of the Lifts.	Diameter of the Plunger, etc.	Load in Pounds (Weight of Water lifted).	Load per square inch on the Piston.	Number of Strokes, and Number of Strokes per Minute.	Actual Horse-power employed.	Consumption of Coal in cwts. (112 lbs.) and lbs. Consum. per Horse-power per hour.	Millions of lbs. lifted one foot high by Consuming a cwt. Coal (112 lbs.).	Average Quantity of Water drawn per minute.	REMARKS. Description of the Pistons—Number of Weeks between the last two dates of Packing. Machinery attached to the Engine. Number of the Boilers and Pressure of Steam per square inch on the Safety-valve.
			ft.	ft.		fms. ft.	ins.	lbs.						gals.	
Carn Brea and Tin- croft, Ld. (Illogan).	Feb. 7 to Mar. 3	80 inches, single, 335 H.P.	8·25	6·66	1p / 1p / 1p / 1p / 2p / 1p / 1p / 1p	33 3 / 34 3 / 34 3 / 34 0 / 64 0 / 25 0 / 19 0 / 20 0	15 / 7 / 14½ / 14 / 14 / 14 / 12	96,717	15·4	192,840 … 5·6	109	2196 3·9	56·6	266	Spring Piston. Drawing 90 fms. perpendicularly, and the remainder diagonally. Main beam over the cylinder. One balance bob at the surface, and three ditto and one angle bob underground. 4 Boilers,—tons *Steam* 35 lbs. per inch.

The most important items which vitiate the reports are—1st, the coal used is now of very varying quality; 2nd, the boilers have varying efficiencies; 3rd, where the pumps work on the incline the load is assumed to be that of a vertical shaft with a depth equal to the *length* of the incline; 4th, the actual length of stroke is more or less uncertain; 5th, the coal used is taken from the invoices of the coal delivered, a deduction for the coal in stock being usually guessed at from an inspection of the heap. These are defects which have been always more or less present in the system. These reports enable us to form a general idea only of the results, but the actual performances of Cornish engines can be determined from the tests now made on engines and boilers separately.

The following table shows the values of foot lbs. of duty per 112 lbs. of coal expressed in lbs. of coal per pump H.P. per hour and in lbs. of steam per I.H.P. per hour, with different rates of evaporation.

Duty of Pumping-Engines.

Duty per 112 lbs. of Coal in Millions.	lbs. of Coal per P.H.P. per hour.	EVAPORATION—6 lbs. of Water per lb. of Coal.		EVAPORATION—8 lbs. of Water per lb. of Coal.		EVAPORATION—10 lbs. of Water per lb. of Coal.	
		lbs. of Steam per P.H.P. per hour.	lbs. of Steam per I.H.P. per hour on a Mechanical Efficiency of 80 %.	lbs. of Steam per P.H.P. per hour.	lbs. of Steam per I.H.P. per hour on a Mechanical Efficiency of 80 %.	lbs. of Steam per P.H.P. per hour.	lbs. of Steam per I.H.P. per hour on a Mechanical Efficiency of 80 %.
40	5·54	33·24	26·6	44·3	35·4	55·4	44·3
50	4·43	26·58	21·26	35·4	28·3	44·3	35·4
60	3·7	22·2	17·7	29·6	23·6	37	29·6
70	3·17	19	15·2	25·3	20·2	31·7	25·3
80	2·77	16·6	13·2	22·1	17·6	27·7	22·1
90	2·46	14·7	11·7	19·6	15·6	24·6	19·6
100	2·21	13·2	10·5	17·7	14·1	22·1	17·7
110	2·02	12·1	9·7	16·1	12·9	20·2	16·1
120	1·85	11·1	8·8	14·8	11·8	18·5	14·8
130	1·7	10·2	8·1	13·6	10·8	17	13·6
140	1·58	9·4	7·5	12·6	10·1	15·8	12·6
150	1·48	8·8	7·1	11·8	9·4	14·8	11·8

Various Types of Pumping-Engines.—The requirements and conditions of mining are so varied that quite different types of engines are required.

Since its introduction the steam pump has been largely employed underground in forcing water to the surface. In a few cases in collieries the boilers also have been placed underground quite near the engines, but the general practice has been to transmit the steam from the surface.

Fly-wheel engines are also used underground; they are usually of the cross compound type, with double-acting plunger pumps.

Underground pumps are also worked by means of hydraulic power transmitted from the surface. This method of pumping is dealt with in a separate chapter. Electricity also is coming into use as a medium of transmission of power to underground pumps, but up to the present time it has not been largely used except for small powers. The chief difficulty has been in the application of the quick-running motor to the slow-running pump, involving high speed gearing, belts or ropes, all of which are practically objectionable. Attempts are now being made to produce a quick-running pump and a slow-running motor, thus minimising the gearing required between the motor and pump. It is in this direction that improvements must be made if electricity is to be applied to heavy pumping.

A large electric installation for underground pumping has recently been installed in a colliery near Bochum in Westphalia, but the author is not able to give the results of its working. The power applied is about 600 H.P., and is probably the largest plant of the kind in use.

The main objection to having the pumping plant underground arises from the circumstance that as surface plant must be erected when sinking the shaft it is usually found convenient to adapt it for the permanent as well as for the temporary work.

Large pump rods take up room in the shaft, but it is a question whether any of the forms of underground steam engines or systems of power transmission are so economical in fuel and upkeep as a well-designed

PLATE III

[To face page 95.

General View of Pumping and Winding Engine Houses and Pit-head Frame, Kachidachi Coal Mine, Japan.

surface engine and pit-work when the power applied is considerable. Surface engines, both non-rotative and rotative, vary very much in design and arrangement.

The old Cornish engine and its pit-work has been superseded by improved forms; but long experience with the Cornish engine has demonstrated the value of working with engines which have a distinct pause after each stroke. Many Cornish engines have worked for half a century with very little cost for maintenance.

Rotative engines are used to actuate heavy pit-work, but when so applied, there is more vibration, more wear and tear involving greater cost of upkeep than there is with non-rotative engines.

Non-rotative engines usually have the same speed of stroke, whatever number of strokes they may make per minute, the variation in the number

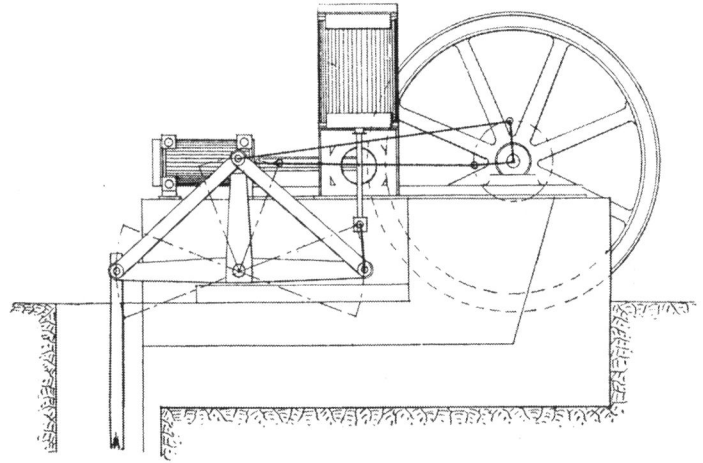

FIG. 84.—Regnier's Pumping Engine, in which an auxiliary engine is employed to help the engine over the deadpoint when moving slowly.

of strokes being governed by the pause between the strokes, but with rotative engines there is a difficulty in running very slowly, and a pump which is being moved very slowly is very frequently inefficient. In such cases a leakage of air, or a leakage of water past the pump valves, gives rise to shocks and to a loss of effect.

In other chapters we have illustrated and described different forms of non-rotative engines. We will now briefly notice some of the types of both rotative and non-rotative engines in use.

When direct-acting rotative engines are employed, very heavy fly-wheels become necessary to insure getting over the centre at slow speeds; attempts have been made to work the engine slowly with a light fly-wheel by the employment of a small auxiliary engine having a crank at right angles to the main crank to assist the engine over the deadpoint, but the system has not found favour in this country, nor does it appear to be a good practicable method of getting over the difficulty. At the end of the stroke of the main crank, when the auxiliary engine comes into full operation, there is no resistance except the frictional resistance of the engine, and the auxiliary engine naturally produces a quick speed just at the time when the actual pause is desirable for the proper action of the pump valves.

Several engines have been built on this plan, such as Regnier's engine (fig. 84).

Kley's Pumping-Engine.—Kley has also endeavoured to overcome the difficulty of the deadpoint by his engine illustrated in fig. 85.

In this engine the object has been to combine the advantages of the

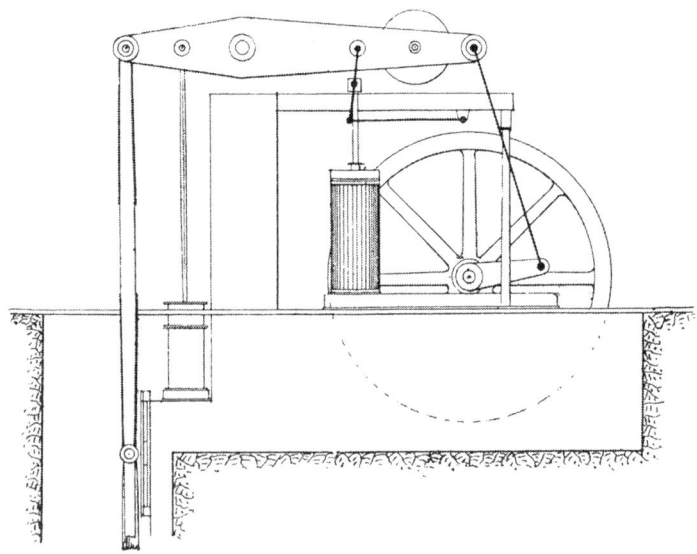

Fig. 85.—Kley's Engine, in which the fly-wheel does not turn completely round when going slowly, but turns backward and forward, the piston not making a full stroke. When running fast the engine works as an ordinary rotative engine.

non-rotative with those of the rotative engine, and to obtain a pause at the end of the stroke with an engine having a crank connecting-rod and fly-wheel. Should the speed of the engine be too great for the pause, then the engine would go on working like an ordinary rotative engine. To secure this double effect a peculiar construction of valve gear was necessary. The engine valves are of the ordinary double-beat type, and are operated by a gear which is a modification of the Cornish gear, having tappets and catches so arranged that the valves are operated whether the crank should happen to go over the centre or not. Should the crank get over the centre, it continues to work in that way ; if not, then the piston makes a shorter stroke, and the engine works as a non-rotative engine. At high speeds the fly-wheel goes continually round : at slow speeds it moves backwards and forwards without turning the centre.

An engine of this kind has been erected at Idria : it was built by E. Skoda of Pilsen, and has been provided for the extension of workings 255 metres below the adit level, as a substitute pending the reconstruction of two existing pumping plants driven by water wheels. It is based on the system of combining a double-acting rotatory engine with a cataract so that it may be worked either continuously or intermittently through a range of speed between half a revolution and eighteen revolutions per minute. The arms of the main beam placed below the cylinders are of unequal length, giving a stroke of 1·5 metres in the shaft and 2 metres on the low-pressure piston, with a further length on the steam side to a total of $6\frac{1}{2}$ metres which carries the balance weight of the pit-work and the connecting rod for the fly-wheel, giving a radius of 1·8 metres for the crank path. The pit-work is of the Rittinger telescopic form, with differential plungers, the total lift of 259 metres being divided over three setts of 55·6 metres, 100 metres and 105·5 metres rise respectively. The main rods, made of Siemens-Martin steel of 32 tons tensile strength with 20 per cent. elongation and 45 per cent. contraction at fracture, are circular in section and tapered in four thicknesses of 135 millimetres, 110 millimetres, 80 millimetres and 75 millimetres. The total weight of the pumps and pit-work is 65·17 tons, of which 50·5 tons are in the moving parts, and the water load is 27·616 tons on the up and 14·02 tons on the down stroke.

The maximum working stresses on the rods correspond to a 10 fold safety factor at the top, 10·3 fold in the middle, and 13·8 fold in the bottom sett.

The load on the up stroke is $27·6 + 50·512 = 78·112$ tons against $50·512 - 14·017 = 36·495$ tons excess power on the down stroke. The latter quantity is nearly all counterbalanced by cast-iron weights amounting to 20·4 tons attached to the inner arm of the beam between the points of attachment of the piston and connecting rod. A portion of the weight is also provided by the steam pistons, which are unusually thick and heavy, that of the high-pressure cylinder weighing 26 cwts. and that of the low-pressure cylinder 66·6 cwts. The fly-wheel, 7·5 metres in diameter, weighs 20·65 tons, and the total weight of the engine and condenser is 137·2 tons. The average working speed on the trials was eight to nine revolutions per minute, corresponding to between 120 I.H.P. and 150 I.H.P., while the steam consumption under the most favourable conditions is 9·74 kilogrammes per H.P. hour. Wood and coal of a low (six-fold) evaporative power are used as fuel. The normal work of the engine is lifting $1\frac{1}{4}$ tons of water per minute to a height of 270 metres when making six revolutions per minute, corresponding to 100 gross and 75 effective H.P. which can be doubled when required.

The results of three years' working give the steam consumption as 13·66 kilogrammes per effective H.P., that of an underground compound condensing engine of 120 H.P. in another shaft being 14·39 kilogrammes.

The total cost of the engine, boiler, pit-work and buildings was £9363, of which amount £3912 are charged to the engine and the travelling-crane in the engine-house, and £2809 to the pumps and pit-work.*

There are other forms of rotative engines for actuating the single rod system of pumps. Two of these are shown in figs. 86 and 87.

Geared engines have been used both for the single and double rod

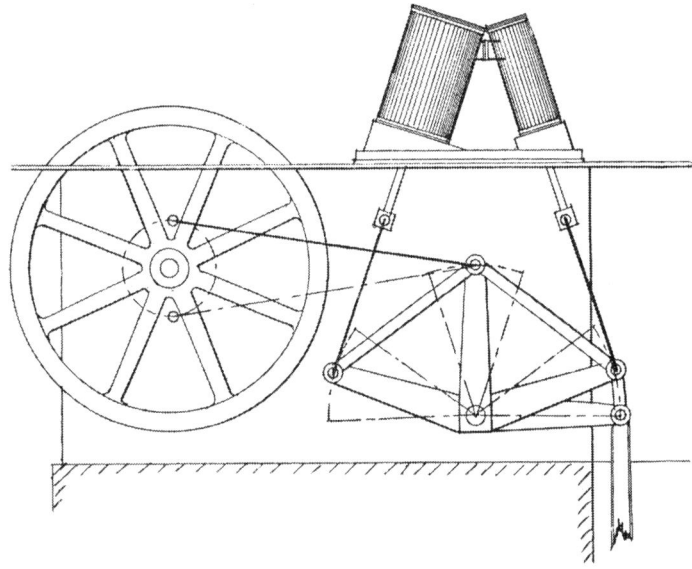

FIG. 86.—American Compound Rotative Pumping-Engine for actuating a single line of pump rods.

system of pumps, but such engines cannot be said to be suitable for heavy pumping.

An example of an American geared engine is given in fig. 88, and the description of another is as follows :—

The pumping machinery used in the iron and copper districts of Michigan usually consists of Cornish plunger pumps which are operated by geared engines, the latter making from three to sixteen strokes for each stroke of the pump.

The largest plant of this type yet erected is that of the Calumet and Hecla copper mine at Calumet, Mich. There are two lines of pumps varying in diameter from 7 in. to 14 in., and with an adjustable stroke varying from 3 ft. to 9 ft. The object of the adjustable stroke is to diminish the

* *Trans. Inst. Civil Engineers* cxxxvii.

capacity of the pumps in the dry season. Each line of pumps is driven from a crank placed on a steel spur-wheel shaft 15 in. in diameter, making ten revolutions per minute. The mortise spur wheels have a diameter of

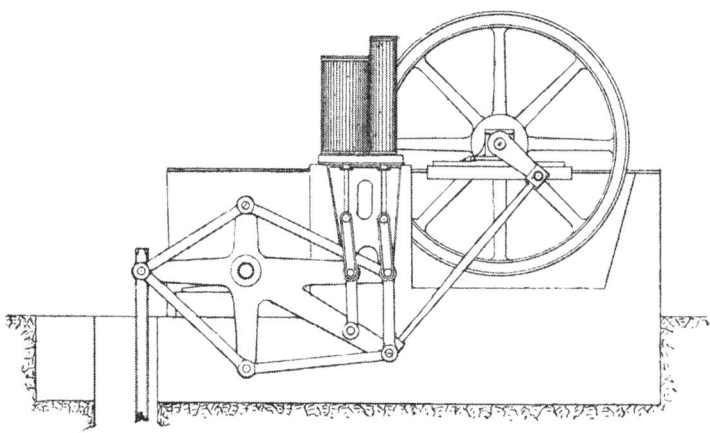

FIG. 87.—American Compound Rotative Engine for actuating a single line of pump rods.

22½ ft. at the pitch line, with two rows of teeth, each 15 in. face. The pitch is 4·72 in. Engaging with the mortise wheels are pinions of gun iron of 4 ft. 6 in. diameter, placed on steel shafts of 12 in. diameter, and making

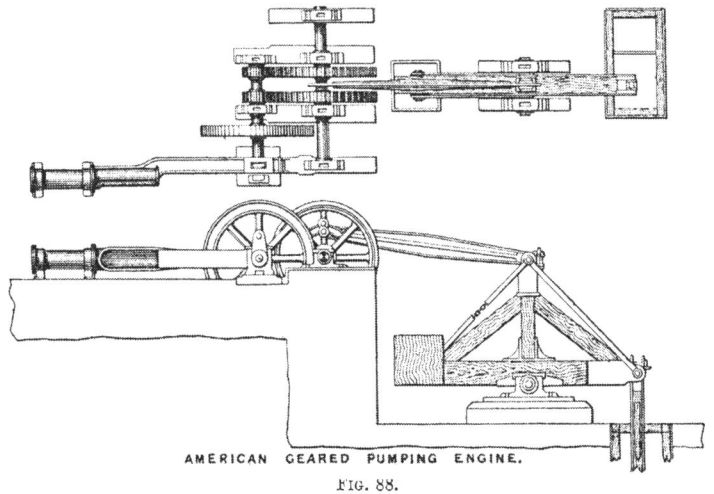

AMERICAN GEARED PUMPING ENGINE.
FIG. 88.

fifty revolutions per minute. The 12-in. pinion shafts are driven through mortise wheels 12 ft. in diameter and 24 in. face by pinions, 3 ft. 9 in. in diameter, which make 160 revolutions per minute. The engine is 4700

H.P., and in addition to driving the pumping machinery does the hoisting and air-compressing for the Calumet mine. In the same building with the mine pump gearing is a duplicate arrangement for operating the man engines. In order to operate the mine pumps and man engine for the Hecla mine, it was necessary to use rocking shafts, which are made of gun-iron and hollow; they are 32 in. in diameter outside, and the thickness of metal is 4½ in. The pump rocking shaft is 39 ft. 4½ in. long over all, in two sections, and weighs 40 tons. Rockers are placed on each end of this shaft; one is connected with a crank on the mortise wheel shaft, and the other with the surface rods that work the pump bobs. These rods are of Norway pine, 12 in. by 12 in. in section and 1000 ft. long. There are two bobs, one above the other, with axes at right angles, each weighing about 25 tons. The connection from the upper bob to the lower has hemispherical pins and brasses to accommodate vibrations in right-angled planes. The slope of the main pump is 39°, and the machinery has been designed to

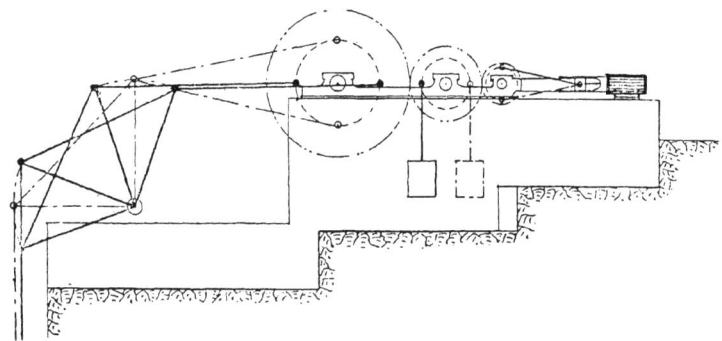

Fig. 89.—Geared Pumping-Engine, having weights hanging on cranks on an inter-mediate shaft, so arranged as to equalise the resistance on the engine.

raise water from a depth of 4000 ft. The pumps are of the usual Cornish plunger type with flap valves. There is an auxiliary engine of the Porter-Allen type for driving the pumps and man engines when the main engine is not working. It makes 160 revolutions per minute, the same as the rope wheels. The seeming complication of this arrangement is due to the fact that it had to be adapted to existing works for increased depths, and to be put in without interfering with the daily operation of the mine.

One of the objectionable features in all geared engines is, that at the time of the turn of the stroke, when the pump offers no resistance, the speed of the engine becomes greatest; the engine runs with a variable load of such a nature that the pumps turn the stroke quickly when a pause would be desirable. This acceleration of speed at the end of a stroke is a defect, to correct which Mr Charles Bridges in America, in 1883, patented a device shown in fig. 89. The intention was to equalise the resistance on the engine by means of balance weights, but the device does not appear to have been put in practice.

Non-rotative pumping engines in mines, when worked on the multiple

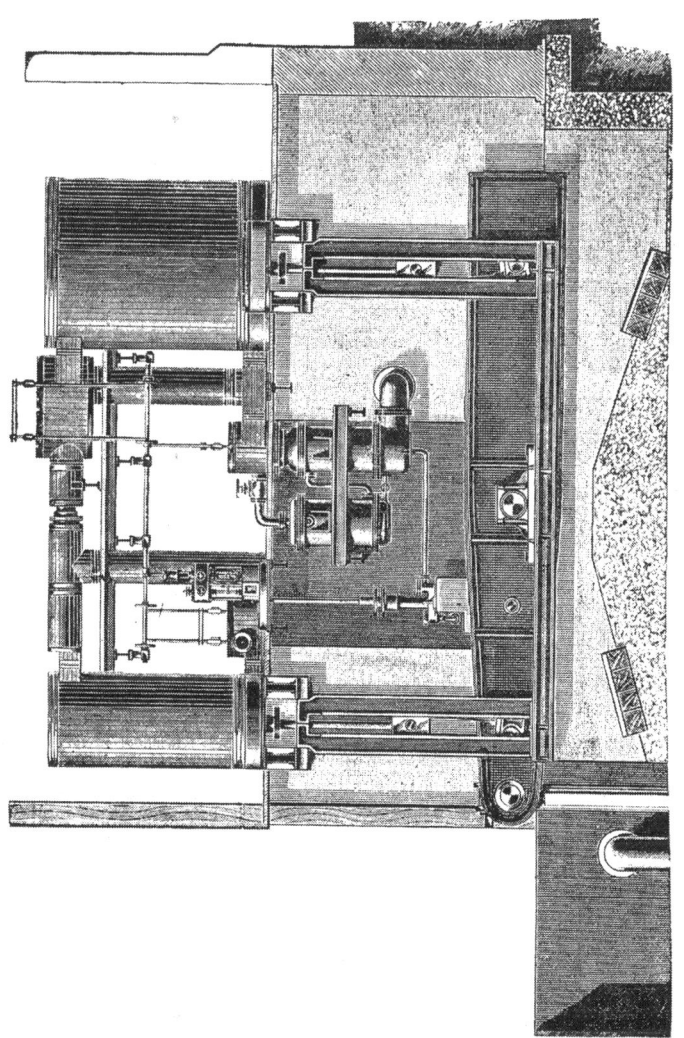

FIG. 90.—Elevation of Compound Cornish Engine—Waihi Mine.

cylinder system, have generally the Woolf system of steam distribution. The receiver system demands a greater energy of mass to enable the expansion of steam to be complete, and, therefore, the strains imposed on the moving parts become very great. It is not expedient to employ the receiver system in either non-rotative or direct-acting rotative engines employed to actuate heavy pit-work. Mines are sometimes drained by means of steam engines placed at the bottom of the pit. When rotative engines are so employed, it is practicable to use the receiver system, provided the fly-wheel is heavy enough. The high- and low-pressure pistons should actuate double-acting pumps direct, and be coupled to the crank shaft with cranks at right angles. When non-rotative engines are employed, the inertia of the water column may be made use of, as explained on page 76, and the most economical result with such engines is obtained with *long stroke* tandem compound or triple engines. Duplex and other short stroke steam pumps lose much by clearance. 1 in. clearance in a 15 in. stroke is nearly 17 per cent., whilst on a 5 ft. stroke it would only be one-fourth of that amount. Pumping-engines when employed for the water supply of towns, etc., are not subject to the same practical considerations as mining engines, because they are generally applied under different conditions and in quite a different way.

Compound Cornish Engine.—In figs. 90 and 91 we give the elevation and plan of an engine designed by the author for the Waihi Mines, New Zealand. Both cylinders have the Cornish cycle of steam distribution.

A section of the cylinders showing steam-valve nozzles and feed-heaters is given in fig. 83 and fully described in Chap. V., page 89. The cylinders are 45 in. and 90 in. in diameter, both having a stroke of 8 ft. The pumps are of the plunger type, 19 in. in diameter, and have a stroke of 10 ft. The total lift is 1000 ft.

The pumps are shown in figs. 123 and 124, Chap. VII. Fig. 90 is an elevation and fig. 91 a plan of the engine. The condenser is of the surface type, and is placed with the air-pump in a water tank outside the engine-house, as shown in fig. 91. The boiler pressure is 120 lbs.

Pumping Shafts.—Where the shaft is entirely used for the pumping plant, the steam cylinders may be placed directly over the pump rods, as in figs. 92 and 93, and then the balanced system of pump rods may be employed, the engine being a compound double-acting engine with valve nozzles, as shown in figs. 94 and 95. Another practical example of such an engine is given in fig. 96.

This form of engine has been frequently applied in German collieries by Mr Wippermann, who has also designed and erected the engine illustrated in fig. 97.

In this engine the parallel motion beams take the place of the triangular rocking levers of fig. 96.

In the collieries of this country the horizontal engine (fig. 64, Chap. IV.) is generally preferred, because it gives more room at the top of the shaft.

In fig. 98 we illustrate a pumping-engine at the Calumet and Hecla

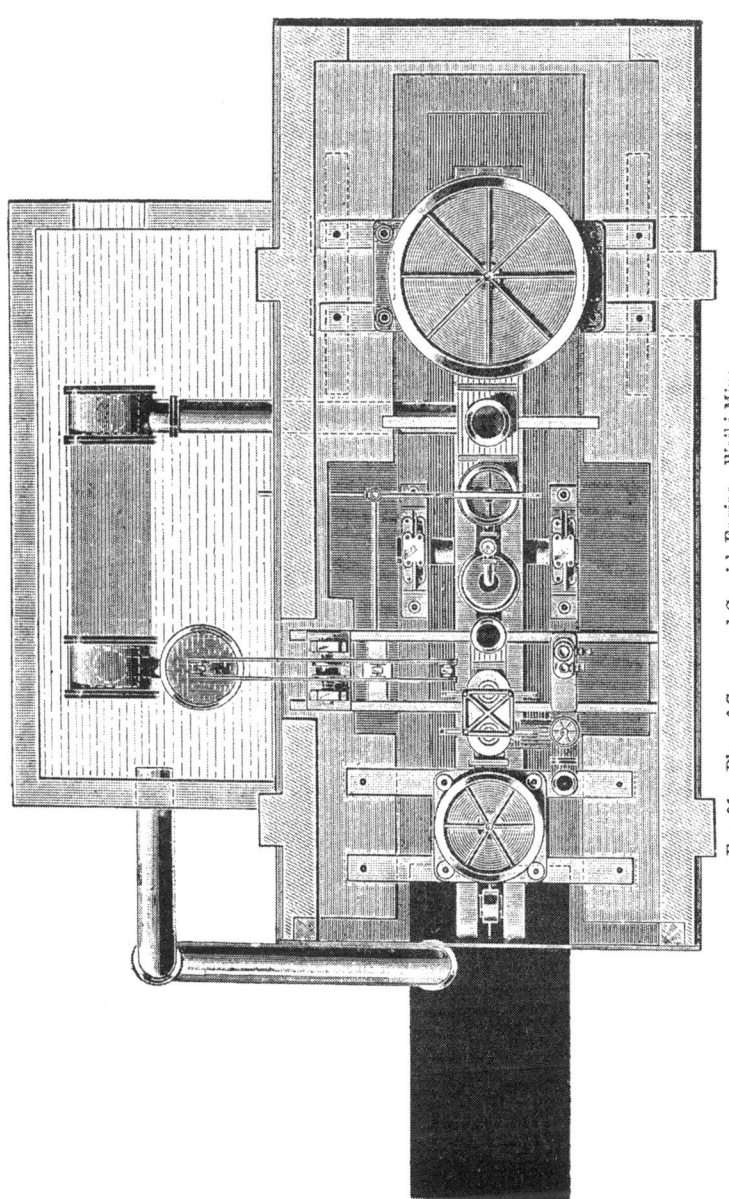

FIG. 91.—Plan of Compound Cornish Engine—Waihi Mine.

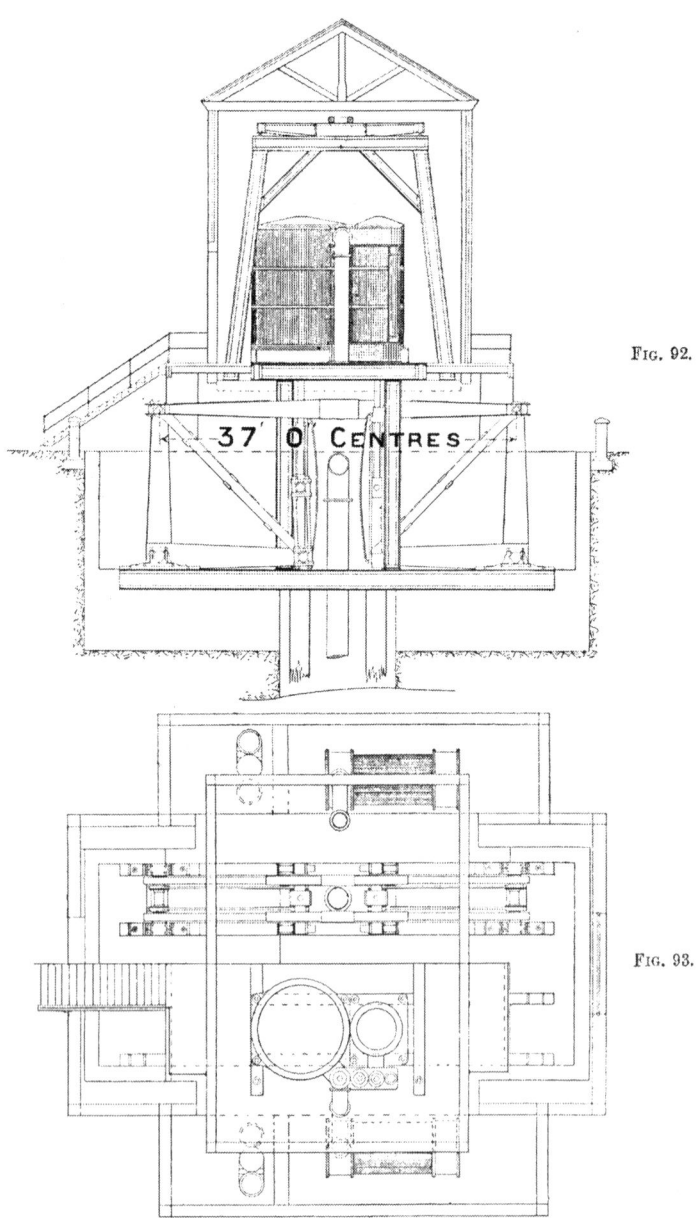

FIG. 92.

37' 0 CENTRES

FIG. 93.

FIGS. 92 and 93.—Elevation and Plan of a Pair of Compound Engines placed
directly over the shaft, each engine actuating a double line of rods.

Mine for pumping water for use in the stamp mills. There are three pumping-engines, two of which have a capacity of 20,000,000 gallons per day, while that of the third is 10,000,000 gallons per day.

The water is elevated between 50 and 60 ft., and is used for treating the stamped rock. Two of the engines are of the inverted compound beam and fly-wheel type (fig. 98).

The largest of the compound engines is named 'Ontario,' and has a vertical low-pressure cylinder 36 in. in diameter, and a high-pressure cylinder 17½ in. in diameter, the stroke of both being 5 ft. These are inverted

FIG. 94.   FIG. 95.

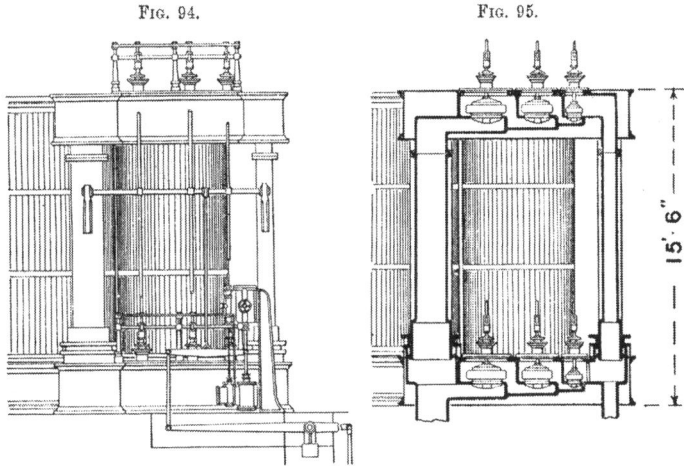

FIGS. 94 and 95.—Valve Gear and Section of Nozzles and Valves of Compound Double-acting Engine with Cylinders over Pump Rods.

Elevation and plan of two engines over one shaft are given in figs. 92 and 93.

over a beam or rocker, and the pistons are connected to opposite ends of the same (fig. 98).

The beam attachment of the main connecting rod is made to a pin placed above and midway between the pins for piston connections. The throw of the crank is 5 ft. There are two differential plunger pumps having upper plungers 20 in. in diameter, and lower plungers 33 in. in diameter, with a stroke of 5 ft. These pumps are vertical, and placed beneath the engine bed-plate, to which they are attached by strong brackets. The pump under the low-pressure cylinder is worked direct from its crosshead by an extension of the piston rod. The other pump is worked by a trunk connection from the opposite end of the beam. The radius of the beam is but 50 in., but the connections to it are made very long by links.

The lower plungers work through sleeves in diaphragms situated in the centre of the pumps.

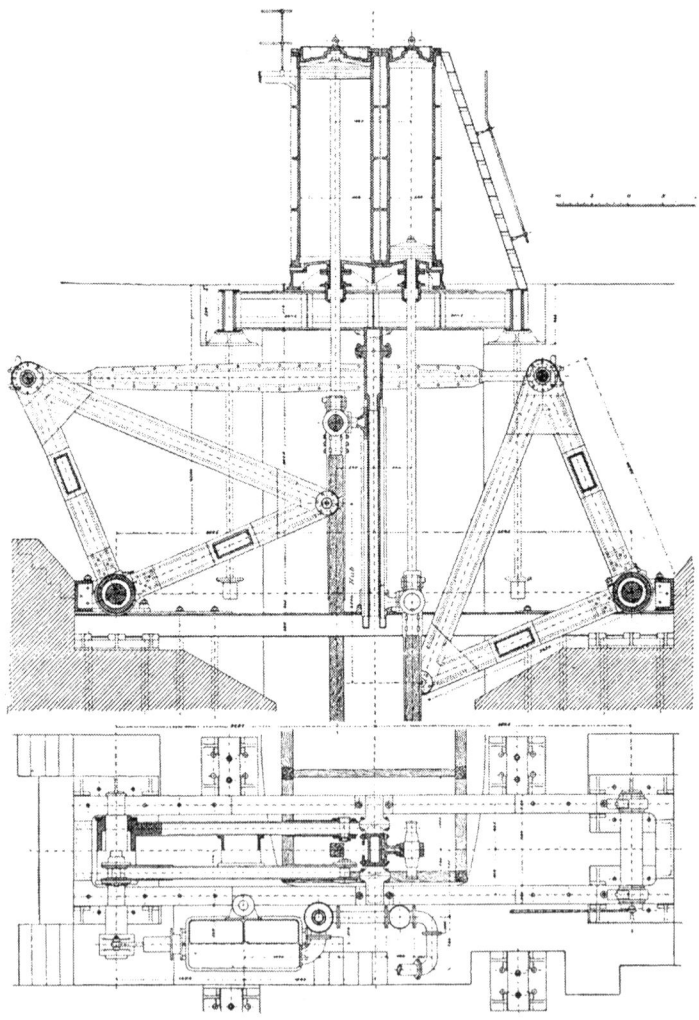

FIG. 96.—Vertical Compound Double-acting Engine with cylinders directly over the shaft, the engine actuating a double line of rods—used in some German collieries.

In these diaphragms the openings for the delivery valves are made. These valves are of brass, faced with rubber, and closed with brass spiral springs. Their diameter, however, is $5\frac{1}{4}$ in., and there are 72 suction and 72 delivery valves for each pump. It will readily be seen that the action

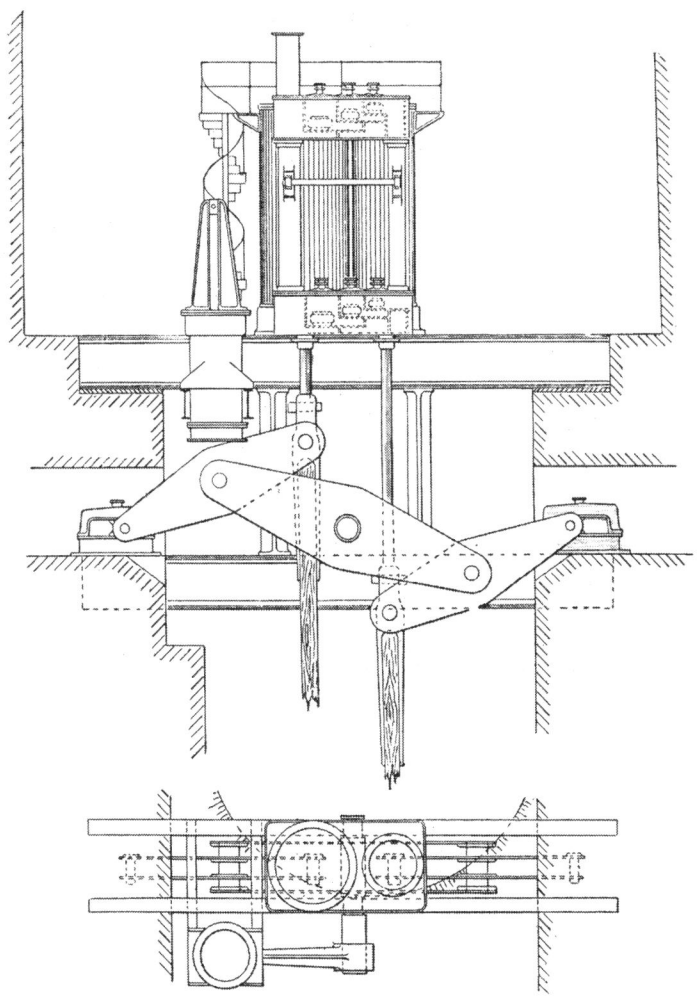

Fig. 97.—Double-acting Compound Differential Engine.

of these pumps is similar to that of the bucket and plunger, each pump having one suction and two deliveries for each revolution of the engine.

The 'Ontario' is designed to run at a maximum speed of 33 revolutions per minute.

In fig. 99 we give an illustration of Barclay's grasshopper beam engine. This engine has been made both with single and with compound cylinders. The engine may be single or double acting.

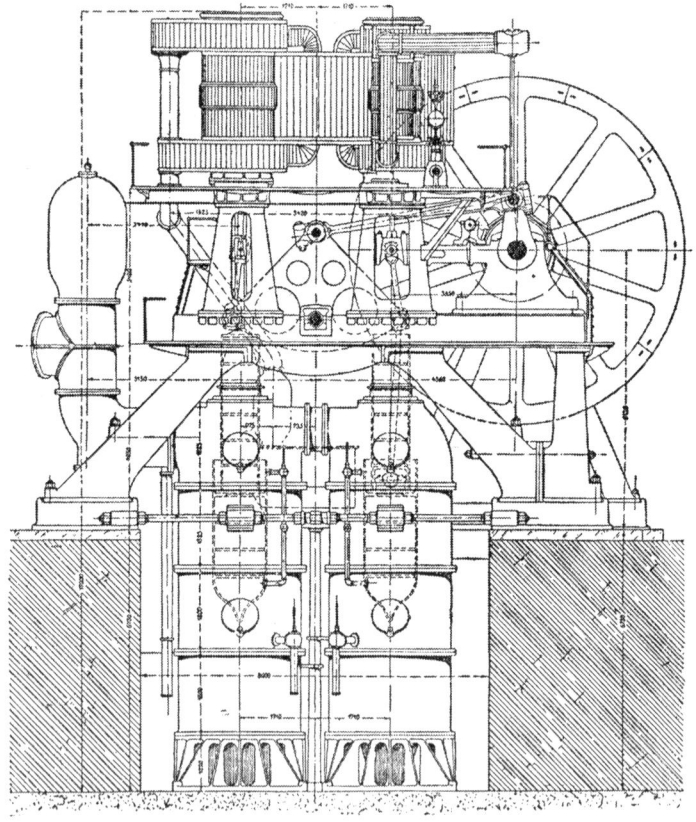

Fig. 98.—Pumping-Engine—Calumet and Hecla Mine.

When of the latter construction, a heavy balance beam is required. The valve gear employed is of the ordinary Cornish type.

Fig. 100 represents a small compound Cornish engine supported on a steel framing.

This engine was designed to be independent of masonry foundation, in order that it might be easily taken down and re-erected at another place with little expense. In fact, it is semi-portable. The connection of the piston rods with the beam is made by means of a triangular connecting rod. The steel framing which carries the engine also forms a support for

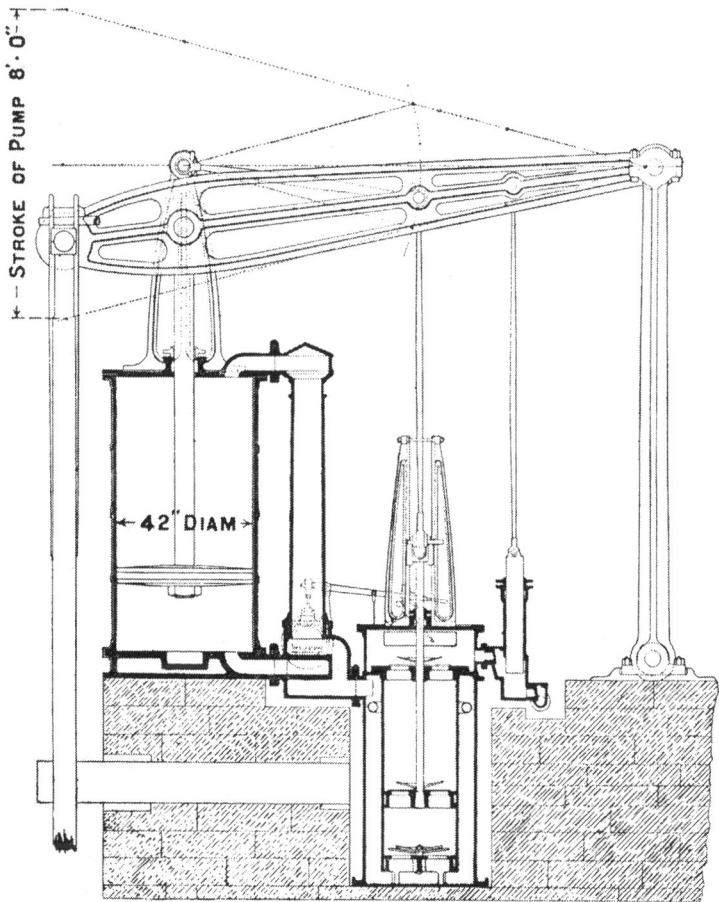

FIG. 99.—Barclay's Grasshopper Beam Engine.

screw gear for the purpose of lowering and lifting the sinking lift in the shaft.

The steam valves are of the ordinary D-slide valve form, actuated by means of the differential gear.

Example of large Pumping-plant.—The pumping-plant illustrated in the folding plate (Plate IV.) and the following engravings is probably the largest pumping-plant erected on one shaft. It was designed by the author for the Müke mine in Japan. The complete pumping-plant is capable of raising 9000 gallons per minute from a depth of 900 feet when the engines are running at six strokes per minute. This equals 58,000 tons of water

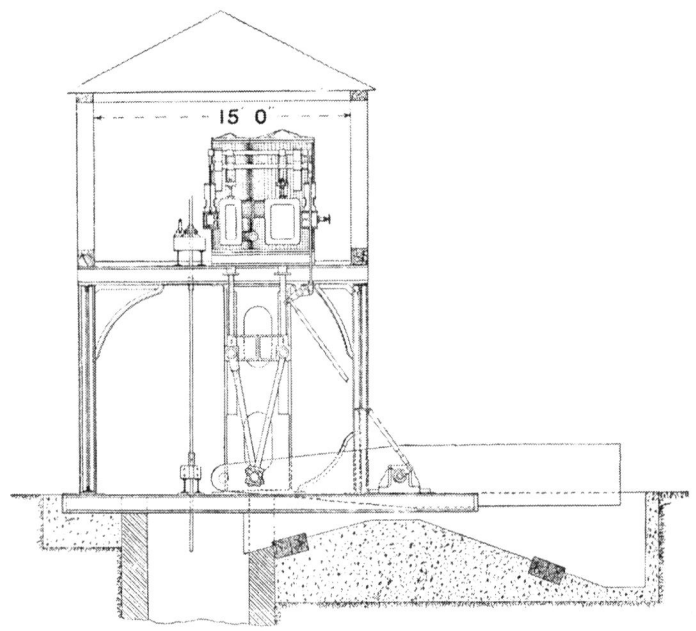

Fig. 100.—Small Compound Cornish Engine with triangular connecting rod attachment to beam showing suspending rods and screws for bucket lifts.

in twenty-four hours, or 13 million gallons. The total horse-power in water lifted is 2500.

The shaft is rectangular, 41 ft. long by 12 ft. wide, and is made to accommodate four sets of twin pit-work and two pairs of winding cages.

Plate IV. shows the general arrangement of the surface plant — the pit-head frame, the pumping-engines, winding-engines, and steam capstans.

A perspective view of the pit-head frame is given in fig. 101. The following illustrations give details of the pump work, the methods employed in sinking the shaft, the arrangements for sinking, permanent pumps, etc.

PLATE IV.]

[To face page 110.

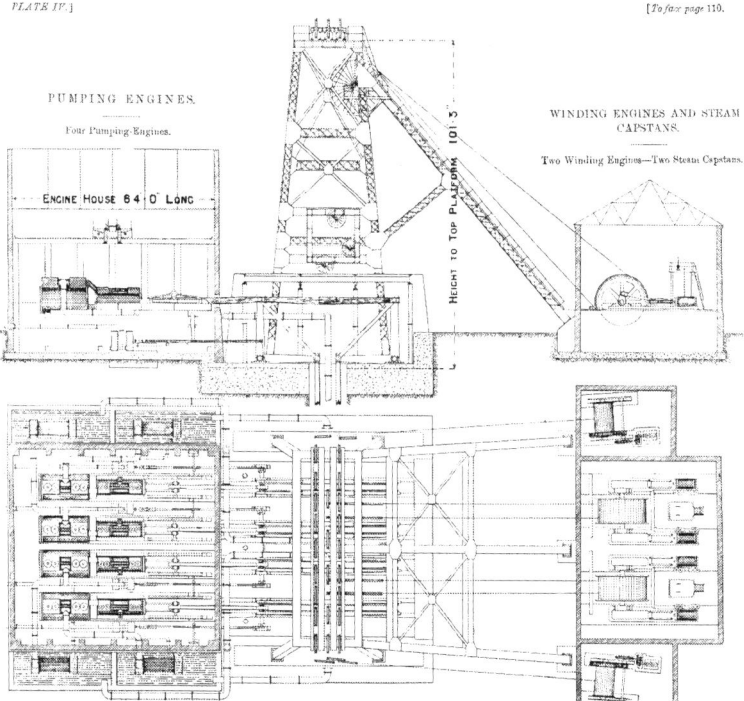

PUMPING ENGINES.

Four Pumping-Engines.

ENGINE HOUSE 64·0″ LONG

HEIGHT TO TOP PLATFORM 101·5′

WINDING ENGINES AND STEAM CAPSTANS.

Two Winding Engines—Two Steam Capstans.

General Arrangement of Pumping and Winding Engines.—Shaft 41 ft. long × 12 ft. wide—Muke Mines, Japan.

Pumping-Engines.—There are four pumping-engines which are arranged side by side in one large house. The surface condensers of the engines are placed in tanks outside the engine house.

The engines and condensers are clearly shown in plan and elevation on the folding plate, whilst the engines themselves are given in detail in figs. 102 to 104; on page 112, fig. 104 shows a section of the steam cylinders, and figs. 102 and 103 the valve gear, etc. The valve gear is of the differential type actuating drop valves. The cylinders are 45 and 90 in. in diameter and have a 12-ft. stroke. A section of the air pump is given

Fig. 101.—Steel Pit-head Frame.

in fig. 105, and that of the condensers in fig 106. The steam passes through the tubes of the condenser, the condenser itself standing in a water tank with the water surrounding the tubes. In fig. 105 it will be seen that the air pump stuffing-box is provided with a pipe communicating with the hot well. The object of this is to introduce water to a lanthorn bush inserted in the packing, thus preventing air leaking into the pump through the stuffing box. Each engine actuates a double line of pump rods by means of a pair of quadrants, as indicated in Plate IV.

Pit-head Frame (fig. 101).—This frame is built of steel. It has a total

height of 101 ft.   Over the winding pulleys are placed girders for
carrying the capstan pulleys.   The winding pulleys are 17 ft. and the
capstan pulleys 5 ft. in diameter.   Each capstan pulley will carry 30 tons.

Fig. 102.   Fig. 103.   Fig. 104.

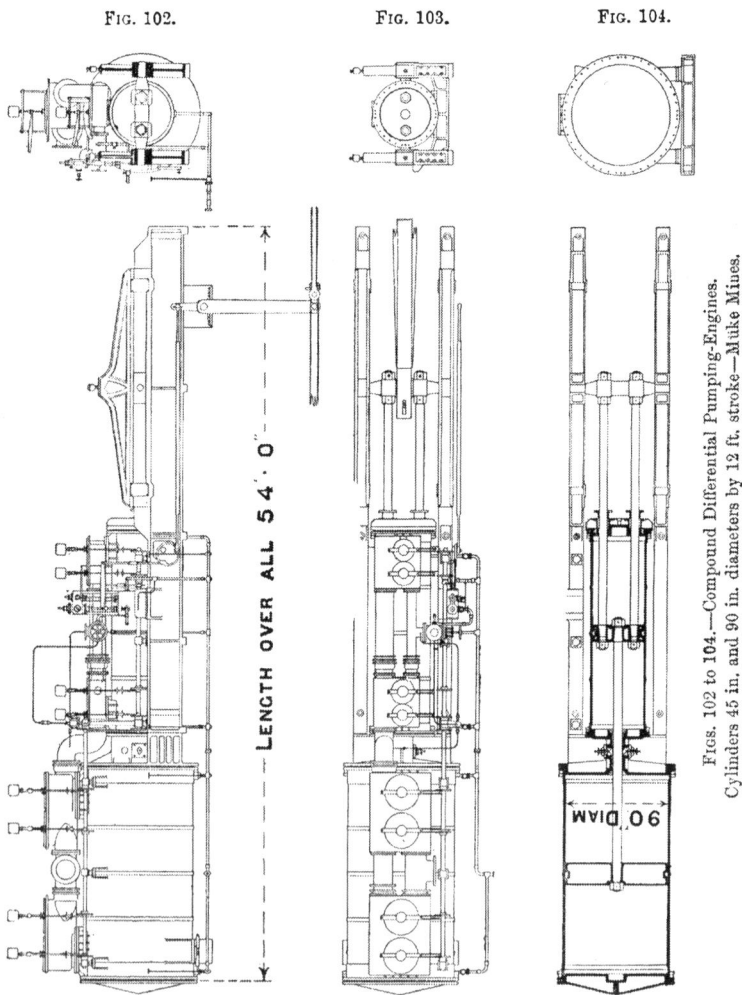

Figs. 102 to 104.—Compound Differential Pumping-Engines.
Cylinders 45 in. and 90 in. diameters by 12 ft. stroke—Muke Mines.

At each end of the winding engine house is placed a 30-ton steam
capstan for the purpose of lifting and lowering the pump work.

Guide pulleys are fixed on each side of the pit-head frame for leading
the capstan rope to the overhead pulleys.

System of Sinking the Shaft.—Each engine is provided with a pair of

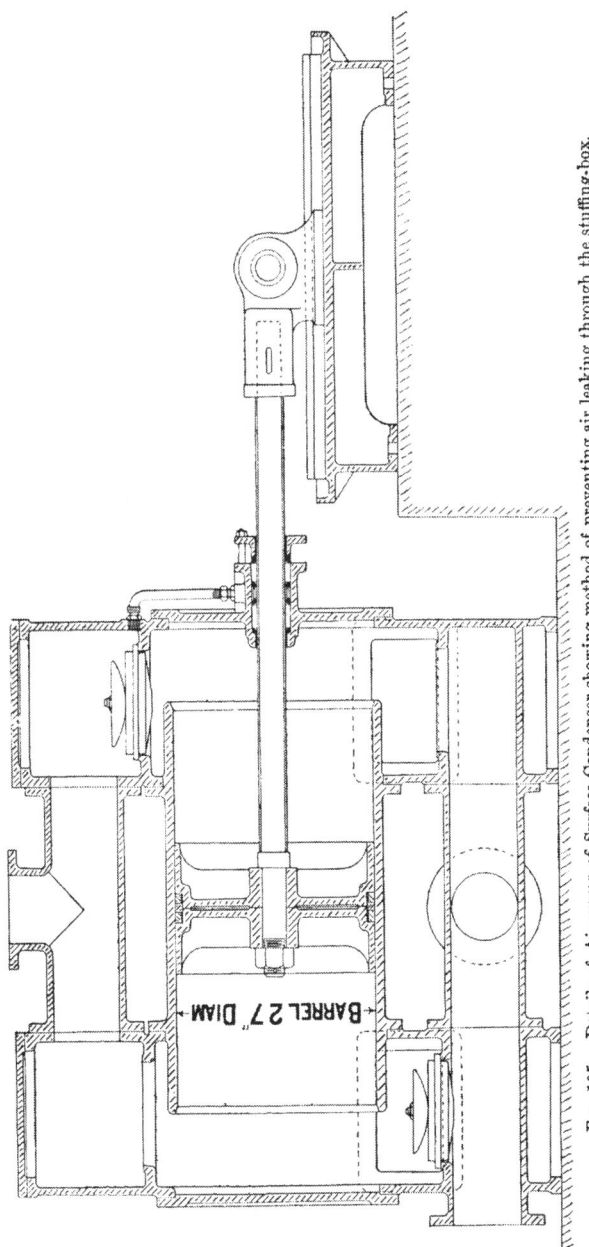

BARREL 27″ DIAM

Fig. 105.—Details of Air-pump of Surface Condenser showing method of preventing air leaking through the stuffing-box.

sinking pumps, each 300 ft. long, for the purpose of sinking in stages of 300 ft.   On the completion of the first stage, plunger pumps are put in to take the place of the sinking pumps, and then the sinking pumps are used for another stage of 300 ft., when the operation of putting in plunger lifts is repeated.

Sinking Lifts.—Each pump is 23 in. in diameter, has a 12-ft. stroke, and is suspended in steel rods as in fig. 107.

That the sinking pumps may stand clear of the plunger lifts, they are worked from a set-off on the main rods, as in figs. 108 and 109, and in detail in fig. 110.   In the shaft above the sinking lifts are placed steel girders, from which the sinking pumps hang in the suspending rods, the top of the suspending rods being provided with screws, as shown in fig. 111.   The screws are made long enough to allow of an addition of a length of pipe; the screws have nuts on which are fixed worm wheels gearing into worms, actuated by means of Pelton wheels taking their power from the water column, as in fig. 112.   The lower girders are for

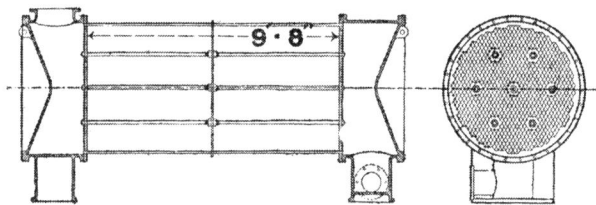

FIG. 106.—Details of Surface Condenser.
2400 square feet of tube surface.   Number of tubes, 954.

taking the weight when it is required to add an additional length of rods.   The Pelton wheels are shown in section in figs. 113 and 114.   It will be seen that there are three wheels on one shaft.   The two outer wheels are made to run in one direction for lifting and the centre wheel in the opposite direction for lowering.   The general arrangement of the Pelton wheels for three sets of sinking lifts is given in fig. 115, and a side view of the screws in fig. 112.

Plunger Lifts.—A plan of three pairs of plunger lifts showing the positions of the sinking lifts is given in fig. 116, and an elevation in fig. 117.   That the plunger lifts may be all in a line, one under the other, steel side rods are carried down from the plunger head and connected to the pump rod directly underneath, as in fig. 117, and also shown in figs. 108 and 109, together with the suspending rods for the sinking lifts.

Pump Rods.—The pump rods are 22 in. square, built up of four 11-in. square timbers, and are guided every 45 ft., as in fig. 118.   The catch-pieces and hanging beams between each plunger set are shown in fig. 119.   The method adopted for connecting the bucket rod to the main rods is illustrated in fig. 110.

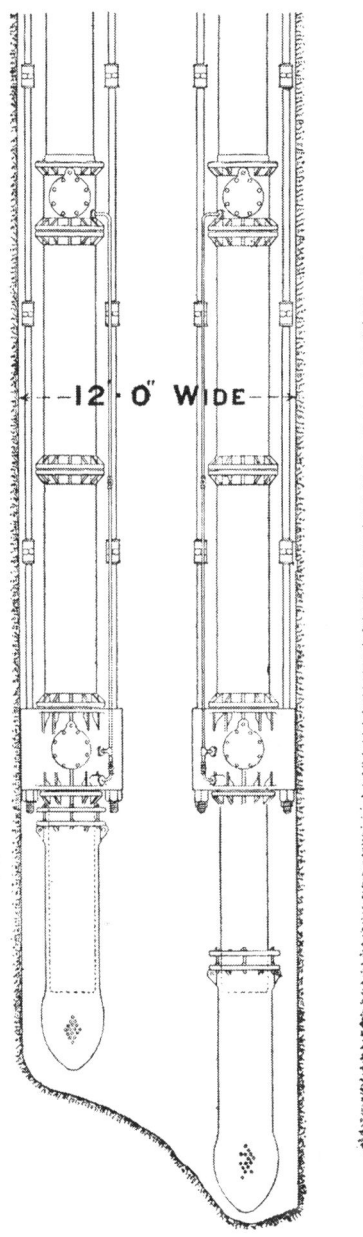

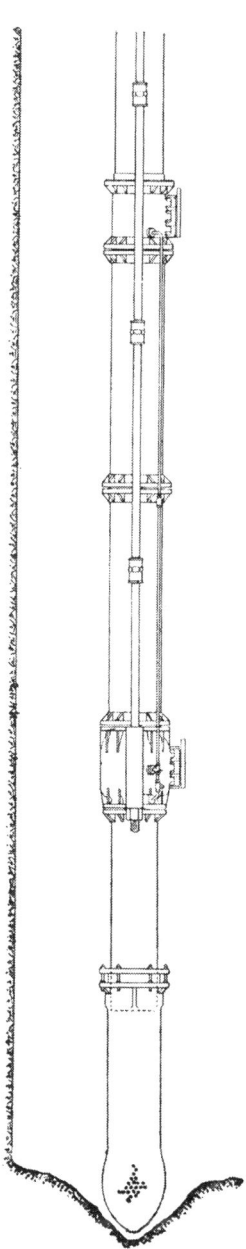

—12' 0" WIDE—

FIG. 107.—Sinking Lifts showing Telescopic Suction Pipe and method of suspending pumps.

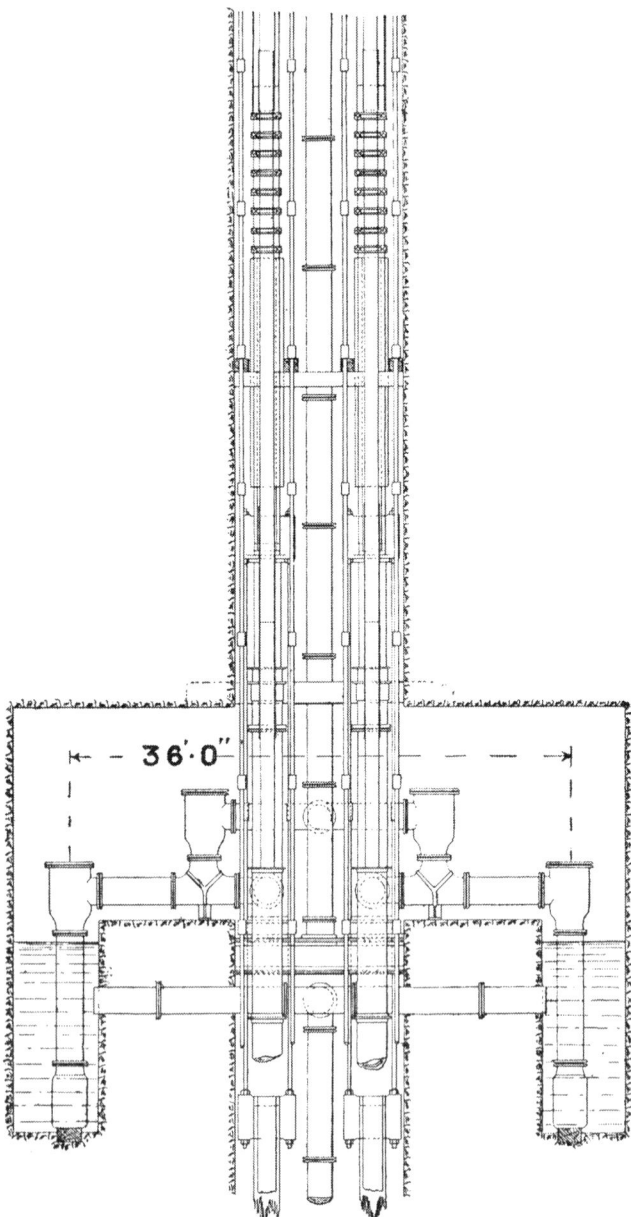

Fig. 108.—General Arrangement of Plunger Lifts showing position of rods for lowering sinking lifts.

36'·0"

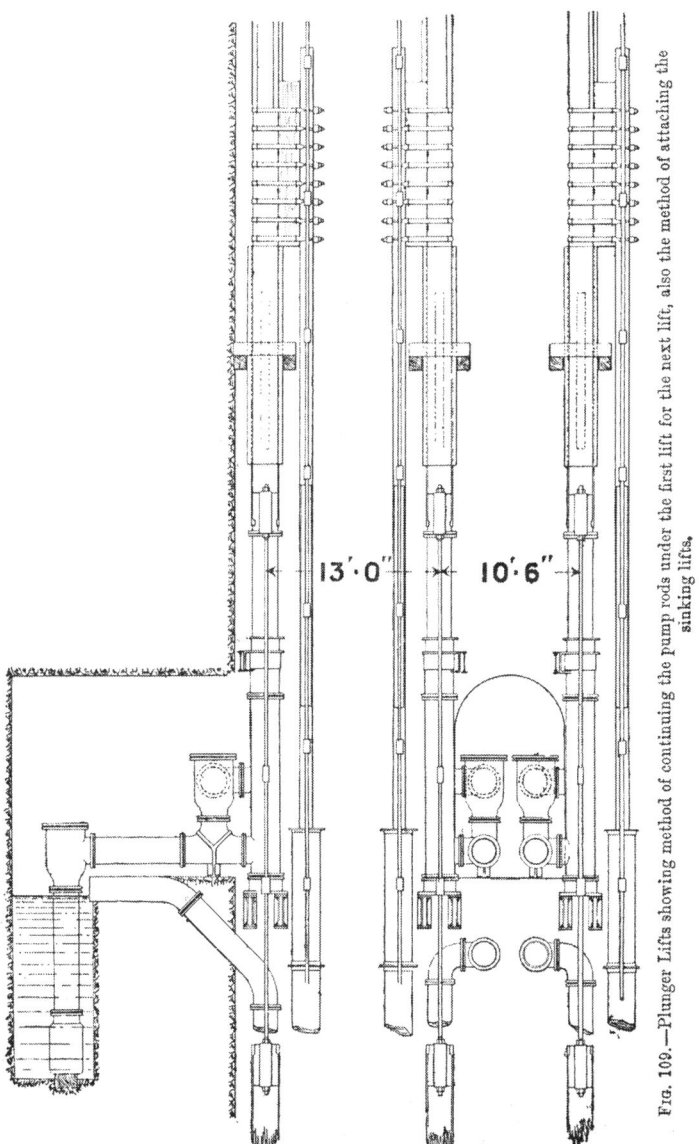

13'·0"   10'·6"

Fig. 109.—Plunger Lifts showing method of continuing the pump rods under the first lift for the next lift, also the method of attaching the sinking lifts.

Underground Engines and Pumps.—There are various kinds of engines and pumps used underground.

In the early days of underground engines, the author had occasion to put a pair of engines and pumps at the bottom of the pit at the Clay Cross Colliery, and an experiment was then made to ascertain the loss from condensation in the steam pipes.

The shaft was 1000 ft. deep, and the boilers were on the surface. The following are the results of the experiment.

Result of experiment to ascertain the loss caused by taking steam down a coal pit:—

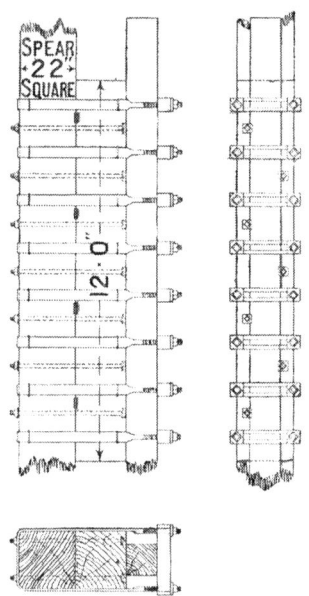

FIG. 110.—Detail of Spear Rods showing connection for rods of sinking lifts.

Diameter of pipes, .	7½ in.	
Vertical height of pipes in pit, . .	915 ft.	
Horizontal length on the surface, .	185 ,,	
Total length, . .	1100 ,,	
Length of pipes with non-conducting cement, . .	1015 ,,	
Length of pipes un-clothed, . .	85 ,,	
Pressure of steam at surface, .	45 lbs. per sq. in.	
Pressure of steam at pit bottom when engines were standing,	46	,,

The engines consisted of a pair of single-cylinder direct-acting engines:—

Diameter of cylinders, ...	20 in.
Length of stroke, ...	36 ,,
Speed of engines, ...	. 10 double strokes per minute.
Water from condensation in the pipes, . . .	. 12 cubic ft. per hour.
Water from condensation when the engines were standing,	. 8 ,, ,,

The steam pipes stood on a strong piece of timber at the pit bottom, and were kept in a vertical position by stays fixed 27 ft. apart vertically.

An expansion joint was provided near the top, and the expansion of the pipes was found to be 18 in., or 2 in. for every 100 ft. in length.

It is clear that a very great loss must occur from long steam pipes however well they may be covered with so-called non-conducting composition.

Underground engines are of various types. Ordinary steam pumps with steam-moved valves, single and compound, but with various combinations of steam cylinders and pumps to suit the various conditions of application, and with various contrivances for giving motion to the valve of the steam cylinder; the object of the design is frequently to produce a compact and cheap pump. These are made single and compound, condensing and non-condensing, and two engines and pumps are often combined on

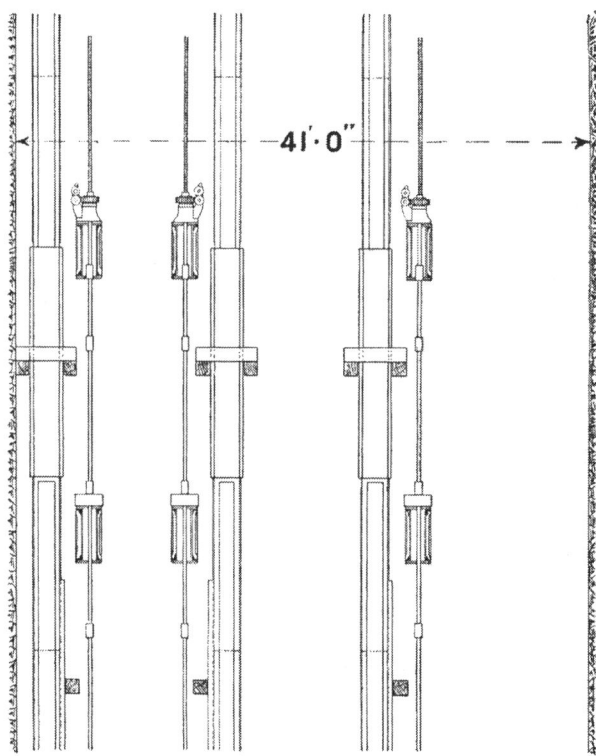

—41'·0"—

Fig. 111.—Detail of Pump Rods showing lowering rods and screws for sinking lifts.

the duplex plan of steam valve motion described in Chap. X., fig. 175. For the purpose of making these pumps cheap and self-contained, they have a short stroke. Where compactness is a great consideration a short stroke pump is useful, but as all non-rotative engines and pumps must have considerable clearance in the steam cylinders, the longer the stroke the less the percentage loss from clearance.

Where the work to be done is considerable and permanent, a long stroke is to be preferred for non-rotative engines. There is no difficulty in arranging long stroke engines on the duplex plan—that is, in having two complete double-acting engines and pumps with the valve motions so arranged that when the two engines are working together one engine commences its stroke before the stroke of the other is completed. At the same time, each engine may have its valve gear so constructed that in the case of accident to one engine the other may be kept at work. This is superior to the usual duplex engine, which consists of two complete engines and pumps mutually dependent on each other. If one engine is out of order, neither can be worked. On the question of compactness the ordinary duplex pump may be desirable for small powers, but for large powers other designs and arrangements are to be preferred.

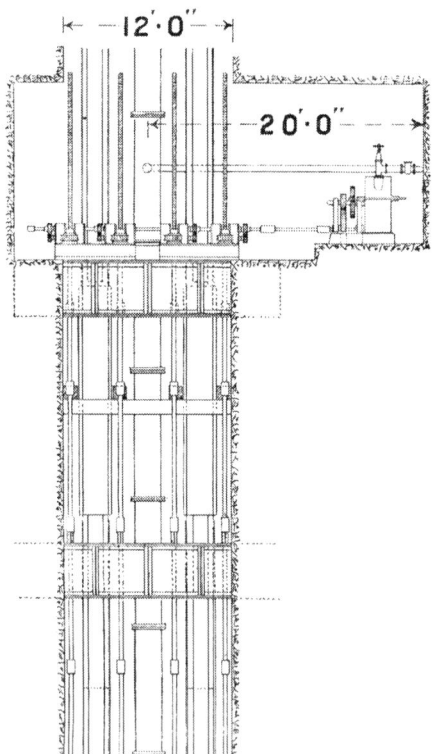

Dip Workings. — The dip or inclined workings in mines are sometimes drained by means of steam pumps, and as the pumps have to be moved from time to time, the short-stroke self-contained pump is convenient.

The author has recently met with a case where all the water of the mine followed the workings to the 'dip.' The shaft was 500 ft. deep and provided with a surface engine and shaft pumps. The workings extended from the bottom of the mine to the dip, a distance of 3000 ft., with an inclination of 1 in 10.

FIG. 112.—Arrangement of Pelton Wheels and Lowering Gear for Sinking Lifts.

Steam pumps of various types, simple and duplex, etc., were used to pump the water from the workings up the incline to the main sump. The total length of steam pipe was about 4000 ft., and the average consumption of coal per pump H.P. for all the steam pumps was 30 lbs., and as the boilers evaporated 6 lbs. of water per lb. of the slack coal used, the consumption of steam per pump H.P. for the system (including all losses of

FIG. 113.   FIG. 114.

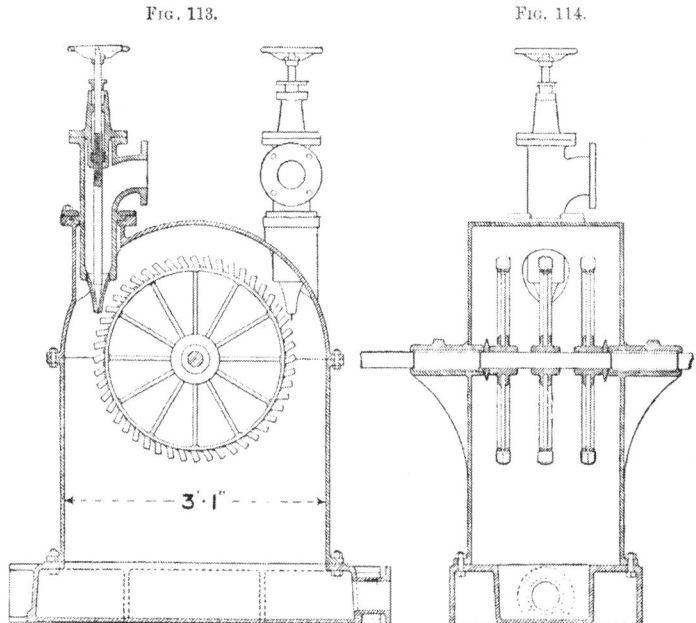

FIGS. 113 and 114.—Details of Pelton Wheels for working worm and screw gear for lowering and lifting sinking lifts.

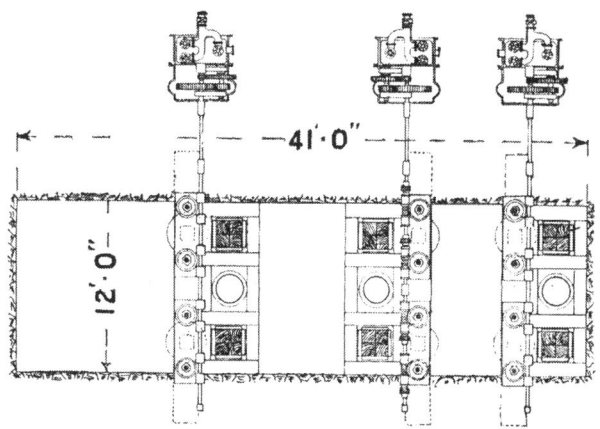

FIG. 115.—Plan showing Pelton Wheels and Lowering Gear for sinking pumps.

condensation in steam pipes, etc., amounted to 180 lbs. The work done was about 200 H.P.

This clearly was a case where hydraulic transmission of power from the surface would effect a great economy.

Rotative Engines Underground.—For permanent work underground various forms of rotative engines are used.

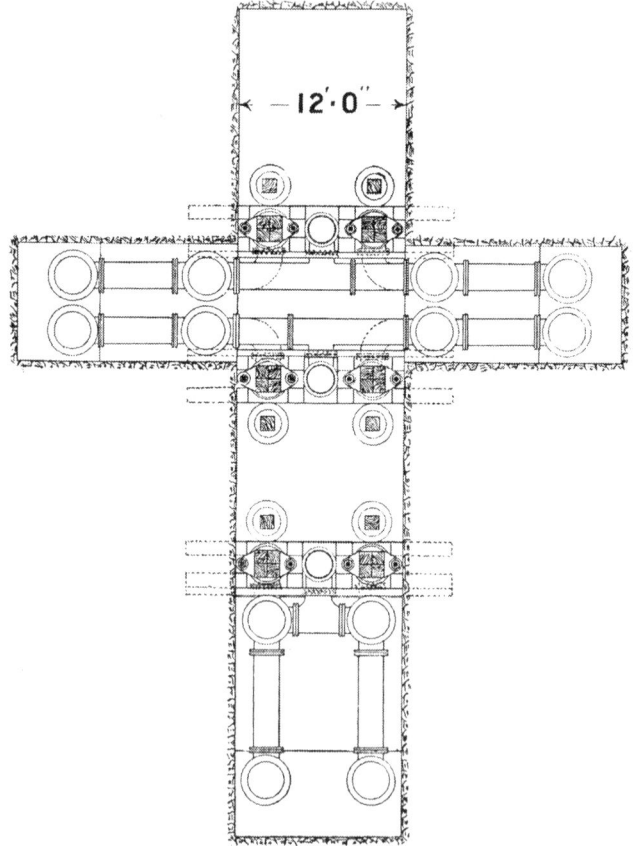

Fig. 116.—Plan of Plunger Lifts showing position of sinking lifts.

The type much used in France and Germany is the ordinary cross compound engine with double-acting plunger pumps.

The following are particulars of examples of such engines:—

Underground Pumping-Engine at the Hugo Mine, Westphalia.—Depth from surface to engine, 652 yards; distance from pit bottom, 280 yards.

The engine is of the horizontal cross compound type, having two cylinders. Attached to the low-pressure piston rod is a double-acting plunger pump, and attached to the high-pressure piston rod by means of a three-armed lever are two plunger pumps employed to raise the water to the feed

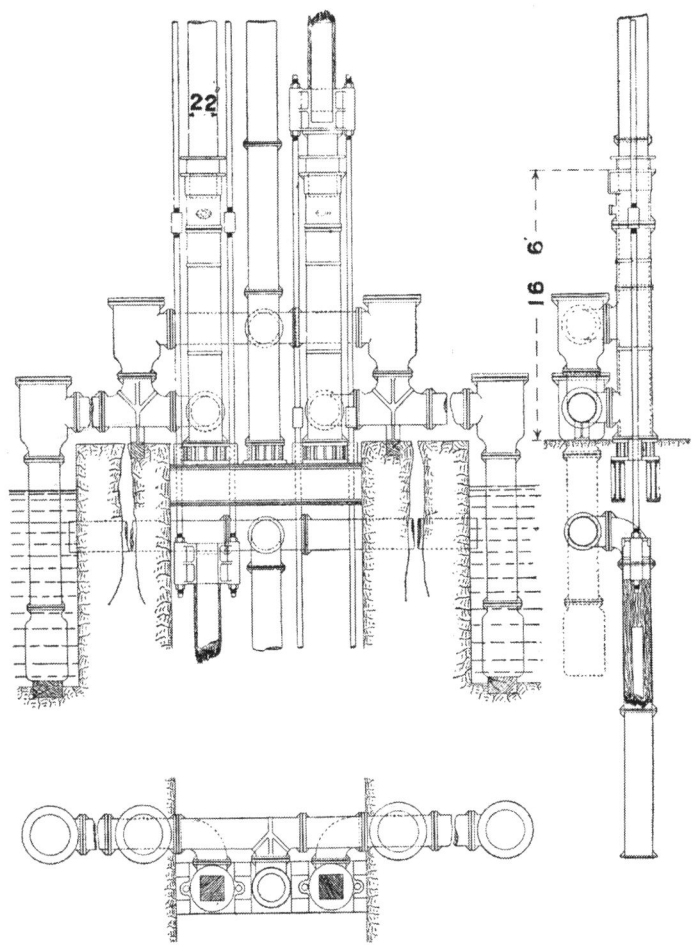

FIG. 117.—Elevations and Plan of Plunger Lifts.

cistern, a height of 45 ft. The steam from the low-pressure cylinder and the jackets is condensed by the water lifted by these pumps. The diameter of the high-pressure cylinder is 27·55 in., and of the low-pressure 45·27 in.; the length of stroke is 47·24 in.; the capacities of the cylinders are in the ratio of 1 to 2·75; the diameter of the main plungers is 7·55 in., and of rod 3·93 in.

The rising main is 9¼ in. inside diameter, with a total length of 1027 yards, the vertical rise being 641 yards. The steam pipe is 9½ in. in internal diameter, 1100 yards long, and is carried down the pit and along the connecting gallery side by side with the rising main. The steam pipe is covered with a layer of 'infusorial' earth ⅛ in. thick, over which is laid 1¹¹₁₆ in. thickness of paper pulp, and that has a covering of jute canvas lapped with galvanized iron wire with a double coating of sheet iron and lead,

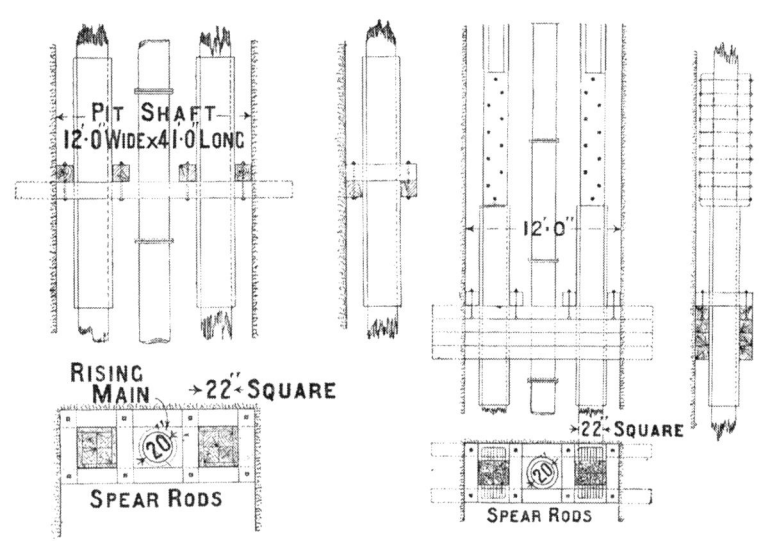

Fig. 118.—Detail of Pump Rods showing method of guiding or staying—Müke Mines.

Fig. 119.—Details of Pump Rods showing catch-pieces, banging beams and rising Main—Müke Mines.

painted with asbestos paint outside. The results obtained from trials continued over three-and-a-half months, showed that the work developed in the engines was 348½ H.P., and that of the pumps 287·55 H.P., or a useful effect of 82½ per cent. The actual discharge of the pump was 98·7 per cent. of the theoretical quantity. The consumption of steam under these conditions was 7782·32 lbs. per hour, of which amount 1741·16 lbs. was condensed and removed by the drain cocks. It appears, therefore, that in spite of the great care taken in protecting the steam pipe by non-conducting coverings, the loss by condensation is very considerable, being 22·4 per cent. of the total steam supplied.

The cost of the two engines and the necessary underground works was as follows :—

Cost of engines, rising mains and steam pipes,	.	.	£12,000 0 0
Buildings and proprietary works underground,	.	.	5,779 15 0
Total, .	.	.	£17,779 15 0

The engine when making 40 revs. per minute lifted 960 tons of water in a shift of eight hours at a total cost of—

Steam,	.	.	.	.	.	£1 14 0
Wages,	.	.	.	.	.	0 6 8
Lubrication,	.	.	.	.	.	0 3 3
Lighting and current repairs,		.	.	.	0 7 1
		Total,	.		.	£2 11 0

for 960 tons of water lifted 1968 ft.

It will be seen that the cost of the whole installation was about £60 per effective H.P., without boilers.*

Combined Direct-acting and Rod-pumping Engines.—At Arsimont, in Belgium, the drainage engines are required to lift a quantity of water varying from 1800 cubic metres per day in summer to 4000 or 5000 metres in wet winter. This has to be raised partly from 203 metres, and partly from 260 metres, but ultimately the lower level will be extended to 350 metres.

To meet these conditions, two compound horizontal engines have been placed at the 260 metre level, each driving four plunger pumps, which force the water to the surface, while a part of the power is transferred by spur gearing from the fly-wheel shafts to a pair of Rittinger telescopic pumps placed in a vertical shaft below the engine chamber, now 17 metres deep, but which will ultimately be carried down 90 metres more to the full depth of 350 metres.

The principal dimensions of the engines and pumps are as follows :—

A.—*Rittinger Pumps in Pit.*

Diameter of plungers,	.	.	.	.	.	0·495 metre.
Length of stroke,	.	.	.	.	.	1·200 ,,
Capacity per stroke,	.	.	.	.	.	0·232 cubic metre.
Diameter of suction pipe,	.	.	.	.	.	0·500 metre.
Diameter of rising pipe,	.	.	.	.	.	0·600 ,,
Velocity of water in rising pipe at 30 revolutions or 10						
strokes per minute,	.	.	.	.	.	0·850 ,,

---

* *Proc. Inst. of Civil Engineers*, vol. cxviii. p. 515.

B.—*Steam Engines.*

Diameter of H.P. cylinder, ...	0·600 metre.
Diameter of L.P. cylinder, ...	1·000 ,,
Stroke of pistons, ...	0·800 ,,
Length of receiver, ...	2·150 ,,
Diameter of receiver, ...	0·400 ,,
Diameter of fly-wheel, ...	3·500 ,,
Diameter of steam pipe, external, ...	0·200 ,,
Number of expansion joints in 250 metres, . . .	3
Diameter of exhaust pipe when working high pressure,	0·095 metre.

C.—*Condenser.*

Diameter of air pump, ...	0·500 ,,
Length of stroke, ...	0·320 ,,
Capacity per revolution, ...	0·080 cubic metre.

D.—*Force-pumps of Engines.*

Height of lift above engines, ...	250 metres.
Height of lift above sump, ...	268 ,,
Diameter of plungers, ...	0·180 ,,
Length of stroke, ...	0·800 ,,
Theoretical discharge of each engine per 24 hours at 30 revolutions per minute, ...	2·733 cubic metres.

E.—*Rising Pipes.*

Diameter of discharge pipes in the engine-room, . .	0·200 metre.
Speed of water per second at 30 revolutions, . .	0·970 ,,
Diameter of large air vessel, ...	0·700 ,,
Height, ...	6·100 ,,
Thickness of casting, ...	0·040 ,,
Diameter of rising pipes in up-cast pit,	0·280, 0·290, and 0·300 ,,
Thickness ,, ,,	0·030, 0·025, and 0·020 ,,
Velocity of water in main from the two engines at 30 revolutions, ...	0·260 ,,

The results obtained at the regular working speed of 30 revolutions were :—

Effective discharge of the pumps, . . .	97½ per cent.
Water lifted by both engines in 24 hours, . .	5·318 cubic metres.
Steam consumption at boilers per useful H.P. per hour as determined from the discharge, . . .	16·2 kilogrammes.
Total condensation per hour = 450 litres, . .	12 per cent.
Pressure of steam in the boilers, . . .	5 atmospheres.
Pressure of steam in the steam drier in engine-room,	4¾ to 5 ,,
Vacuum in condenser, ...	0·760 metre.*

On a High Lift Underground Pumping-Engine.—The engine has a single horizontal cylinder 0·75 metre in diameter, and a 0·80 metre stroke, provided with Meyer expansion gear adjustable by hand. The initial steam pressure ranges from 2½ to 3¾ atmospheres effective, and the cut-off from $\frac{3}{16}$ to ¼ of the stroke. The pumps—two single-acting plungers 110 millimetres in diameter, placed back to back—are in line with the steam pistons, and are connected with it by a cross-head, which also drives two small and heavy fly-wheels placed behind the steam cylinder.

* *Proc. Inst. Civil Engineers*, vol. cvii. p. 515.

The slide-valve is driven by an eccentric on the fly-wheel and the expansion valve from the cross-head. The cast-iron framing is exceedingly solid, the dimensions having been calculated on the supposition that it is a beam supported at the ends, and loaded uniformly through its length in order to render it independent of movement on the ground.

The condenser, with its air-pump of 0·260 metre diameter, is placed on one side of the engine. The whole of the water to be lifted passes through the condenser, and is delivered under pressure to the suction valve of the main pump.

The useful effect realised by the engine and pump combined varies from 71 to 77 per cent. of the power expended, as determined by experiments made in September 1888, when the pump valves had been at work five months without being changed. The principal results of these trials are contained in the following table :—

No. of Experiment.	1.	2.	3.	4.	5.
Date (1888).	Sept. 3.	Sept. 3.	Sept. 4.	Sept. 4.	Sept. 4.
	Centigrade	Centigrade	Centigrade	Centigrade	Centigrade
Temperature of water in pump,	33°	35°	29°	29°	29°
Temperature in hot-well of condenser, ...	47° to 49°	50° to 52°	47° to 48°	48°	49°
No. of revolutions, . . .	48	44	52	56	68
Theoretical discharge per hour, cubic metres, . . .	43·77	40·12	47·44	51·07	62·01
Measured discharge, . .	39·08	35·65	43·17	48·51	59·55
Duty of pumps, per cent., .	89·3	89	91	95	96
Work in water lifted H.P., .	83·4	76	91·24	103·48	127·24
Indicated H.P., . . .	117·5	109	126·99	134·28	167·24
Useful effect, per cent., . .	71	70	71·8	77	76*

* *Proc. Inst. Civil Engineers,* vol. cvii. p. 513.

CHAPTER VII.

PIT-WORK.

THE term 'pit-work' is employed to designate the pumps in the shaft and all that appertains thereto—pump rods, guides, shaft timbering, etc.

Pump rods are usually of red pine having a tensile strength of from 12,000 to 14,000 lbs. per sq. in. After making all allowances for the weaknesses not generally detected in large baulks of timber, and giving a proper factor of safety over and above the maximum strains to which such rods are exposed, the working load should not exceed 500 to 600 lbs. per sq. in. in main rods. Higher strains are employed when there is a good practical reason for keeping down the size, as in bucket pumps where the size of the rod is limited by the size of the pump. Iron or steel rods are used for

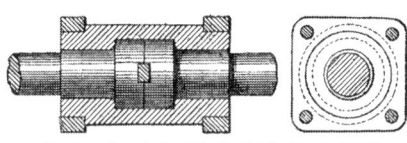

COUPLINGS OF WROUGHT IRON PUMP RODS,

FIG. 120.

bucket lifts, sinking lifts, etc., but the main rod of a large pumping-engine should be of wood.

Iron and steel rods have been used both in the solid and in built-up forms, but such rods have not generally found favour. The joints are apt to work loose, and the practical difficulties of effecting repairs is very great. Fig. 120 represents a form of wrought-iron or steel-rod coupling used in Germany, and also in a few cases in this country. Pump rods built up of girder sections, sometimes of the box form, have been used, but it has been found that there is a great difficulty in keeping the joints from working loose. Iron and steel rods of all forms, where they are subject to alternating tensile and compression strains, are subject to greater vibrations and derangement than wooden rods. The rods should be heavy enough to overcome the resistance of the pump so that there may be no thrust on them from the engine during the down stroke. When the double rod or balanced system of pump rods is employed, it is usual to make each rod at least 20 per cent. heavier than the water load.

When single rods are used, as with the Cornish engine, the surplus

Engine here.

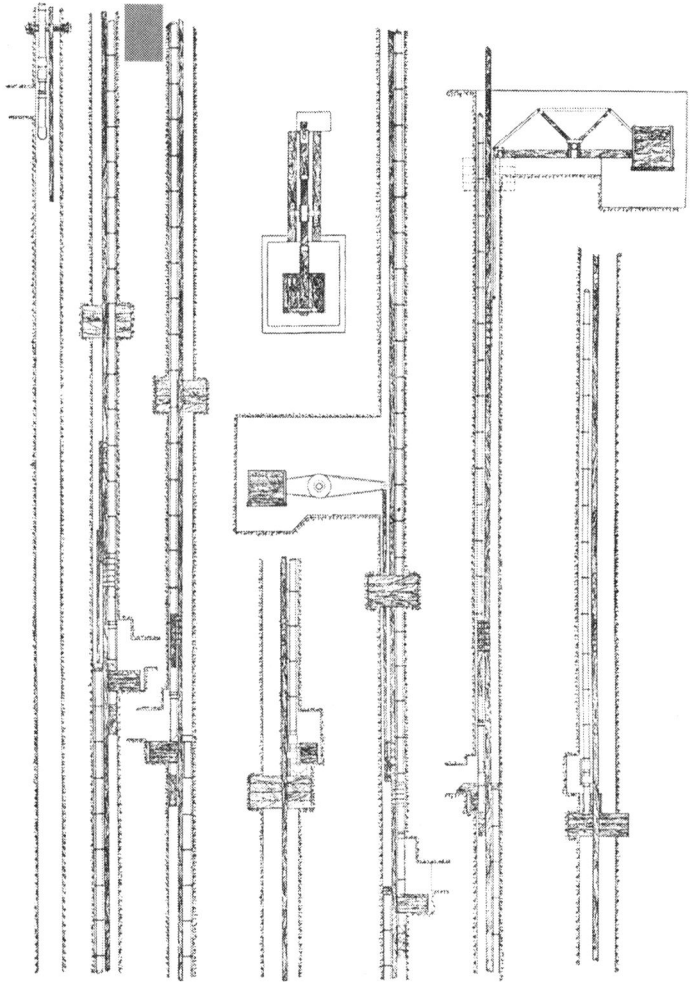

FIG. 121.—Cornish Pit-work showing front and side elevations of pump work, rods, and balance bobs for the entire depth of shaft.

weight of the rod is often much greater; the surplus is then taken up by balance bobs, one being placed on the surface where nice adjustments can

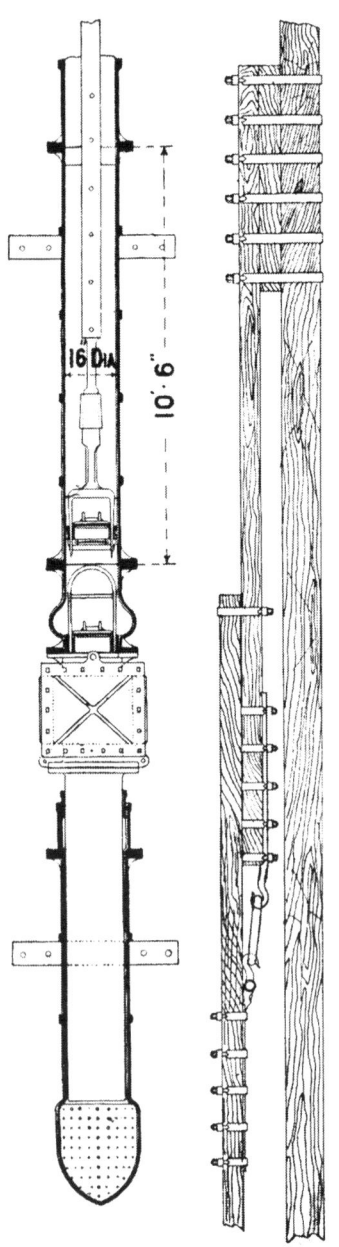

FIG. 122.—Details of Cornish buckets or sinking lifts

easily be made by putting on or taking off weight. In a great length of rods, other balance bobs, placed underground, are sometimes found necessary. An example of the Cornish system of pit-work is given in fig. 121. In this system the sinking lift is used for a depth of 30 to 40 fathoms; then a plunger pump is put in at that depth, and the sinking lift used for a further depth; and so on. A detail of the sinking pump is given in fig. 122, and of the plunger pump in figs. 123 and 124.

For a lift of not more than 40 fathoms leather-faced flap valves may be used both in the sinking and plunger pumps, but much higher lifts have been made possible by alterations in the form of the pumps and the use of double-beat and other types of valves in the place of the old Cornish type.

The jointing of the rods should have careful attention. Splice jointing is not to be recommended. There is nothing better or more easily constructed than butt joints.

The ends should be carefully 'squared,' and the bolt holes for the strapping plates bored with a little 'draw,' so that when the bolts are all tightened up the ends of the rods may be drawn together.

Butt joints are easily tightened by wedging should they become loose.

There is less vibration with wooden rods than there is with iron or steel; the wood being so much larger in section for a given strength is in a better proportion for the function it has to perform, viz., that of lifting the water by

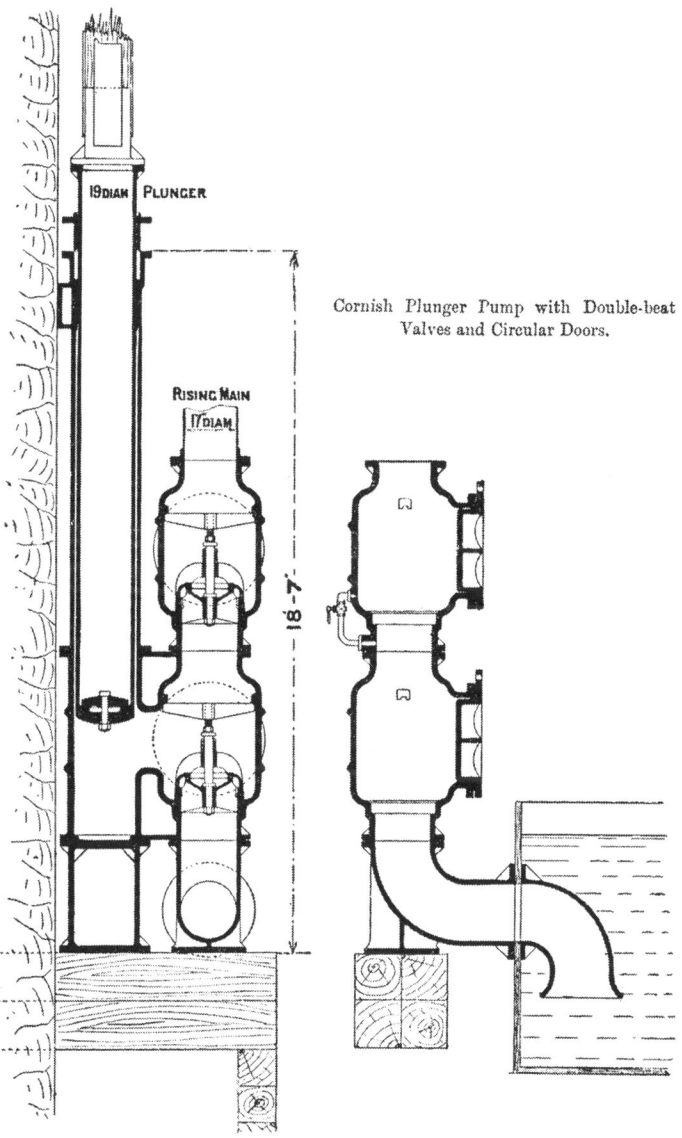

19 DIAM PLUNGER

RISING MAIN

17 DIAM

Cornish Plunger Pump with Double-beat
Valves and Circular Doors.

18′-7″

FIG. 123.   FIG. 124.

its own weight, and in working subject to alternate compression and
tensile strains.   During the down stroke the engine should not put any
thrust whatever on the rod.   If it does,
then unpleasant buckling and vibration
will occur.   The rod should hang on
the pin of the beam or quadrant with
considerable weight during its descent,
therefore it must be heavier than is required to merely overcome the resistance
of the plunger.

Wood rods require less staying than
iron or steel.   It is usual to make each
length of rod 50 ft., placing a stay in the
middle of each length.   Each joint is
made with four strapping plates of mild
steel.   It was the custom to make the
plates of hammered iron and to taper
them so that they might have the
greatest strength where most needed,
viz., in the centre, but parallel plates
of mild steel are now usually employed.
Rubbing-pieces of hard wood should be
fastened on the four sides of each rod
where the guides occur.

Attempts have been made to form
the pump rod into a rising main, thereby
economising space in the shaft.

A pump constructed on this plan
(shown in fig. 125) has been erected in
the north of England, but we have had
no information as to its efficiency.   We
fear, however, that the practical difficulties of applying the system satisfac-
torily to heavy pumping will be found
to be very great.

In inclined shafts rods are carried
on iron rollers.   An example of pit-work
in an inclined shaft is given in figs. 126
and 127.   The machinery includes a
pumping engine, a man engine, and a
capstan engine,—the relative positions
of the engines at the head of the shaft
being shown in plan in fig. 127.   The

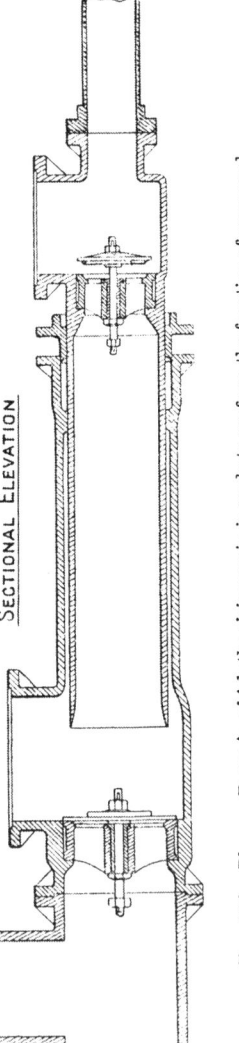

SECTIONAL ELEVATION

FIG. 125.—Plunger Pump in which the rising main is made to perform the function of pump rod.

first 423 ft. of the shaft is vertical, and the remainder inclined at an angle
of about 60 degrees with the horizontal.

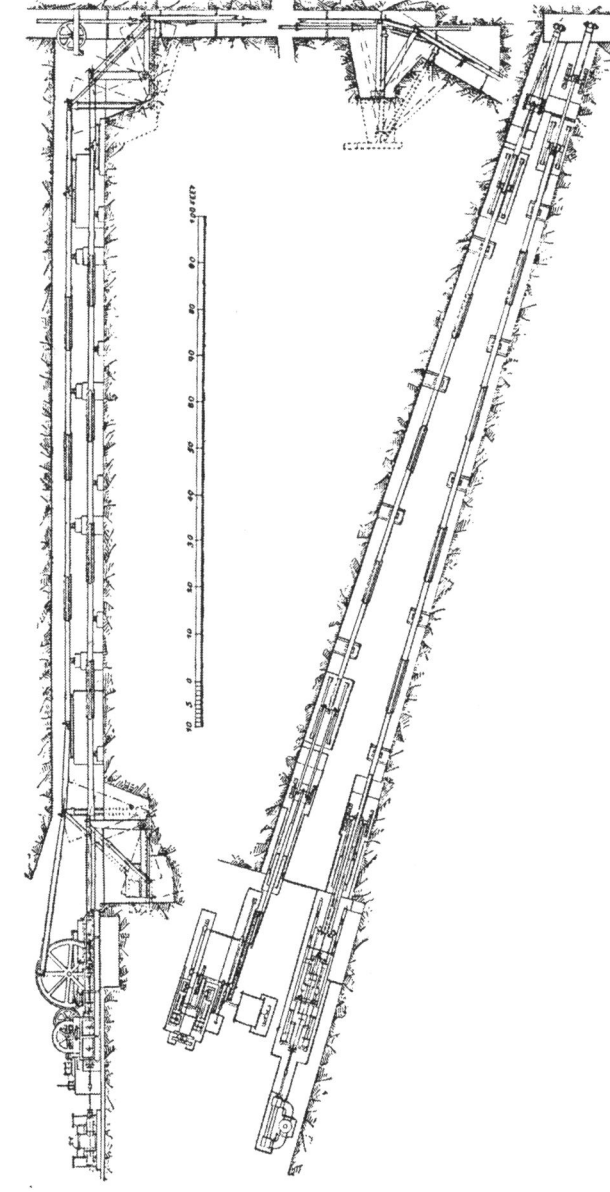

Fig. 126.

Fig. 127.

Figs. 126 and 127.—Compound Horizontal Engine with horizontal, vertical, and inclined pump rods, also man and capstan engine combined. Plan and elevation.

The man engine and the capstan are driven by an engine having a pair of cylinders 11 in. in diameter and a 20-in. stroke. On the crank-shaft of this pair of engines is a pinion gearing into a spur wheel on a counter shaft, this counter shaft also carrying a pinion gearing into a spur wheel on the capstan drum shaft, and a spur wheel gearing into a large spur wheel carrying a crank pin, from which the rod of the man engine is operated. The capstan drum is 6 ft. in diameter and 5 ft. 3 in. long, running at the same speed as the man engine. It is adapted for lifting a load of 5 tons.

The man engine has a single line of rods balanced as shown in figs. 126 and 127. The rods have a stroke of 12 ft., and make five strokes per minute. In ascending or descending the men step alternately from the stands carried by the rods to fixed stands at the side of the shaft.

It will be noticed that the balance quadrant is placed next the engine, so that the horizontal spears are in tension ; this is also the case with the pumps. The rods are made to run on rollers. Both the capstan and the man engine can be thrown out of gear.

The pumping machinery consists of a compound differential pumping-engine, actuating three plunger pumps in a line, one below the other.

The engine has cylinders 26 in. and 46 in. in diameter respectively, with an 8-ft. stroke, making eight double strokes per minute, at which speed the pumps will raise 150 gallons per minute. It is worked with steam at 50 lbs. pressure.

The total vertical height of 2211 ft. is divided into four lifts, the bottom pump being a bucket lift. The plunger pumps are of the construction shown in fig. 128. These pumps have plungers $8\frac{1}{2}$ in. in diameter, covered with gun metal. The valves are double-beat, and are also of gun metal.

A large example of inclined pit-work is to be found at the Lindal Moor Mines. The shaft is inclined, but quite straight, having a vertical height of 600 ft. in an incline 1000 ft. long. The pump rod is 24 in. square, the section being built up of four 12-in. baulks. It runs on rollers, and at the bottom, attached to the end of the rod, is a plunger 30 in. in diameter, having a stroke of 10 ft. The vertical height of the lift is 600 ft.

Height of Lift.—As before intimated the possible height of the plunger lift has been greatly increased by improved construction of pump.

The Cornish H piece is very weak in form, and the flap valve is unsuited to heavy pressures.

The modern form of pump is that illustrated in Chap. VI., fig. 117. To accommodate this form of pump it is necessary to make an excavation in the side of the shaft to receive the valve boxes and the suction tank. The only working parts of the pump standing in the shaft are the plunger and plunger barrel.

If the pit has already been sunk, it is easy to determine what length the lift should be, because it is quite practicable to use lifts as high as

800 to 1000 ft. with safety. Lifts of over 800 ft. have been largely in use for many years. If, however, the pit has yet to be sunk, then it is a question the consideration of which must be governed by the special circumstances of the case. The system of permanent pumping may be determined by the selection of the system of sinking. In the Cornish system a plunger is needed every 40 fathoms, but by employing two sinking pumps, one in advance of the other, a depth of 80 fathoms can be accomplished before a plunger is put in; greater depths can be covered by specially designed bucket lifts; such lifts have been used for 300 ft., so that two sinking lifts would cover a depth of 600 ft. Temporary

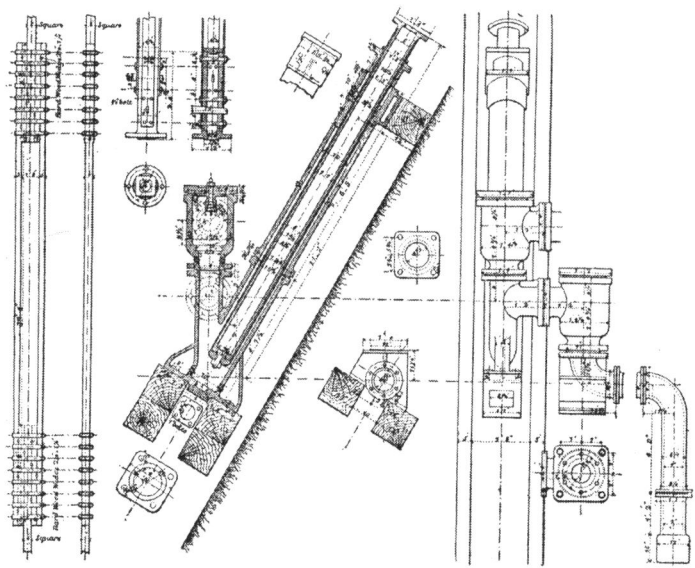

Fig. 128.—Details of Plunger Lift, Rods, and Connections, as fixed in the inclined shaft indicated in figs. 126 and 127.

steam pumps are often placed in recesses in the side of the shaft during the sinking operations, and retained there for emergencies after the completion of the pit. Steam pumps are also suspended in the shaft to take the place of sinking lifts. These are questions for the consideration of the engineer having a knowledge of the local circumstances.

Sinking pumps are suspended by rods, chains, wire rope, blocks, or by wooden rods. For heavy pump work, steel rods and lowering screws, as illustrated in Chap. VI., should be employed. A powerful steam capstan is also required for lifting and lowering the pump work and putting it in position. It is advisable to have the permanent pit-head gear erected before the sinking is commenced.

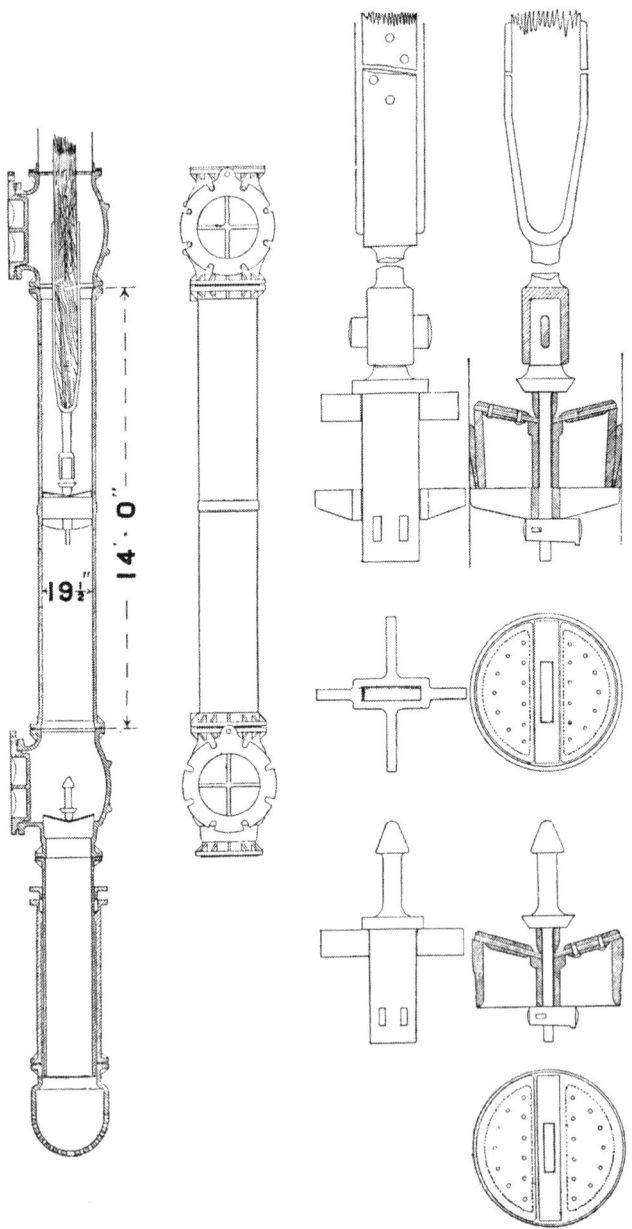

Fig. 129.—Modern Bucket Lift for Sinking Purposes.

An example of large modern sinking and permanent pumping plant is fully described and illustrated in Chap. VI.

Bucket Lifts for Sinking Purposes.—The Cornish bucket lift is illus-

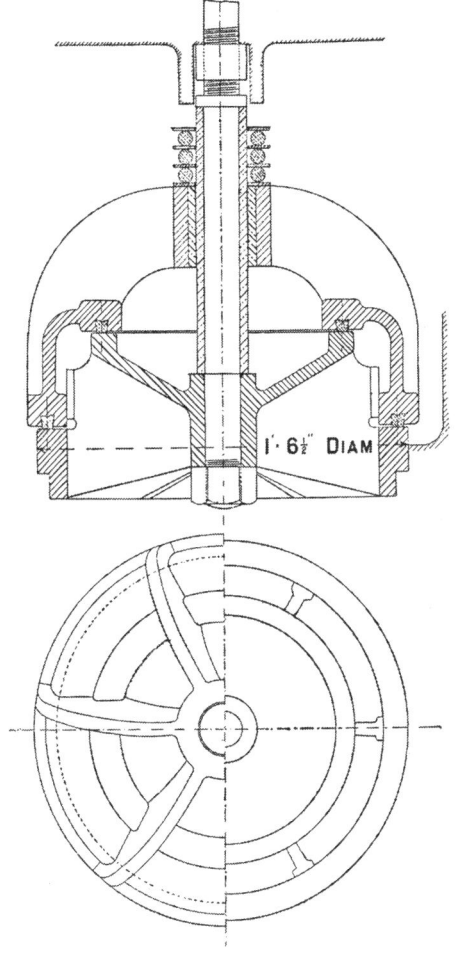

I·6½" DIAM

FIG. 130.—Double Beat Pump Valve as used in plunger pumps of mining pumps.

trated in fig. 122 ; it is used for depths of 30 to 40 fathoms. The suction pipe is provided with a sliding wind bore, the bottom of which should be made very strong to resist the effects of blasting. Wooden plugs are driven into the upper holes, where necessary, to keep down the level of the water in which the men work. Another device is that of carrying down the suction pipe inside the 'wind bore' nearly to the bottom. The wind bore is suspended by a pair of blocks, and lowered as the ground is excavated. When the sinking lift itself is suspended in lowering screws, the sliding wind bore is sometimes dispensed with, but it is generally useful to have it. The working barrel should be made much longer than that necessary for the stroke of the pump, it may then be lowered some distance whilst at work without the bucket coming out of the barrel.

The suction valve box is provided with a door-piece of sufficient size for easy examination and removal of the valve and seating; while above the working barrel a similar door-piece is provided for examination and removal of the bucket. Originally these door-pieces were made square with flat doors, and for small pumps and light pressures the design is not particularly objectionable, but the modern form of door-piece is globular with a circular door (fig. 129), thus securing greater strength.

Where clack doors are used, a safety clack piece is often provided above the door-piece, the valve of which can be drawn by means of fishing tackle from above, for use in the event of the water rising so fast as to prevent the pitman from putting the door in place, as shown in the Cornish bucket lift (fig. 122).

The old plan was to secure the valve seatings in their places by making them taper, and winding thick cord around them, or 'gearing' them with leather like that of a bucket. Taper seatings have often given great trouble by seating themselves so firmly that rods and screws have been required for the purpose of drawing them. Sometimes the seating loop breaks under such strain, and the lift has to be taken out bodily; to meet such contingencies the supports of the lift should be so placed as to allow the lift to be raised bodily by screws or capstan. For heavy pumping it is not advisable to have taper seatings, but to construct them in the manner shown in fig. 50, Chap. III. This seating can be drawn by means of fishing tackle attached to a wire rope. Various kinds of fishing tackle are in use.

For heavy pumping a stronger form of valve and box than that shown in connection with the Cornish pit-work (fig. 122) is required; double-beat valves (fig. 130) are then used. The single flap and double flap or butterfly valves are less liable to become permanently 'gagged' by bits of wood, stone, etc. Special care is needed to exclude gags when double-beat valves are employed. With small sinking pumps flexible suction pipes are used, and sometimes several suction pipes are attached to the same pump, so that more than one pool in the shaft bottom may be commanded at the same time.

Pulsometers are also used as pilots to sinking lifts.

For heavy pumping it is not generally advisable to have bucket-door-pieces of the size necessary for removing the bucket and valve seating, because of the great weight and space occupied. The lifts should be so arranged, when possible, that head-room is provided sufficient for drawing the rods bodily out of the lift by the steam capstan, and then a fishing tackle lowered by a rope is used to bring up the valve and seating.

CHAPTER VIII.

Bailing Tanks.—There are various ways of dealing with water in sinking shafts, and where the quantity to be dealt with is small the difficulties are not serious, but the problem how to deal with large quantities economically and expeditiously is a matter of some difficulty.

The actual sinking of the shaft has to be considered in connection with the pumping appliances. The space available for the pumping arrangements is often very limited after sufficient space has been allowed for the mining operations. The shaft may have to be lined with tubbing; these and other circumstances, general and local, require consideration.

When the quantity of water is small, it is frequently dealt with by the winding engine. The kibbles for raising water (fig. 131) have a valve in the bottom which may be lifted by a hand lever for discharging; or the valve may be opened by lowering the kibble on to the discharging trough on the top of the shaft. An ingenious form of bailing tank (fig. 132) has been used. The tank b is a cylinder with a piston a, and a foot valve, the piston being attached to the winding rope, and the tank lowered into the water at the bottom of the shaft. The piston descends to the bottom of the cylinder when the tank has been lowered, and rests on the bottom of the shaft. On the tank being drawn up the piston first rises till it comes to a stop at the top of the tank. During the rise of the piston, water has been drawn in through the foot valve, and the tank is drawn up to the surface with the water in it. The water is then discharged by lowering the tank on to the discharge launder, where a provision is made for lifting the foot valve. Thus the filling and the emptying of the tank is done automatically. Large bailing tanks are used with special arrangements for filling them, such as pulsometers, steam pumps, etc.

Bailing tanks may be used to raise the water to the surface, or to a point in the shaft where pumps have been provided. The water in the upper strata may be caught by ring dams or other methods, and pumps placed there to prevent the water falling to the bottom of the shaft. The water thus caught may be dealt with by temporary or permanent pumps, the method adopted being governed by the local and special circumstances of the case.

Bailing tanks are sometimes hung in the shaft, into which winding tanks are lowered by the winding engine. The suspended tanks are provided with filling appliances, such as pulsometers, steam pumps, etc. The arrangements for discharging the winding tanks or kibbles are various; a running platform which is run in over the shaft and under the tank may

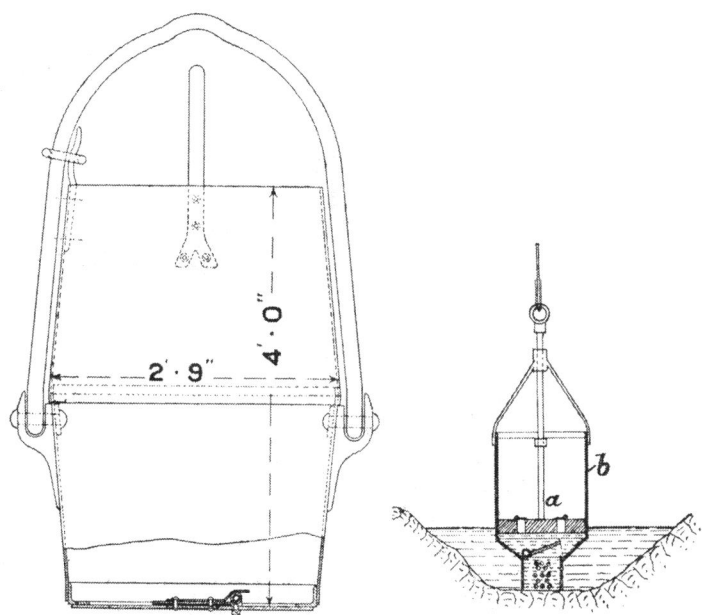

Fig. 131.—Mining Kibble with valve in bottom for use in winding water.

Fig. 132.—Cylindrical Tank with piston and valves for bailing water.

be used. The platform forms a discharging launder for the water. This is a convenient and safe arrangement.

Tipping appliances have also been used.

Professor Galloway has designed special tank appliances for winding water, used in unwatering a pit in South Wales. The mechanism and the results obtained are described in detail in the *Transactions of the Federated Institution of Mining Engineers*, vol. xiii., from which we have derived the following information.

There are two water tanks in the positions of ordinary winding cages provided with wire rope guides. Automatic appliances fill and discharge the tanks. One of the tanks is shown at A (fig. 133).

In the bottom of the tank are fixed two flap valves, opening inwards, and attached by rods to the levers D D. The filling of the tank is effected by simply lowering it into the water in the shaft. On the tank reaching the surface, the levers D D come into contact with levers C C (fig. 134), to which are attached weights sufficiently heavy to open the valves. The tank in this position is ready to discharge its water into two shoots, one on each side of the shaft. To conduct the water to these shoots, the bottom of the tank under the valves is provided with inclined shoots,

Fig. 133.　　　　　　Fig. 134.　　　　　　Fig. 135.

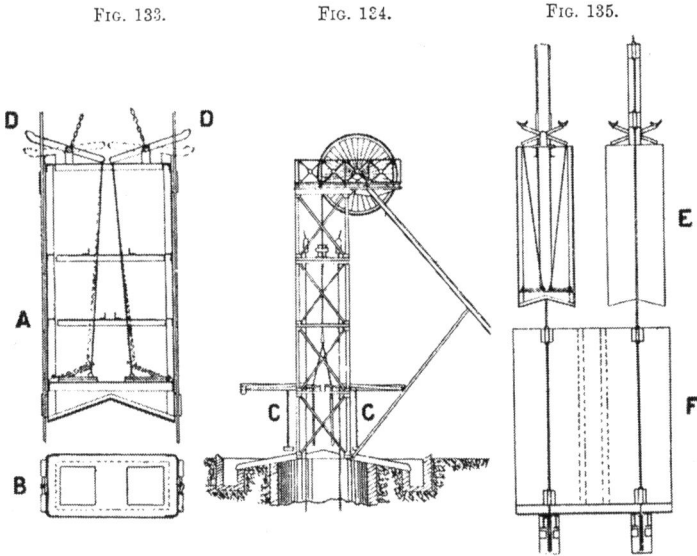

Figs. 133-135.—Galloway's Tank Appliances for winding water.

which direct the water from the tank to the shoots at the side of the shaft, the water leaping over the short intervening space.

Fig. 135 represents a method of using the same tanks for winding water from a tank suspended in the shaft.

The tanks with all their mountings weighed about 3 tons each, and the weight of water carried by each of them was nearly $4\frac{1}{2}$ tons. The top of each tank was covered by a plate with a rectangular hole in it, just large enough for a man to pass through, and the position of this hole was naturally such that the rods which connected the levers to the valves passed through it. When the tanks arrived at the surface they were full to the very brim so long as when lowered they were dipping under the surface of the water in the pit. The necessary labour consisted of three enginemen working continuously in shifts of eight hours each. In addition to the enginemen, two banksmen working in shifts of twelve hours each, and one engine-cleaner by day, were employed.

The principal work of the tanksmen consisted in counting the number of tanks drawn, observing when the ropes required to be lengthened, assisting in lengthening them, and keeping an eye upon the operations generally. The maximum number of tanks drawn was 94 per hour. Allowing 9 tanks per hour for contingencies, that is, taking 85 per hour as an average for 24 hours, the quantity and cost per ton would be somewhat as follows, when the depth of the water from the surface was 600 ft.

Quantity of water raised in 24 hours = 2,040,000 gallons = 9107 tons ; H.P. 257 ; coal consumed at 10 lb. per H.P. per hour, 27 tons 10 cwts. ; duty of 1 cwt. (112 lbs.) of coal, 22,254,196 foot pounds.

COST PER TON OF WATER RAISED 600 FT.

	£	s.	d.
3 Enginemen at 5s. 6d., ...	0	16	6
2 Banksmen at 4s. 2d., ...	0	8	4
1 Engine-cleaner at 3s. 6d., ...	0	3	6
27½ tons of coal at 6s. 6d., ...	8	18	9
Stores and light, ...	1	0	0
Ropes, at 0·05d. per ton raised, ...	1	17	6
Total, . .	13	4	7

Or 0·347d. per ton, or 0·259d. per 1,000,000 ft. pounds.

It will be remarked that the duty does not much exceed one-third of that of a fair Cornish pumping-engine, and reached only to about one-fifth of that of a very high-class pumping-engine of the most approved type ; and, therefore, one would not be justified in resorting to this method of raising water for a prolonged period of time. It had, however, the advantage of easy application and low first cost. The Cornish engine and pumps requisite for the same work would have cost £5000 or £6000, whereas the two tanks and their accessories cost only £140.

Pit Sinking.—Assuming that all the water in the shaft or pit has to be delivered from the bottom to the surface (except such water as may be kept out by lining or tubbing), we will proceed to consider the methods of dealing with large quantities of water during the sinking of the shaft. Winding the water is impracticable ; it must be done by a system or systems of pumping.

The Cornish system is described in another chapter, but Cornish pit-work involves the Cornish system of permanent pumping in short stages, a system useful in many cases for metalliferous mining, but not desirable when long lifts are practicable.

In the Cornish system the sinking lift seldom exceeds 200 ft., so that the plungers are about 200 ft. apart ; a shaft 1200 ft. deep would have six plunger pumps, each with a 200 ft. lift.

The construction of the Cornish pit-work is such that lifts of much more than 200 ft. are not practicable.

The valves are not adapted to high pressures, nor are the valve boxes,

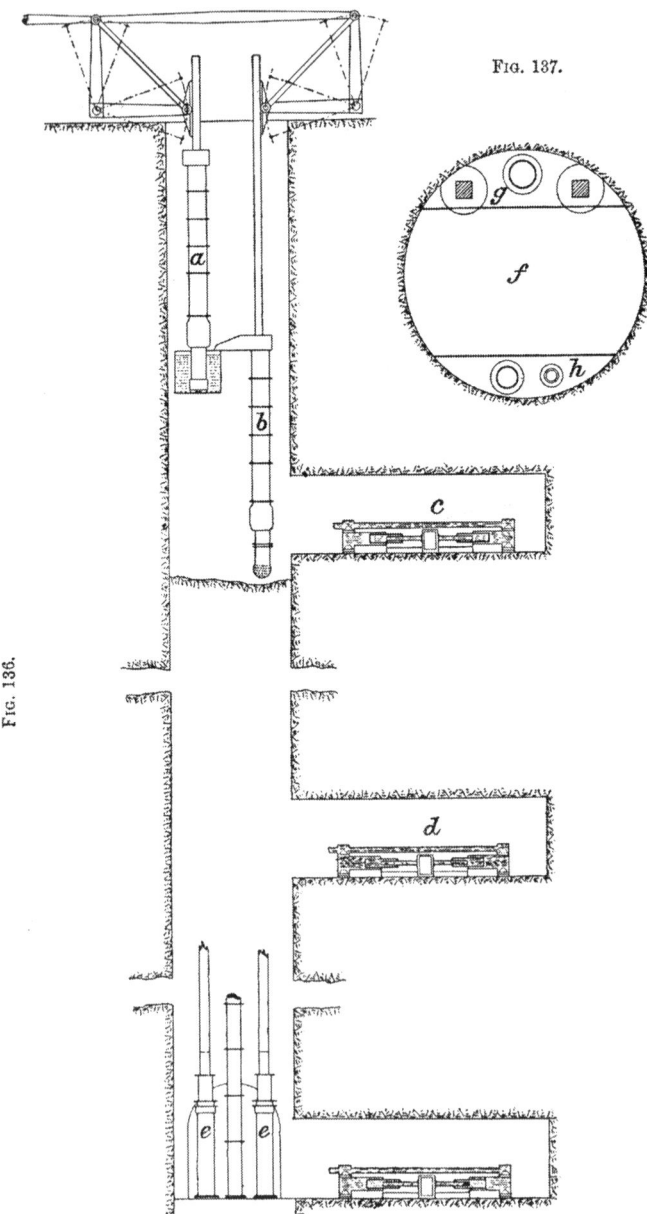

Fig. 137.

Fig. 136.

Figs. 136 and 137.—Illustration of a Method of Pumping during shaft sinking, and the installation of permanent pumping plant.

or 'H pieces,' as they are termed. The distribution of metal in the design is bad, and even for light pressures the weights are excessive. The author claims to have initiated and introduced the system of pit-work now adopted for high lifts.

Valve boxes are made circular and spherical instead of square and cubical, thereby obtaining a better distribution of metal, but that necessitated quite a new arrangement of pit-work.

Sinking pumps of large size are now used for lifts which do not exceed 300 ft., and plunger pumps with pump rods for lifts of 1000 ft. and more.

We will now consider the problem of sinking a shaft and providing it with permanent pumping plant.

Take the case of a colliery shaft to be sunk to a depth of from 900 to 1200 ft. In permanent pumping this is not too great a depth for one lift either with underground, or surface engines and pump rods. The question is how shall we proceed to sink the shaft so that when it is completed it may be equipped with modern and approved permanent pumping plant. It could be done by the use of the Cornish system of sinking in short stages, putting in plunger pumps at each stage, and, on the completion of the shaft, erecting the permanent plant in another part of the shaft, and abolishing the plant used in sinking.

A heavily watered mine should be provided with a duplicate pumping plant. A convenient arrangement is to have a surface engine with pump rods and plunger pumps capable of pumping all the water, and underground pumps equal to the full work, as reserve power. The underground pumps may be either steam or hydraulic. The arrangements for sinking the shaft might then be such as are indicated in figs. 136 and 137. Let it be required to sink the shaft to a depth of 1200 ft. Sink to 400 ft. with the sinking pumps a and b, then put in the underground pump c. The sinking pumps may then be used for the next 400 ft., when the second underground pump d would be put in, and so on for the next stage. On the completion of the shaft, plunger pumps might be put in as at e, and rods carried up to the surface to complete the permanent surface pumping plant, the underground plant pumping the water during that time. The sinking pumps a and b may be lowered by means of rods and screws, as described in detail in Chap. VI.

To avoid constantly lowering the sinking lifts, pilot pumps may be used, the lift pumps being kept supplied by means of tanks.

Where the shaft is very heavily watered the sinking pumps a and b may be used side by side for a depth of 300 ft., and the remaining 100 ft. or more covered by the use of pilot pumps.

Let fig. 137 be a plan of the shaft; then the surface engine would actuate pump rods in the space g, and the space h would serve for the accommodation of the steam or hydraulic pipes and rising main for the underground engines, the space f being reserved for the winding cages.

K

Sinking pumps are sometimes suspended by means of spear rods and rope blocks, as illustrated in fig. 138.   Steam pumps are also employed for

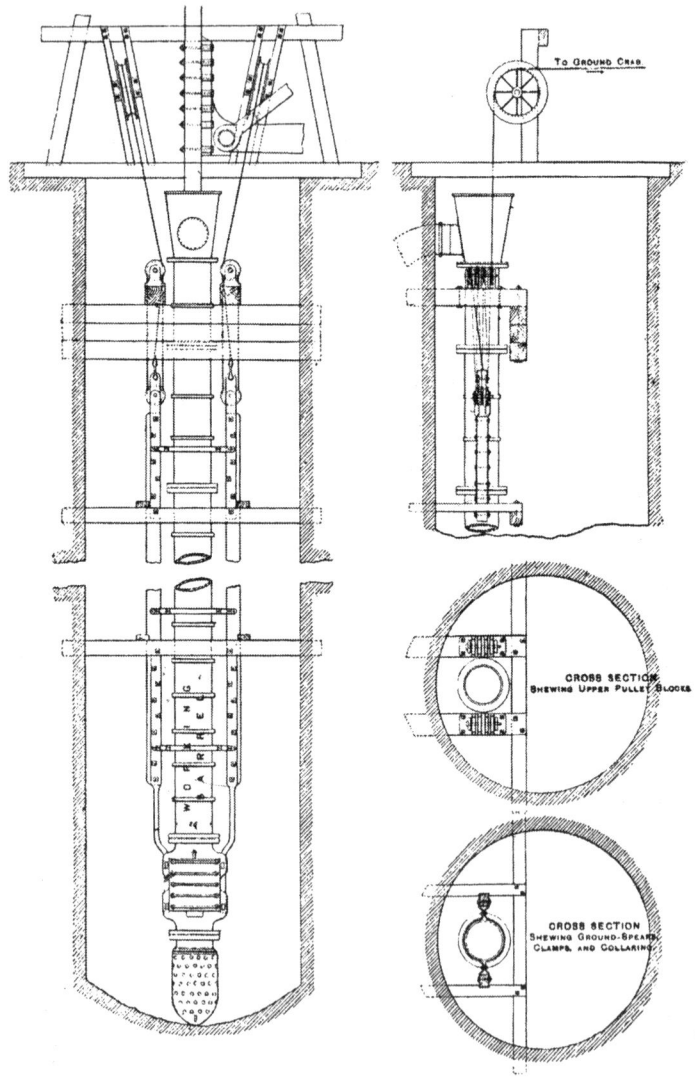

Fig. 138.—Sinking Bucket Pump suspended by means of rope blocks.

sinking purposes ; sections of a pump of this class are shown in figs. 139 and 140.

Pulsometers are also used; an illustration of the ordinary form is given in fig. 141.

The pulsometer is a self-acting steam pump without pistons. It is

Fig. 139.   Fig. 140.

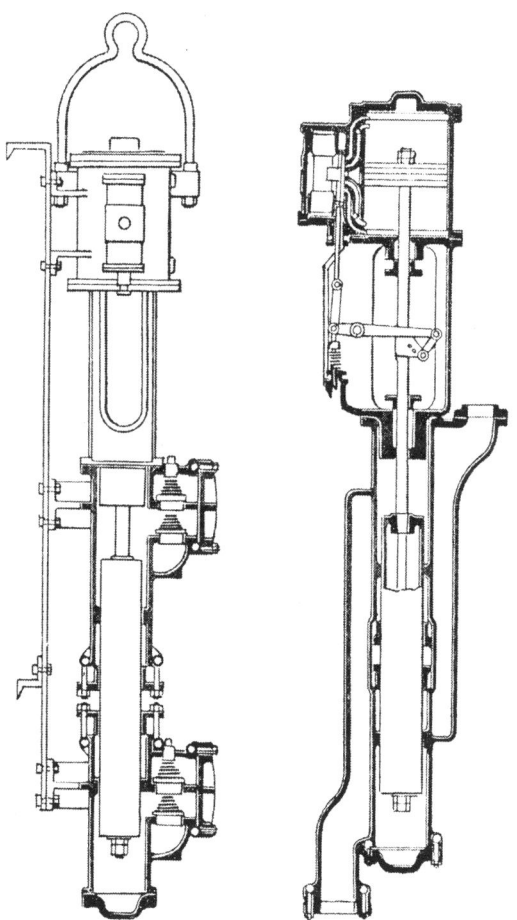

FIGS. 139 and 140.—Cameron Sinking Pump.

the old Savery engine, fully described and illustrated in Chap. I., made automatic and portable.

Because of its portability and handiness it is useful for all kinds of temporary work. The consumption of steam for work done is very great.

Referring to the illustration, fig. 141, A A are two vessels alternately filled with water and discharged. E E are the suction and F F the delivery valves. K is the steam pipe from the boiler, and C the suction pipe for the water.

Steam is assumed to be coming the direction of the arrows into one of the vessels marked A and to be forcing the water through the delivery valve F into the delivery pipe; at the same time water is coming from the suction pipe into the other vessel A, a partial vacuum having been obtained by the condensation of the steam. On the complete, or nearly complete,

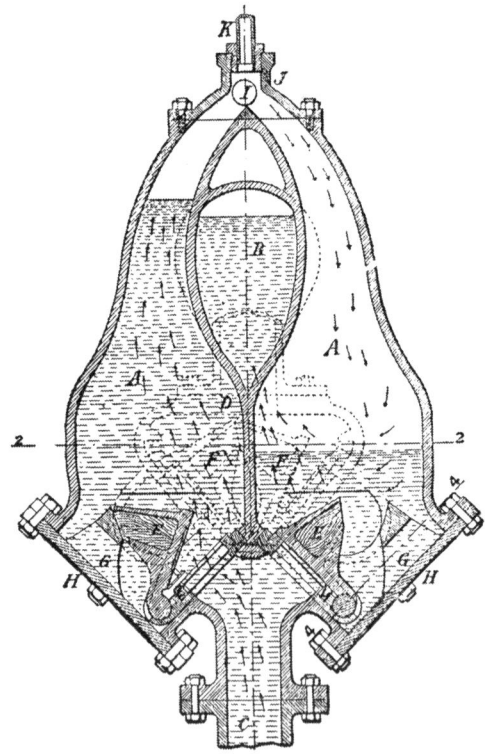

Fig. 141.—Section of Pulsometer.

filling of this vessel, the valve I is forced over so as to close the communication with the vessel which has just been discharged and admitting steam to the vessel which has just been filled. The action has thus been reversed, and the operation goes on automatically, one vessel filling whilst the other is emptying.

The choice of plant and appliances is largely influenced by the magnitude of the work, and by local circumstances.

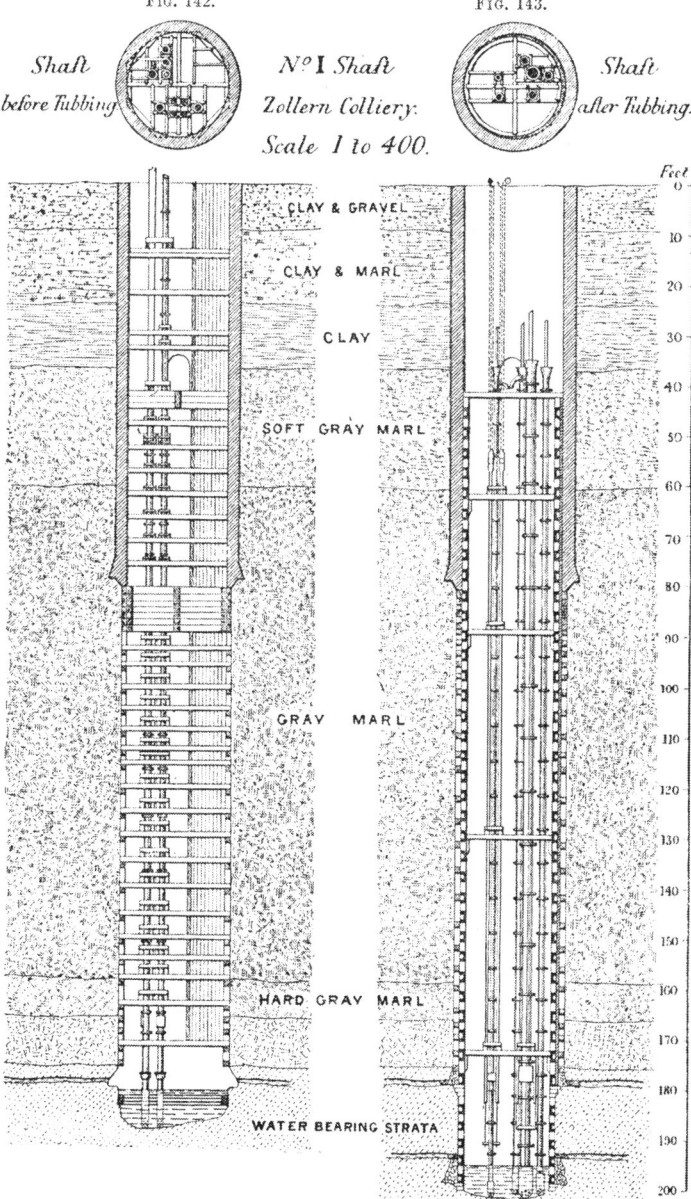

FIG. 142.

FIG. 143.

Shaft before Tubbing

Nº I Shaft Zollern Colliery. Scale 1 to 400.

Shaft after Tubbing.

CLAY & GRAVEL

CLAY & MARL

CLAY

SOFT GRAY MARL

GRAY MARL

HARD GRAY MARL

WATER BEARING STRATA

Feet
0
10
20
30
40
50
60
70
80
90
100
110
120
130
140
150
160
170
180
190
200

FIGS. 142 and 143.—Sections of Shafts—Zollern Colliery.

As an example of a heavily watered pit, and the difficulties met with in dealing with the water, we have taken the following description of the methods employed at the Zollern Colliery, Westphalia, from a paper by

FIG. 144.   FIG. 145.

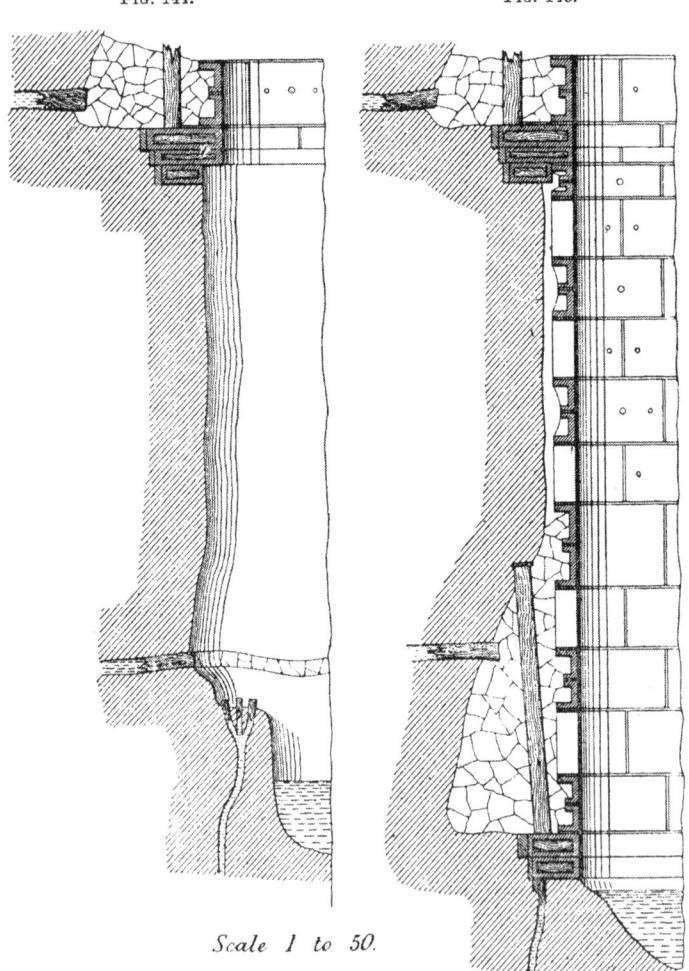

Scale 1 to 50.

FIGS. 144 and 145.—Sections of Shaft Tubbing—Zollern Colliery.

Messrs T. and W. Mulvaney in the *Proceedings of the Institution of Mechanical Engineers*, 1882.

"In this colliery, situated near Kirchlinde, and containing 3000 acres, two shafts were commenced by the original company, much nearer to the southern outcrop of the marl formation than in the collieries previously described. In consequence, the thickness of the marl is only 351 ft. to the surface of the coal measures, which at this point is only about 49 ft. under the level of the sea.

"This position of the shafts had the disadvantage of being nearer to the supplies of water from streams on the surface, from one of which the writers subsequently discovered water flowing regularly into the marl, which is more fissured and broken near the outcrop.

"Though the site, for the reason above stated, and also as regards its position in the colliery itself, was not well chosen, yet the plans of the colliery buildings, boilers, machinery, etc., were on a very large scale, and the boilers and engines powerful and good. The former company are believed to have commenced work about 1856, by sinking two large round shafts, 24 to 25 ft. in diameter, intended for brick walls of great thickness, as at that time applied by German mining engineers for damming back the water. These shafts were sunk to the level of the first water feeder, which was met at about 182 ft. from the surface, or about 139 ft. 6 in. below an adit which had been constructed for carrying off the water from the pumps.

"The general section (fig. 142), and the enlargement of the bottom (figs. 144 and 146), show clearly the condition in which the writers found both shafts as sunk down to feeder No. 1; and figs. 145 and 147 show the manner in which they finished them, down to the feeder No. 2 in shaft No. I., and to the feeder No. 1 in shaft No. II. This latter shaft, as mentioned below, they subsequently completed down to its present depth of 943 ft., for coalwork, pumping, and ventilation.

"It will be seen from the plan in fig. 142 that in shaft No. I. the German engineers had ten sets of pumps firmly built into the shaft, with an enormous mass of timber framing; according to the system of that time the wind-bores were movable, or telescopic, so that they could be removed on firing shots or changing; and the pumps were lengthened by common pump pipes (each one lachter, or 6 ft. 10 in. in length) added on below in the shaft. Thus the space, even in these shafts of such great dimensions, was so encumbered with timber as to render sinking, even with moderate quantities of water, a very slow, expensive, and difficult operation. The writers have little doubt that when the feeder No. 1 was first met with, and even before it was widened out by the constant flow of water, it must have yielded 600 cubic ft. per minute. Under such circumstances, and with the inability in some of the pumps to change either buckets or clacks for packing, at the surface, it is only wonderful that the engineers succeeded, even in course of time, by continuous pumping and partially exhausting the feeder, in sinking the sump, and in preparing, as shown at bottom in the section (fig. 142), the foundation for the great walling below the first feeder.

"In April 1867 the writers, having acquired the colliery, commenced preparations for recovering shaft No. I. They encountered great difficulties in the commencement; but by hanging in one large set of pumps, 32 in. diameter, they so far lowered the water as to enable them to take out the

FIG. 146.

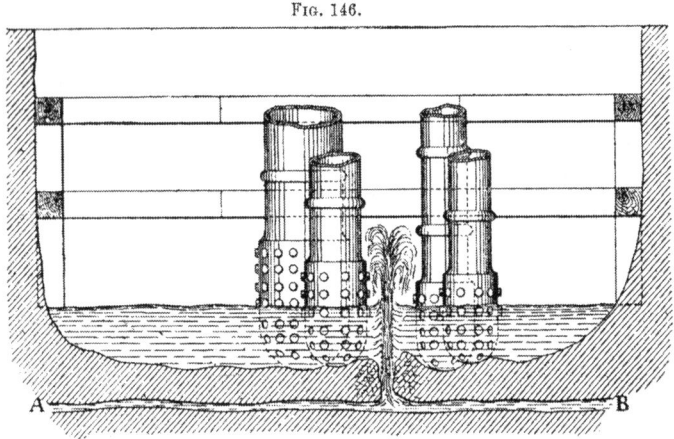

FIG. 147.

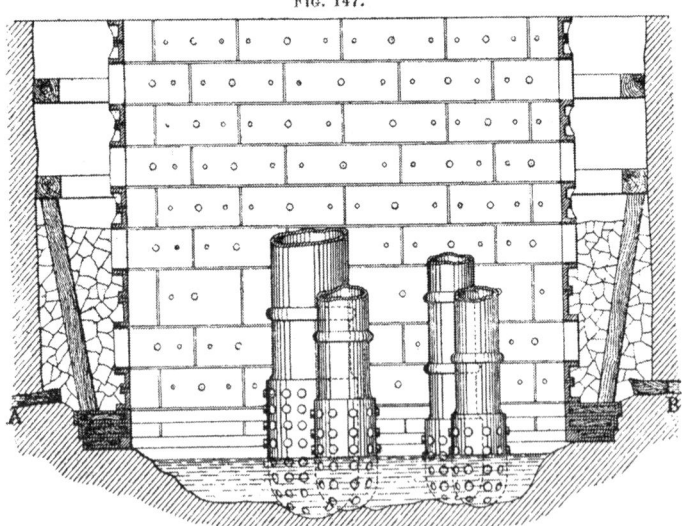

FIGS. 146 and 147.—Section of Shaft showing suction pipes of sinking lifts—
Zollern Colliery.

German pumps and timber, and then to hang in other large sets of 18 in., 19 in., and 20 in. diameter, as hereinafter mentioned, and as shown in fig. 143; and by 30th November 1867, after wedging off part of the supply of water coming from the horizontal cleft or fissure, they were enabled to commence cutting out the foundation for the wedging cribs, designed for the tubbing of a shaft 17 ft. 6 in. diameter. They adopted this dimension as that most suitable, according to the extensive experience they had obtained in the opening out of such large coal-fields, where the coal formation with its numerous beds is likely to reach 2500 or 3000 ft. depth below the surface.

"This size of shaft compelled the writers to cast, at their 'Vulcan' Iron Works on the Rhine, tubbing and wedging cribs of proportionately increased dimensions and thickness, and of improved construction with greater width of flanges. This caused some delay; but though over 600 cubic ft. of water per minute was being pumped from the first feeder in shaft No. I. (this water coming through the crevices left by the rough wedging in the fissure), the tubbing of the upper lift above this first fissure was set and completed up to the level of the delivering drift by 6th March 1868, as shown in fig. 144.

"On further sinking shaft No. I. to the second feeder, it was found that there was such a connection between the feeders in the two shafts that it was advisable and economical, both as to time, cost, and application of steam power, to pump the water and carry on the operations in both shafts together.

"Accordingly a direct-acting 72-in. cylinder engine of 11-ft. stroke was erected over shaft No. II.; a Cornish beam-engine, 84-in. cylinder and about 10-ft. stroke, at shaft No. I.; and lastly, a horizontal winding-engine of 42-in. cylinder and 6-ft. stroke, between the two shafts, working with lay spears and quadrants into both shafts. By this means the writers were enabled to work together nine sets of pumps, as follows:—

2 sets of 32 inches diameter.
2 ,, 21 ,, ,,
2 ,, 19 ,, ,,
2 ,, 18 ,, ,,
1 ,, 16 ,, ,,

"With these they pumped out of the two shafts 1200 cubic ft. per minute of the water which escaped through the wedging of the natural fissures in the marl, and completed the closure of the tubbing for the first feeder in shaft No. II. on 9th August 1868, as shown in fig. 147; and for the second feeder in shaft No. I. on 15th November 1868, as shown in fig. 145.

"In the winter of 1868 further operations were suspended, owing to the want of a railway, which had been long previously agreed upon by the Cologne and Minden Railway Company, but not constructed. This suspension continued for the same causes until 1870, when, to ascertain what further feeders were to be expected in sinking shaft No. II., the writers had a boring carefully made a little to the west thereof. This boring gave very precisely the positions and probable dimensions of the water-bearing fissures, which they might expect to meet in sinking the shaft to the coal measures; and these indications, upon the subsequent sinking, proved to be correct.

"The railway company still failing to construct the branch railway, the writers were at last forced to construct, at their company's expense, a horse

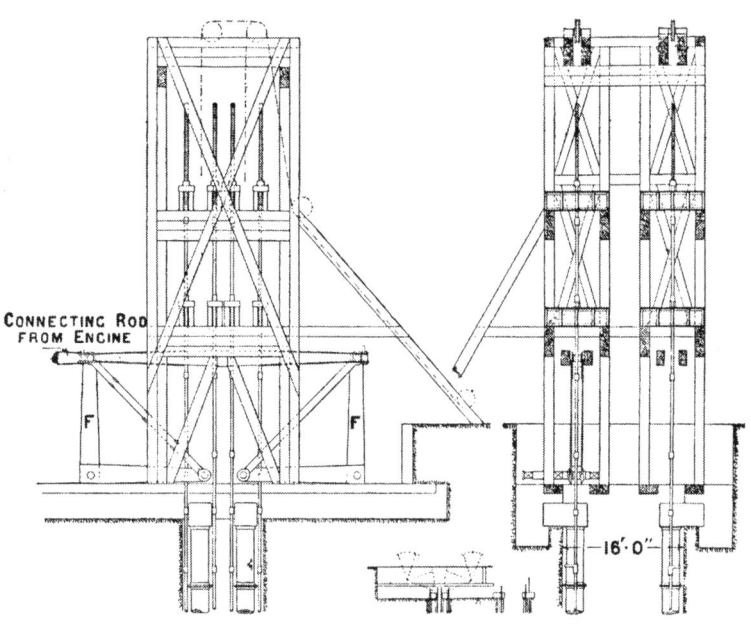

CONNECTING ROD
FROM ENGINE

F   F

16' 0"

Fig. 148.—Unwatering Flooded Colliery—Pit-head Frame for suspending bucket lifts in shaft.

tramway along the public road from Zollern to Hansa Colliery, and sinking operations were recommenced in shaft No. II. in August 1871. Four more feeders, Nos. 2 to 5, were sunk through and tubbed off successfully to the depth of 292 ft. before the end of March 1872, and all water was so completely shut out that the remainder of the shaft-sinking and walling to the net diameter of 17 ft. 6 in. was carried out, first to the coal measures at the depth of 357 ft., and then to the present bottom at 943 ft., without any pumps whatsoever in the shaft.

"The quantities of water shut out by the several lifts of tubbing were carefully measured as follows * :—

Down to, and inclusive of, the first four feeders, . .	1160 cub. ft. per min.	
The fifth feeder, ...	150 ,, ,,	
The sixth or lowest feeder, ...	100 ,, ,,	
Total . .	1410 ,, ,,	

This total is a quantity of water rarely met with in shaft-sinking; but it

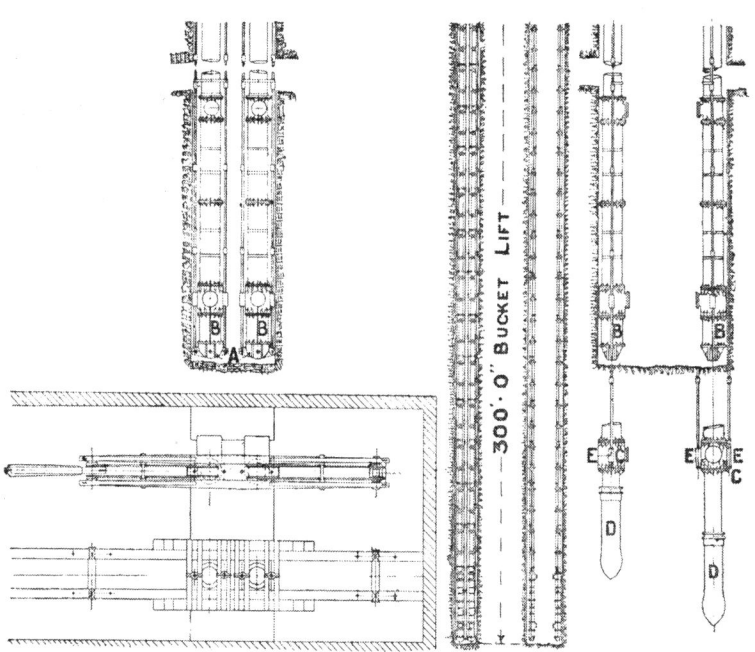

Fig. 149.—Unwatering Flooded Colliery—Suspending Bucket Lifts showing plan of shaft and pumping quadrants.

was shut out so effectually in this large shaft as to enable it to be completed to the bottom without pumps.

"The connection to a railway was not made until 1879, up to which time only preparatory work could be carried on in opening out the colliery from the shaft; but since then the pit has been continually worked, producing now about 750 tons of coal per day, and with very little water to

* In all cases the quantities of water were actually measured at an overflow weir, in a watercourse constructed for the purpose.

pump. It is worked entirely through the single shaft No. II.; but at an early period a second shaft must be completed."

Unwatering a Flooded Colliery.—The following illustrations (figs. 148 and 149) represent the plant used by the author for unwatering a flooded colliery and sinking it deeper, the plant being so designed that when completed it formed the permanent pumping plant of the mine.

A general view of the surface of the mine after the completion of the

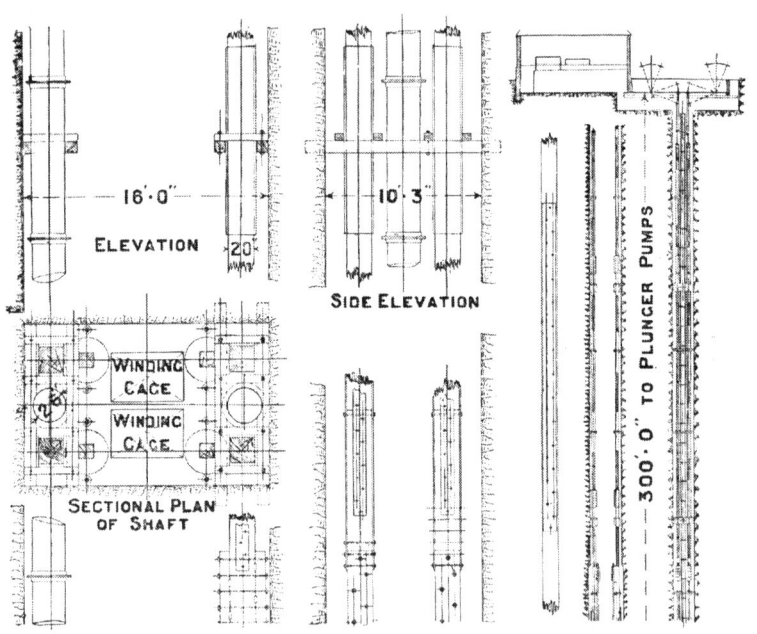

Fig. 150.—Unwatering Flooded Colliery—General Arrangement of Plunger Lifts showing position of sinking lifts for deepening the shaft.

plant is given in the frontispiece, Plate III., Chap. VI., the engine-house and pit-head frame being clearly shown.

The colliery was formerly drained by means of underground steam pumps involving a very great consumption of steam, and causing great inconvenience by the excessive heat in the workings. A sudden inburst of water overtaxed the power of the pumps, and the colliery became flooded. The shaft was 300 ft. deep, and at that level the workings existed; it was, however, required that the shaft, after recovery, should be sunk to a depth of 120 ft. more.

The problem presented was that of getting out the water and deepening the shaft with plant which might remain as permanent plant. It was decided that four 24-in. pumps should be provided to cope with the water in unwatering the mine, and it was evident that bucket lifts were the only suitable pumps under the circumstances, but bucket lifts had not

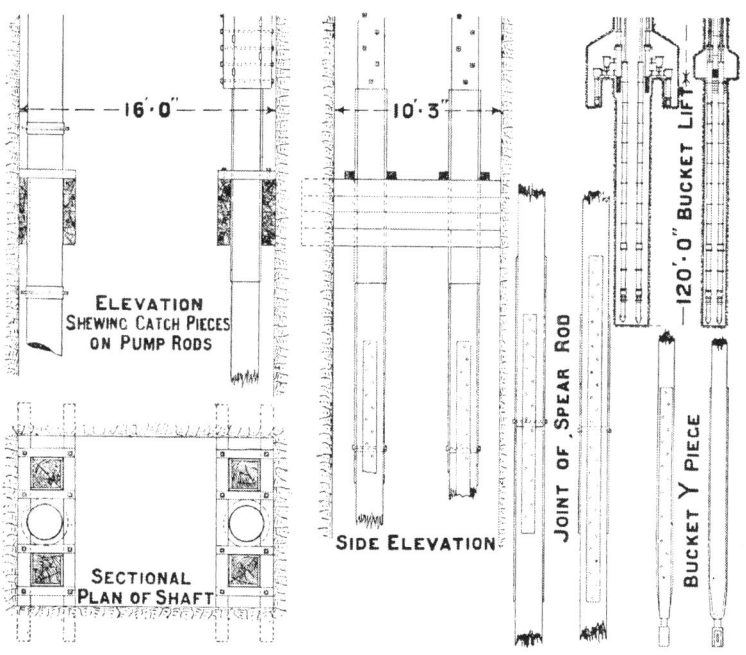

FIG. 151.—Unwatering Flooded Colliery—Plan of Shaft, with details of pump rods.

been used of that size for such a high lift as 300 ft.; hence, if used for that lift, the design required special attention.

The plan adopted was that of having four 24-in. bucket lifts each 300 ft. long, and each of the four capable of reaching to the bottom of the shaft.

The pumps were suspended from a strong pit-head frame shown in fig. 148 by means of rods and screws, and were lowered into the shaft through the water to the full depth, and whilst so suspended were connected by means of quadrants F F to two compound differential engines, a pair of pumps to each engine. The engines had cylinders 45 in. and

76 in. in diameter and a 10-ft. stroke, the pumps having the same stroke as the engine.

The water was pumped away till the bottom of the shaft was reached.

It was then found that one of the engines with its pair of pumps was capable of pumping all the water coming from the feeders.

Plunger pumps had been provided for the permanent pumping. Whilst the water was kept down by the one engine, a pair of plunger pumps (figs. 150 and 151) were put in at the bottom and coupled up by means of wooden rods to the other engine. When that was done and the plungers put to work, the other engine was stopped, the bucket lifts taken out and plungers put in their place, thus completing the permanent plant, which consisted of two pumping-engines, each provided with plunger lifts.

The permanent plant was now extended to the bottom of the shaft. The shaft had then to be deepened by 120 ft. more. The bucket lifts were now available for the purpose. The general arrangement of the pumps and the details of rods, etc., are clearly shown in figs. 149 to 151, and need no further description.

It may be observed that the rising main is larger than necessary, but as a matter of economy the rising mains of the bucket pumps were utilised as rising mains for the plunger lifts.

It will be seen in fig. 149 that the suspending rods for the bucket lifts were attached to a steel beam centred under the snore pipes B B.

The object of this arrangement was to insure strength, not trusting to a cast-iron connection, and at the same time to insure equal tension on the suspending rods.

After the unwatering was accomplished, and the lifts became short sinking lifts, telescopic snore pipes D D were made to take the place of snore pipes B B, and the suspending rods were attached to lugs E E already provided on the sides of the suction valve boxes. Doors for the convenience of examining the suction valves and removing gags are shown at C C.

Explanation of the Principle.—Hydraulic transmission of power in mines is becoming largely employed, especially for pumping. For that purpose it is very conveniently applied, as the natural speed of water-pressure engines is about equal to that of the pump; the engine may therefore be connected directly to the pump rod without the intervention of any mechanism. Where a natural head is available, then water-pressure engines, such as are illustrated in Chap. XI., may be employed, but the power may be obtained from a prime mover such as a steam engine, and it is that form of transmission which we now propose to discuss.

A simple form of hydraulic transmission is illustrated by the diagram (fig. 152).

The steam engine A gives motion to the piston of the water cylinder B, the ends of which cylinder are connected by pipes to the ends of a similar water cylinder C. The piston of the cylinder C is attached to the rod of the pump D. If the cylinders B and C with their connecting pipes are filled with water, and the piston of the cylinder B is moved backward and forward by means of the steam engine A, then as water is practically incompressible, the piston of the cylinder C will partake of the same motion. The piston C will in its turn communicate its motion to the piston of the pump D. By this arrangement it will be seen that the piston of the pump partakes of the same motion as the piston of the steam engine, one stroke of the steam engine producing one stroke in the pump. The connecting pipes between the cylinders B and C may be of considerable length, so that the steam engine may be on the surface of the mine, whilst the pump is at the bottom. When it is advisable to run the engine at a higher speed than that of the pump, gearing is introduced between the engine A and the cylinder B. This is known in this country as Moore's system, but the principle is an old one.

It will be seen that any leakage of water from the stuffing-boxes, pipes, or pistons of the cylinders B and C, or any leakage in the connecting pipes themselves, will alter the relative positions of the said pistons, with the result that the pistons may strike the covers with such force as to

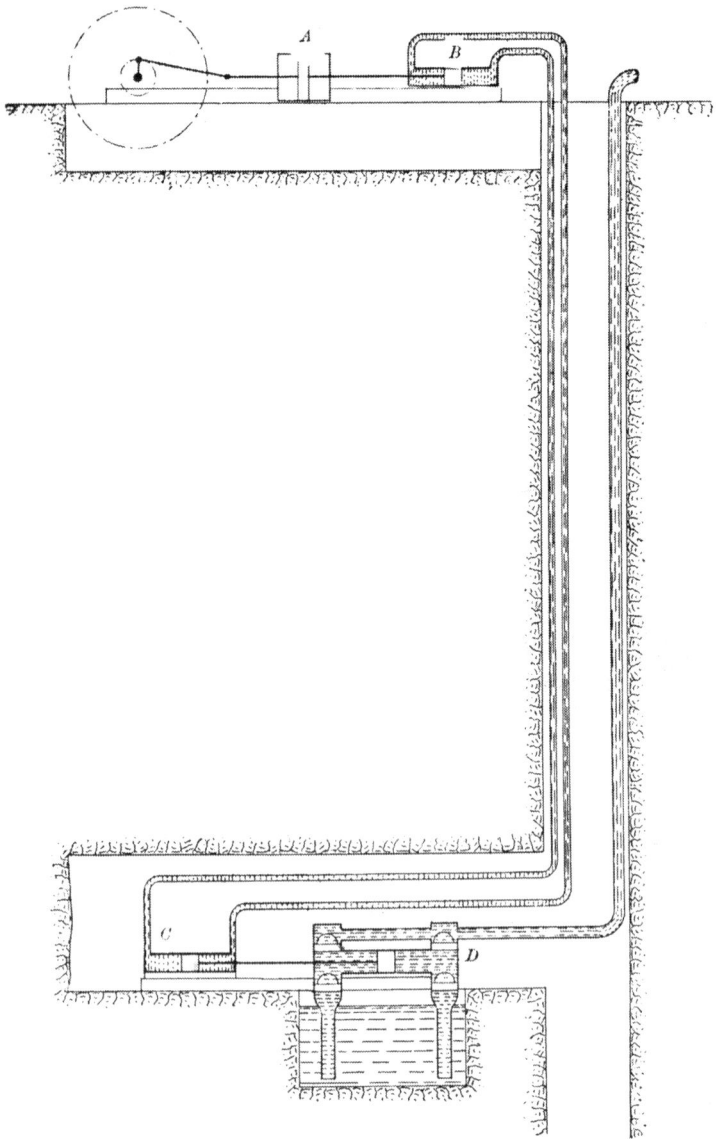

FIG. 152.—Hydraulic Pumping—Water Rod System of transmitting power from the surface to a hydraulic pump at the bottom of the mine.

cause a breakdown of the plant. It therefore becomes necessary to introduce a device by means of which the relative positions of the pistons may be maintained. That is done by an arrangement of tappets and valves so designed that when either piston moves beyond its normal stroke a valve is opened, allowing some water to pass from one side of the piston to the other. It will be readily seen that the system is only applicable to the working of one pump, and that when applied for mining purposes there are three pipes in the shaft—two power pipes and one rising main from the pump.

In practice, small power pipes working under high pressures are employed, and the power cylinders and pumps, instead of having pistons, are provided with plungers. The diagram is only intended to illustrate the principle.

Its Application at Saarbrücken.—This system was adopted in a mine at Saarbrücken about twenty-five years ago.

The mine was 306 yds. deep below the surface. The motor consisted of a double-acting horizontal high-pressure engine having a cylinder 53 in. in diameter, and a stroke of 61·5 in. The piston-rod was continued through both covers, and connected at each end with a plunger 9 in. in diameter. The hydraulic engine underground consisted of four pressure cylinders having rams arranged in pairs, and connected by a crosshead.

Each plunger was 6 in. in diameter, with a stroke of 66 in. Between each pair was placed a pump plunger. The pressure column consisted of cast-iron pipes 5 in. in internal diameter and $1\frac{1}{2}$ in. thick, put together in 10-ft. lengths.

The working pressure at the hydraulic engine was 1150 lbs. per sq. in. For the sake of regularity and safety in working a special safety and filling valve was placed on each pressure pump at the surface, consisting of a chamber with an auxiliary feed valve, a safety valve, and an air-escape valve, which were in connection with a closed cistern on a higher level, kept constantly filled with water. In setting the engine to work, difficulty was at first experienced in adjusting the relation of the stroke of the pumps underground to those above. To effect this a magnetic telegraph apparatus was used, and signals were transmitted, until the two engines were in proper relative positions.

When first put to work the speed did not exceed six double strokes per minute, the speed of the surface and underground engines being, of course, both the same, but after some time the speed was increased to from ten to twelve double strokes per minute, and experiments were then made to determine the efficiency.

It was found that 25 per cent. of the work was absorbed in friction, and that the consumption of coal was at the rate of 22 lbs. per H.P. per hour.

The effective work of the pumps when making 10 strokes per minute was 100 H.P.

It appears that the engine was worked non-expansively, which partly accounts for the high consumption of coal.

The difficulty of keeping the plant in proper working adjustment, and the impracticability of working more than one pump from the surface engine, have been against the system, and although the plant was erected twenty-five years ago, the application of the system has not increased to any great extent.

The total cost of the plant was £5550, or, including erection, £9117, which is at the rate of £55·5 per H.P. for plant alone, or, including erection, £91 per H.P.

Working Results of the Hydraulic-rod Pumping-Engine at Saar-brücken.—This is a sequel to a former memoir on the same subject, giving an account of the results obtained by the engine in continuous work. The speed has been increased from six double strokes per minute to twelve without difficulty.

In order to observe the working pressure upon the different parts of the engine, five pressure gauges were used, placed as follows:—One at the bottom of the rising pipe, and two on each of the rod tubes, one at the bottom and the other at the surface below the point of attachment of the press pumps. The readings of these gauges have been plotted in the manner of indicator diagrams. These diagrams are supposed to be correct within a margin of 1 or 2 atmospheres, and are the best obtainable in the absence of an indicator capable of working under pressures of 80 atmospheres and above. In the rising pipe the pressure was very steady, and varied only between 26 and 28 atmospheres, the change being sensible only at the beginning and ending of the stroke. In the rod tubes underground the variation in pressure was marked by extreme regularity and precision. When the pressure piston was making the return stroke, the gauge indicated uniformly 35 atmospheres; it then rose during the cataract pause with a vibratory movement to about 78 atmospheres, which remained the normal pressure to the end of the working stroke, when it fell with corresponding oscillation back to 35 atmospheres. There was little difference in the maximum pressure between the north and south rods, the latter indicating up to 80 atmospheres, this difference arising from the load not being quite uniformly distributed between the two pumps. The oscillations of the gauges at the surface were somewhat greater; near the end of the return stroke the pressure rose to about 15 atmospheres, increased during the cataract pause to 30 atmospheres, and on the commencement of the forcing action of the plunger suddenly rose to 50. This pressure was maintained regularly till close upon the end of the stroke, when it fell rapidly through 30, 15, 10 atmospheres to 0. In some cases even a negative pressure was observed when, through loss of water in the tubes, a partial vacuum was formed, owing to the waste not being made up quickly enough by the auxiliary feed valves. The velocity of the column of water in the pressure tube was 7·6 ft., and in the rising pipes 3·6 ft. per second.*

Armstrong's Method.—The ordinary system of hydraulic power transmission invented by Armstrong for working dock machinery consists of pumping water by means of a steam or other engine under high pressure into pipes from which the water is taken to work the machinery requiring the power. To compensate for the varying demand for power, Armstrong applied what is termed an 'accumulator,' which consists of a large weighted plunger in a vertical cylinder in constant communication with the power pipes. The rise of the weighted plunger or accumulator was also used to regulate the speed of the steam engine employed to pump the water under the accumulator plunger, in such a way that when the demand for water ceased, and the accumulator became fully charged, the power engine was stopped by the closing of the steam stop valve. When the demand for power recurred, the falling of the accumulator would open the steam valve and the engine would again start into motion.

Davey's Modification.—A modification of this system is peculiarly applicable to mines where the pumps are situated some distance from the bottom of the shaft, and also where the ordinary surface pumping-engine is not applicable, or where it is undesirable or uneconomical to take steam down into the mine. The following illustrations (figs. 153 and 154) represent hydraulic plant, designed by the author for a colliery where the ordinary systems of pumping were, owing to local circumstances, not applicable.

The general arrangement consists of a compound condensing steam engine on the surface employed to pump water under a pressure of 1400 ft., and thereby made to actuate hydraulic pumps underground. The water is conveyed by a power pipe to the hydraulic pumps in the mine, and is returned to the surface, where it is used over and over again. By this arrangement clean water is used to avoid excessive wear and tear of the hydraulic engines, which would take place if they were worked with the dirty mine water. The rising main from the pumps is taken up the shaft and made to discharge through an adit level. As the power water is used over and over again, a little oil may be put in it for the purpose of lubricating the working parts of the power pumps and hydraulic engines.

The oil should be free from acid, otherwise the steel and iron working surfaces will become pitted.

In this system of pumping, nothing is required in the nature of an accumulator of power, as the work done is constant.

All that is required is a constant pressure in the power pipes, and that is secured in this case by the employment of a steam accumulator or regulator, shown in section in fig. 155, and also in elevation in fig 154. The steam accumulator or regulator consists of a plain steam cylinder

* *Proceedings Inst. of Civil Engineers,* vol. xliii.

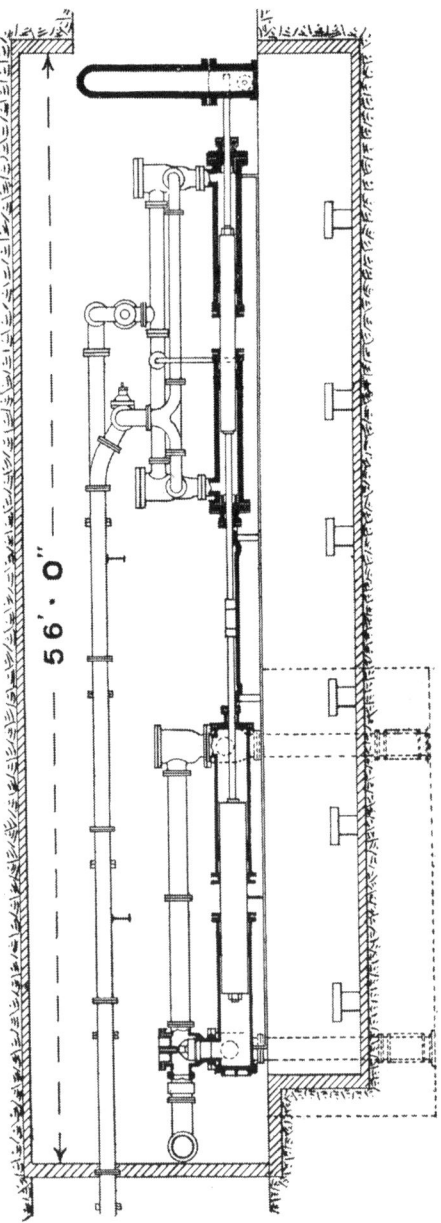

56' 0"

Fig. 153.—Underground Hydraulic Pumps.

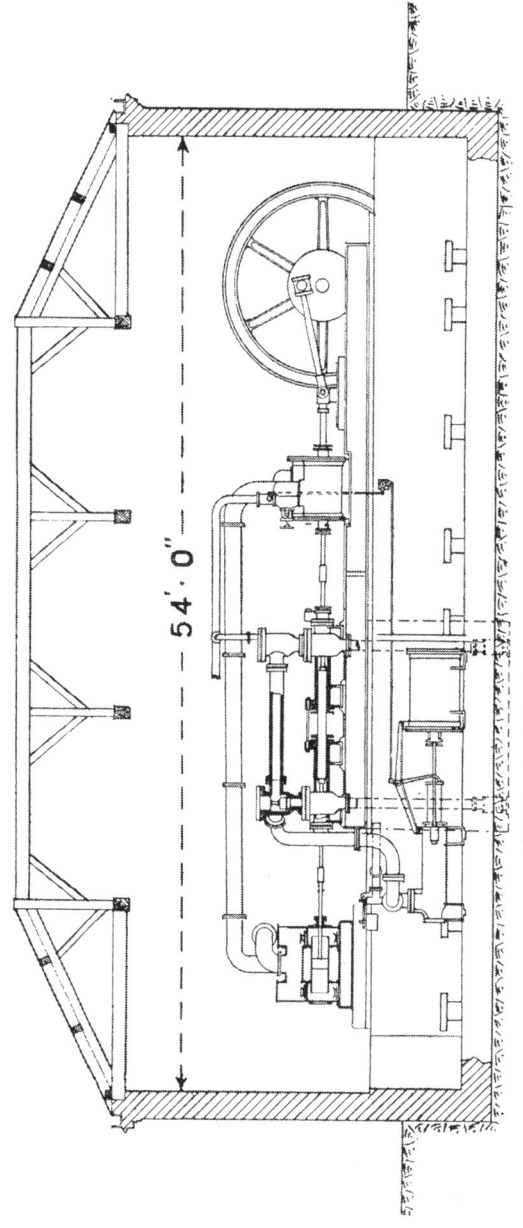

54' 0"

FIG. 154.—Compound Power Engines for hydraulic power transmission.

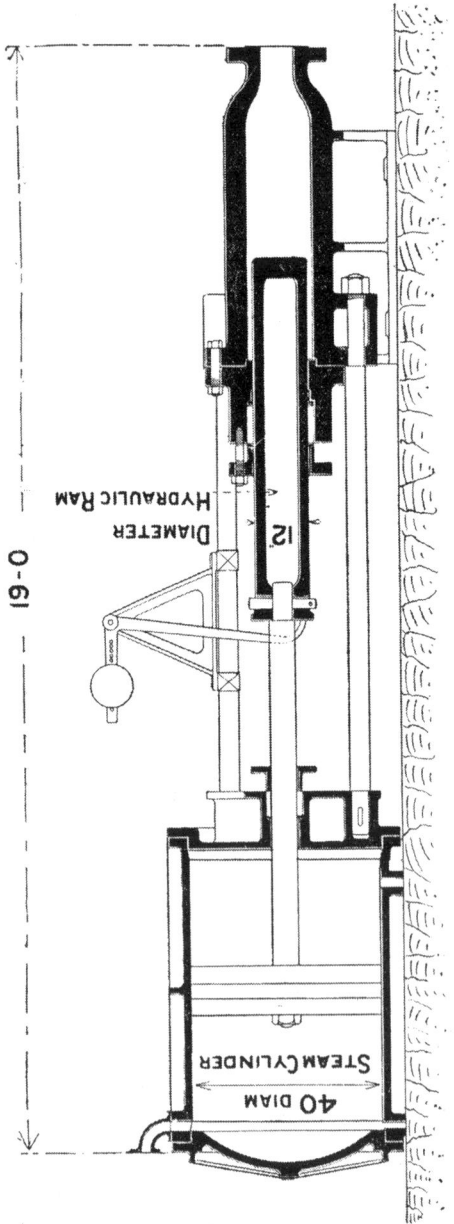

DIAMETER
HYDRAULIC RAM

12

19·0

STEAM CYLINDER

40 DIAM

FIG. 155.—Hydraulic Transmission of Power—Steam Accumulator or Regulator.

having a piston, the rod of which is attached to a plunger working in a hydraulic cylinder in communication with the power pipes. The front end of the steam piston is in communication with the atmosphere; the back end is connected by means of a pipe to the main steam pipe, so that the boiler pressure is maintained at the back of the steam piston, and the relative areas of the steam piston and hydraulic plunger are such as to maintain the required pressure in the power pipes.

The steam condensed in the cylinder is automatically drawn off by means of a steam trap. Variations in the supply to, and the draught from, the delivery pipe are compensated for, and a constant pressure maintained by the movement of the steam accumulator plunger. In ordinary working the variations are small if care be taken to so design the power pumps and power engines that the former supply and the latter take nearly a constant stream. That is secured by making the power pumps double-acting, coupled by means of cranks at right angles, or by the use of three plungers, and by making the hydraulic engines on the duplex plan. Fig. 154 is an elevation of the engines, power pumps, and accumulator, partly in section.

The following are the chief dimensions:—

High pressure cylinder,	30 in. diameter.
Low pressure cylinder,	50 in. diameter.
Length of stroke,	3 ft.
Number of strokes per minute,	23.
Diameter of power pump plungers,	7½ in.
Stroke of pumps,	3 ft.
Pressure at the power pumps,	1400 ft.
Diameter of steam cylinder of accumulator,	40 in.
Diameter of accumulator plunger,	12 in.
Length of stroke,	4 ft.

The hydraulic pumps have power plungers 10¾ in. in diameter, and pump plungers 18¼ in. in diameter, with a stroke of 6 ft., all double-acting.

On the completion of this plant, a careful experiment was made as to the efficiency of the whole system.

During the trial the work done by the pump in raising water represented 124 H.P., whilst the steam engine indicated 229 H.P., thus giving a total efficiency of 54½ per cent. This includes the small amount of work done by the engine in lifting the condensing water to a cooling pond. The cost of the machinery and pipes was £5430, and the total cost of the whole plant erected at Marseilles, including all the shaft work, excavation of chamber for hydraulic pumps, buildings, cooling pond for condensing water, and all other items, came to £8630.

If the engine is worked to its full power, it will indicate about 300 H.P., so that the total cost of the installation comes to £29 per I.H.P., or just over £50 per actual H.P. in work done.

M. Eugéne Biver, the engineer of the colliery, carried out a very exhaustive set of experiments with this plant, the results of which were published in *Le Génie Civil*, 1885, p. 81. An abstract of his report, taken from the *Transactions of the Inst. of Civil Engineers*, is here given.

" This pumping-plant erected at Shuillier, Fuveau, by the Société de Charbonnages des Bouches-du-Rhône, was designed to complete the draining of the Castellare-Leonie section.

" The question to be solved was :—Given a shaft already encumbered with mining plant, with two forcing pumps 20 in. and 16½ in. in diameter, with wooden pump spears, beams, guides, ladders, etc., to double the power by means of apparatus occupying as little space as possible, giving a higher dynamic efficiency, and in no way interfering with the existing machinery, which was to preserve the galleries from being inundated during the installation of the new engine.

" Other circumstances added to the difficulty of the problem. At the pit-mouth, on the surface, was a Bull engine with its cylinder placed vertically over the pit, also a platform and a winding engine. In short, the new installation was not to interfere in any way with the existing machinery, or with the future deepening of the pit.

" The first scheme proposed was the installation at the bottom of the pit of a direct-acting rotative engine, taking steam from boilers on the surface, or from a battery of boilers near the engine in the mine itself. In the first instance, the result must have been considerable condensation and heating of the pit ; in the second, defective installation of the boilers.

" Each case required a room beyond the reach of the water, but this condition could not be fulfilled, and it was necessary that the pumps should be capable of working under water. Messrs Hathorn, Davey & Co. being consulted, proposed the transmission of power to the pumps by water, with pressure engine on the surface, and hydraulic pumping-engines at the bottom of the pit.

" Like installations had been adopted under similar conditions at the mines of Mansfeld and at Saarbrücken.

" The installation may be defined as follows :—An economically working compound steam engine, forcing water at 42 atmospheres by means of two double-acting plunger pumps. At the bottom of the pit, duplicate hydraulic pumps, both worked from the surface engine pressure water pipe, and a return pipe, also a delivery column from the hydraulic pumps to the draining gallery of the mine. The top of the pit is 292·805 metres (950 ft.) above the sea-level. The draining gallery is about 60 metres lower, or 232·820 metres above the sea-level ; the situation of the hydraulic pumps, 140 metres. The total elevation of the water is therefore 232·820 – 140 = 92·820 metres, or in round numbers 305 ft.

" The work to be done was raising 7700 litres (1694 gallons) per minute.

$$\frac{93 \times 7700}{60 \times 70} = 159·10 \text{ H.P.}$$

" The motive power engine is of the compound type, having two horizontal cylinders with an intermediate receiver and cranks at 90°. The motor is calculated to supply 2357·7 litres of water per minute to the hydraulic engines at the bottom under an effective pressure of 42 atmospheres. The high-pressure cylinder has a diameter of 0·762 metre (30 in.), the low-pressure cylinder a diameter of 1·270 metres (50 in.), with a stroke of 0·914 metre (36 in.). The pistons are connected to a double rod, the front rod ending with the crosshead of the engine, the back rod connected to the plunger rod of one of the forcing-pumps. The front rods are 0·120 metre (4·72 in.) in diameter, and the back rods 0·080 metre (3·15 in.).

" The forcing-pumps at the surface are two in number, and exactly alike. The plunger of each is attached to the piston of the motor. The effective surface of the plunger is equal to the difference of the area of the plunger and the rod—that is, S = 234·752 centimetres.

" The capacity created by a full stroke is—

$$0·023475 \times 2 \times 0·914 \quad = \quad 42·9 \text{ litres.}$$
$$\text{For both plungers} \quad = \quad 85·8 \text{ ,,}$$

" The water necessary for the normal feed to the hydraulic engines must be 2357·7 litres (519 gallons) per minute.

" The number of complete strokes per minute with the steam pistons will be $\dfrac{2357·7}{85·8} = 27·5$, corresponding to 27·5 revolutions of the fly-wheel. Between the two engine foundations, and in a line with the fly-wheel, is the steam-accumulator or regulator, in direct communication with the forcing main. The two steam cylinders, though neither of them is steam jacketed, are joined by a cast-iron pipe 10 in. in diameter, forming the intermediate reservoir, and connecting the valve chest of the large cylinder with the exhaust of the small cylinder. The capacity of the reservoir thus formed is 640 litres.

" The relative volumes formed by the pistons in the two cylinders is 1–2·8. The cut-off in the large cylinder is fixed, and corresponds to an angle of 110°.

" The low-pressure diagram shows that the cut-off is equal to 1·09. It may therefore be neglected. The small cylinder is provided with an expansion gear variable by hand. The speed of the engine on the surface must be regulated to suit the speed of the hydraulic pump underground.

" As the work underground is continuous it must therefore be continuous on the surface. There must be no reservoir of work, such as is characteristic of the Armstrong system. An accumulator with such a reservoir of work would not allow the engine on the surface to determine the speed of the hydraulic pumps underground ; that it should be so was a necessary condition, because the hydraulic pumps have sometimes to work under water. To render the action constant the apparatus called a ' regulator ' is introduced. In similar installations in America, the Davey apparatus includes not a regulator, but a simple reservoir of air. This gives excellent results in working, but has the following disadvantages :—

  1st. Its size must be considerable.

  2nd. It must have a large and considerable supply of air. The density increases with the pressure.

  3rd. The bursting of this large reservoir, containing a pressure of 50 atmospheres, would cause a great explosion.

"The steam regulator is composed of two cylinders and of two pistons mounted on one rod. In the first cylinder the back surface only of the piston is subject to the action of steam.

"The motive water in the pressure pipe enters the second cylinder, and acts on the piston in an inverse direction to the steam. The areas Ω and ω of the pistons are calculated, of course, to satisfy the equation $f\Omega = F\omega$, f and F being the respective pressures of steam and of the water in the pipe. This apparatus is very sensitive. It serves as regulator for the compound motive power engine. The piston rod of the regulator is connected directly to the throttle-valve of the engine. In case of the breaking of the pressure column of water, this safety apparatus prevents all further · accident. The steam regulator offers, from a practical point of view, other advantages. The effective pressure of the water in the supply pipes may be varied by simply adjusting a reducing valve in the steam pipe supplying the steam accumulator.

"The engines at the bottom of the pit are twin engines, motors and pumps working with a pressure of $42+15$ atmospheres, and with the return water at a pressure of 15 atmospheres (corresponding to the difference of the level between the forcing pumps on the surface and the engines at the bottom—that is, 153 metres). The hydraulic pumps raise the water from 140 metres to a height of 232·82 metres to the draining gallery. The delivery per minute is 7700 litres—that is, 462 cubic metres (101,640 gallons) per hour. The stroke of the pumps is 1·829 metres (6 ft.). The diameter of each pump plunger is 0·463 metre (18¼ in.), the diameter of the rod being 0·102 metre (4 in.). To deliver 7700 litres per minute, the two pumps must therefore make 6·4 complete strokes. The corresponding speed of the piston per second is 0·39 metre (15·6 in.). The difference of area is $0·050364^2 - 0·035200^2 = 0·015164^2$ metre.

"*Trial of the Engines.*—The diagrams taken at the time of the trial— 29th August 1884—show considerable variation in the pressure of water at different points of the circuit.

"On the surface the variations are more felt in the pump worked by the piston of the small cylinder than in those which are worked by the large cylinder. According to the diagrams the maximum oscillations correspond to variations of 35 to 54 kilogrammes per square centimetre (500 to 700 lbs. per sq. in.).

"At the time of the trial, the water pumped from the mine was not considerable, the fly-wheel of the compound engine making only twenty-two revolutions per minute, corresponding to a delivery to the pressure-pumps of 1887·6 litres per minute.

"Each full stroke of the hydraulic engines at the bottom consumes 368·40 litres of motive water; the number, then, of double strokes of the underground pumps must be

$$\frac{1887 \cdot 60}{368 \cdot 40} = 5 \cdot 12.$$

"Five strokes only were recorded; the remainder was lost by the stuffing boxes, distribution valves, etc. The corresponding delivery of the pump was—

$$1201 \cdot 80 \; l \times 5 = 6009 \text{ litres per minute.}$$

"The effective work in water raised was then

$$100 \cdot 15 \text{ kilogrammes} \times 93 \text{ metres} =$$
$$9313 \cdot 95 \text{ kilogramme metres} = 124 \cdot 20 \text{ H.P.}$$

"According to the diagrams the total work done by the steam pistons was

$$Tm = 229 \cdot 20 \text{ H.P.}$$

Since the useful work was

$$Tu = 124 \cdot 20 \text{ H.P.}$$
$$\frac{Tu}{Tm} = 0 \cdot 542 = 54 \cdot 2 \text{ per cent.}$$

"The work of the air pump in raising the condensing water to the cooling pond (an amount of work which slightly increases the efficiency) is not taken into consideration.

"The Directors of the Société de Charbonnages des Bouches-du-Rhône intend to make use of the installation for working, independently of the pumps described, some underground hydraulic motors.

"It will be only necessary to increase the number of strokes of the compound engine on the surface to provide a larger volume of water under pressure, from which it will be possible to work additional pumps in the 'dip,' and also hauling engines.

"For hauling, rotative hydraulic motors, with two cylinders, having cranks coupled at right angles and valves worked by means of eccentrics, will be employed.

"The cost of the installation was as follows :—

	Francs.	
Surface engines, compound engine, and forcing pumps free on board in London, ...	37,036	70
Hydraulic engines at the bottom transport, marine insurance,	37,415	95
Loading and unloading, ...	13,313	26
Cast-iron pipes (61,997 kilogrammes), and transport, .	35,782	12
	123,548	03
Total cost of whole installation, .	215,771	20*

Or say, £8628."

* Proc. Inst. Civil Engineers, vol. lxxxiv. p. 513.

Later Improvements. — Since this plant was erected, considerable improvements have been made by the adoption of higher pressures, and

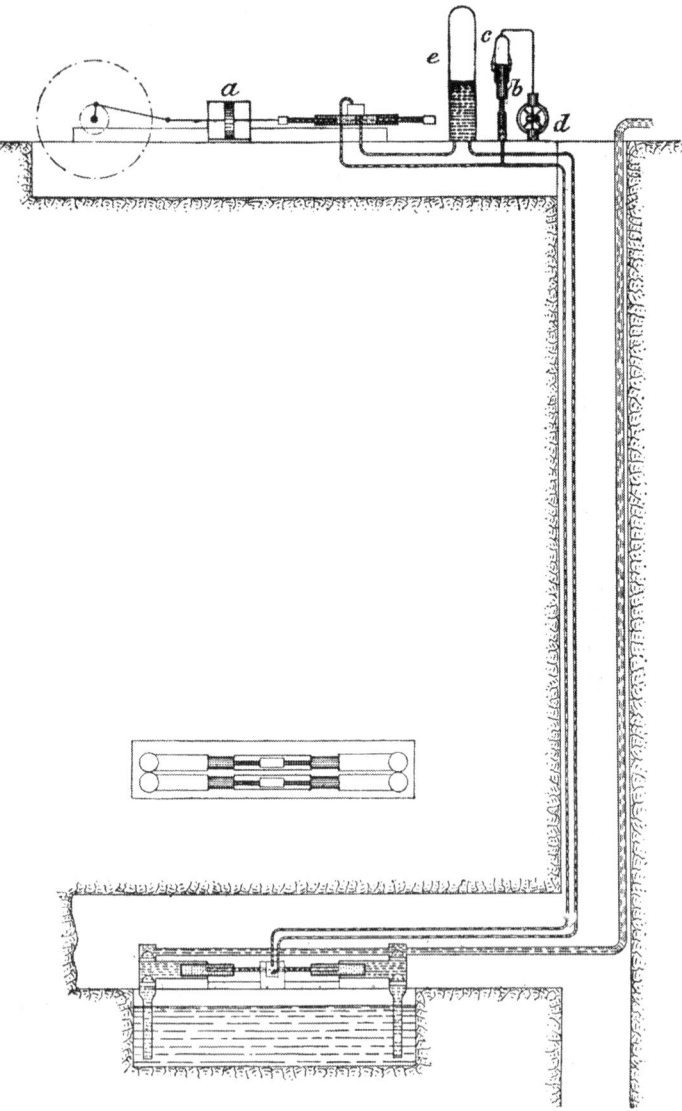

Fig 156.—Hydraulic Pumping System of transmitting power from the surface to hydraulic pumps situated at the bottom of the mine.

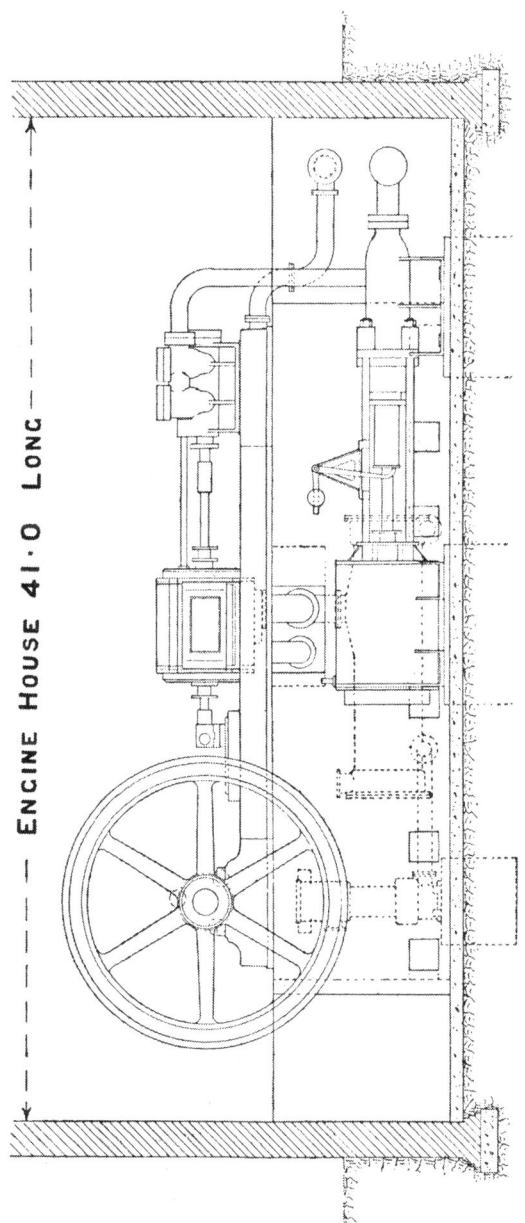

ENGINE HOUSE 41·0 LONG

FIG. 157.—Longitudinal Elevation of Triple-expansion Power Engines for hydraulic power transmission, showing steam accumulator.

improvements in details have enabled higher speeds to be obtained both in the power engines and hydraulic pumps.

By the use of higher pressure the loss from friction in the pipes is

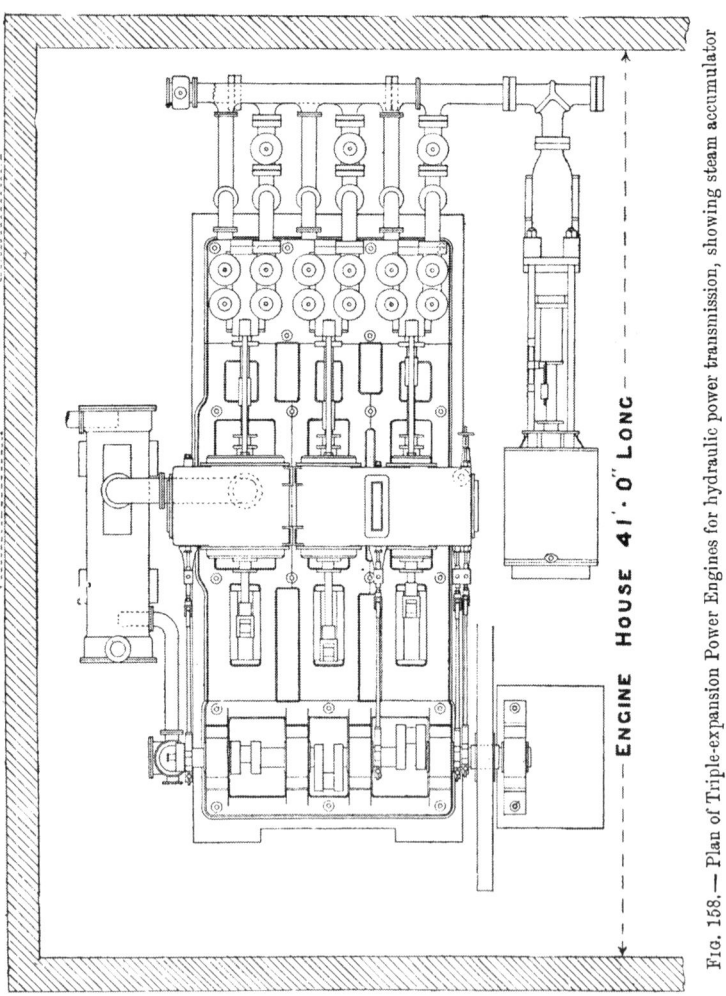

FIG. 158.— Plan of Triple-expansion Power Engines for hydraulic power transmission, showing steam accumulator and surface condenser.

ENGINE HOUSE 41'·0" LONG

reduced. For a given velocity the friction is not materially increased by increased pressure, therefore with higher pressures the friction becomes a smaller percentage of the total power produced. Pressures of 4000 lbs. per sq. in. at the power pumps have been employed, which involves, of course, a greater pressure at the bottom of the mine. It must, however, be remembered that as the power water is returned to the power engines, the hydraulic engines actuating the pump, work with a back pressure.

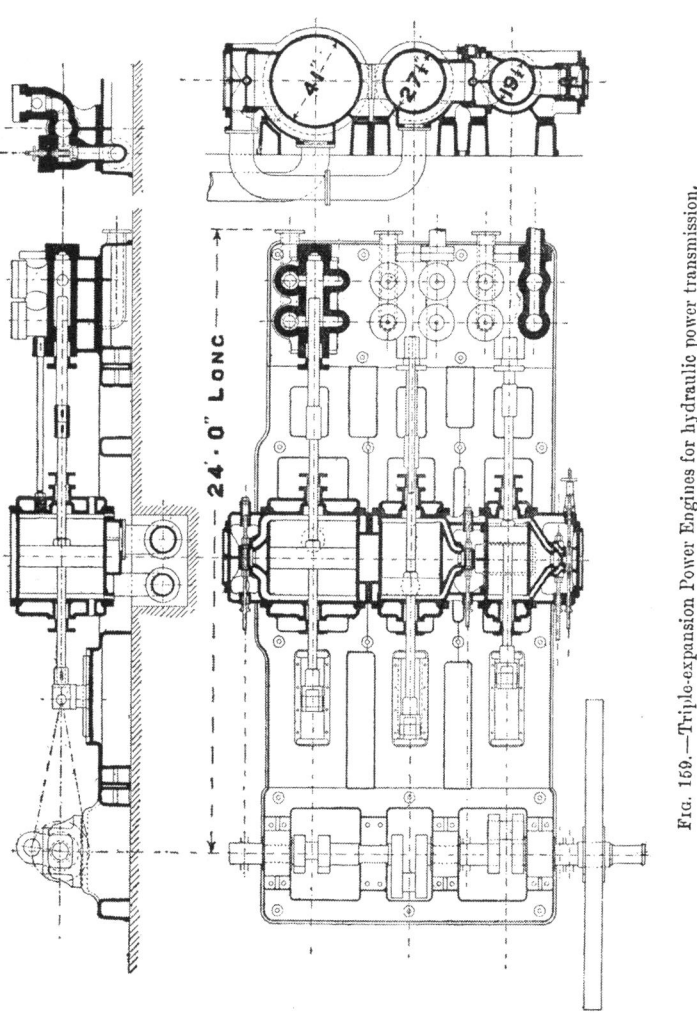

FIG. 169.—Triple-expansion Power Engines for hydraulic power transmission.

The effective pressure is therefore that produced by the power pumps *minus* the friction of the water in the supply and return pipes.

The higher the pressure, the greater the loss from leakage through the valves of the power and hydraulic engines, and the employment of very high pressures such as 4000 lbs. per sq. in. makes it necessary to construct the power valve boxes of wrought steel, and

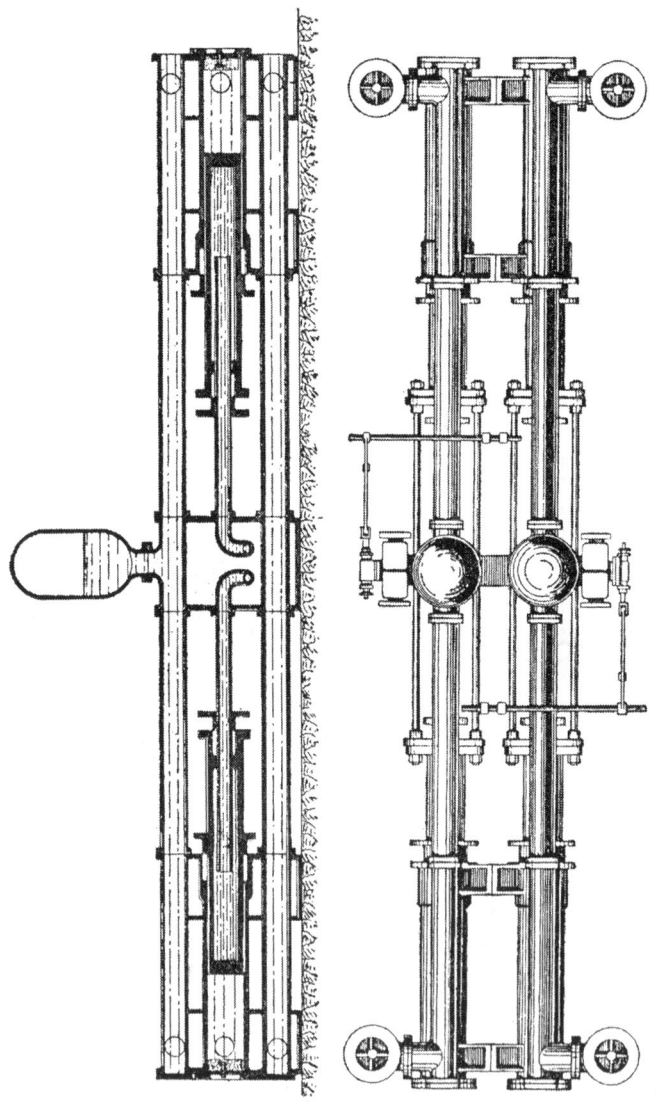

Fig. 160.—Duplex Hydraulic Pump.

the workmanship must be of the highest class. With pressures up to
1000 lbs., ordinary materials and ordinary valve arrangements may be used.

The gross power transmitted through a 6-in. pipe with 1000 lbs. pressure would be 254 H.P. In the ordinary application of hydraulic pumping in mines the efficiency of well-designed plant should be 60 per cent.—that is to say, the work done in pump H.P. should equal 60 per cent. of the I.H.P. of the steam engine. The actual H.P. transmitted through a 6-in. pipe is thus 152.

In two or three Westphalian collieries there has recently been erected hydraulic pumping plant, illustrated in fig. 156. A compound steam engine a on the surface pumps water at from 3000 to 4000 lbs. per sq. in. under the pressure of air in a regulator b. The regulator consists of a plunger in communication with the power pipe, having an air cylinder c above it. In the air cylinder the required pressure is preserved by means of the air pump d. The power water is returned to an air vessel e, from which it enters the suction pipes of the power pumps. The power water is clean, and has some oil added to it.

The plant at 'Pluto' Colliery is thus described :—

On the surface is a twin steam engine with cylinders 870 mm. in diameter, having a stroke of 1080 mm., by which four rams, each 80 mm. in diameter, are worked.

These rams, which are supplied with water containing some vaseline oil, force the water into an accumulator up to a pressure of 200 atmospheres ; from this the water is led through pipes with an internal diameter of 65 mm. and a thickness of 10 mm. of metal to the hydraulic pumping-engine, standing at a depth of 505 metres below the surface. This is a double-action twin engine without revolving parts. The driving gear is so constructed that in the water column there is no pause in motion, whereby, notwithstanding the 250 atmospheres pressure, all blows are completely avoided. The motor pressure water is forced back to the reservoir at the surface through pipes 75 mm. in diameter. The rising main pipes have a diameter of 109 mm.

The hydraulic pumping-engine forces 2·5 cubic metres of water per minute to the surface from 505 metres depth.

Hydraulic pumps are subject to shocks if proper care is not taken in the design. Low-pressure systems suffer more from shocks than high-pressure ones, because the weight of water employed is larger in proportion to the pressure. High-pressure systems with weighted accumulators suffer more from shocks than those provided with steam or air accumulators, because of the inertia of the accumulator weight.

A body of water at rest in a pipe exerts at any point a pressure proportional to the head or vertical height of the column above that point.

Let

h = the head of water in ft. above any point,
p = the pressure in lbs. per sq. in. at that point ;
then $p = 0.43\ h$.

M

A column of water in motion in a pipe has an acquired energy expressed in ft.-lbs. by

$$\frac{W v^2}{2g} = \frac{W v^2}{64\cdot4},$$

in which W is the weight in lbs. and v the velocity in ft. per second.

If the velocity of a given weight of water is not altered, its acquired energy is not increased or diminished. But a column of water moving with an accelerating velocity exerts a less pressure than the same column at rest; and a column of water whose velocity is being retarded exerts a greater pressure than the same column at rest.

In high-pressure systems the energy due to velocity alone is small compared with the total energy, especially with steam or air accumulators.

We have already described fully the system of hydraulic power transmission in mines and the application of the steam regulator or accumulator. The foregoing figs. (157, 158, and 159) represent modern power plant recently designed by the author for a colliery.

In fig. 160 is illustrated the general construction of the duplex hydraulic pumps employed underground. In this particular instance the working pressure at the surface is 1000 lbs. per sq. in.

The hydraulic pumps are placed in heavily watered 'dip' workings with 360 ft. lift to the sump at the bottom of the shaft, from which the water is pumped to the surface by means of compound differential engines. The shaft is 500 ft. deep, so that the vertical height from the hydraulic pumps to the surface is 860 ft. Below the main hydraulic pumps are placed pilot hydraulic pumps for the purpose of following the coal workings into the 'dip.'

CHAPTER X.

ALMOST all types of steam valves are used in pumping-engines, and the forms of valve gears are numerous.

The Cornish gear is illustrated in figs. 161 and 162. The Cornish engine is provided with what is known as 'equilibrium valves,' three in number—the admission, the equilibrium, and the exhaust—and each valve is actuated from a separate valve arbor or shaft. The admission or steam valve is closed by the cut-off tappet A, which is adjustable for different portions of the stroke. The valve is held closed by means of the catch B until it is released by the operation of the cataract rod C. The other valves are closed at the end of the stroke by tappets on another plug rod clearly shown in fig. 161, and are kept closed by catches.

On each gear arbor is a tappet arm shown in the details between figs. 161 and 162. The upper arbor actuates the steam valve, the middle one the equilibrium, and the lower arbor the exhaust valve; a handle is also provided on each arbor, that the valve may be commanded by hand in starting the engine. The plug rods which carry the tappets are clearly shown in fig. 161. The tappets are made of hard wood, faced with leather.

During the steam or indoor stroke the long tappet A on the left hand plug rod closes the steam valve by striking the arm on the upper gear arbor. At the completion of this stroke the lower tappet on the right hand plug rod strikes the arm on the bottom arbor, and closes the exhaust and the injection valve. The cataract now releases a weight on the rod D of the middle or equilibrium arbor, and the weight falls, opening the equilibrium valve. The piston now being in equilibrium, makes its up stroke, towards the completion of which the upper tappet on the right hand plug rod strikes the arm on the middle arbor and closes the equilibrium valve. Steam is now cushioned above the piston, and the engine comes to rest, remaining in that condition till the cataract releases the top and bottom arbors, when the weights attached to them suddenly cause the opening of the steam, exhaust and injection valves.

By adjustments on the cataract rod C, the exhaust and injection valves are caused to open a little before the steam valve by releasing the lower arbor a little before the top one. The steam valve tappet A is made long that it may close the steam valve early in the stroke.

To this tappet a long screw is attached, whereby it may be moved up

Fɪɢ. 161.　　　　　　　　　Fɪɢ. 162.

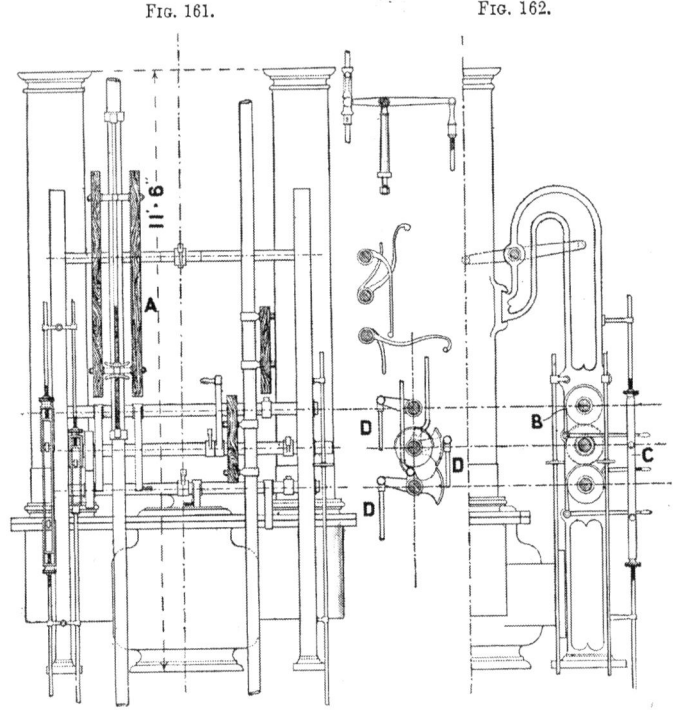

Front Elevation.　　　　　　Side Elevation.

Fɪɢs. 161 and 162.—The Cornish Engine Valve Gear—an illustration of the engine is given on p. 81.

A, Expansion Tappet for closing the steam valve, adjustable for cutting off at from $\frac{1}{4}$ to $\frac{3}{4}$ the stroke.

B, Catch for holding the valve closed, until released by the cataract rod C.

D, D, D, Rods carrying weights for opening the valves when released by the cataracts A.

and down on the plug rod to adjust the degree of expansion. Usually it is so adjusted that the cut-off takes place at about one-third of the stroke or a little later, but the steam is throttled up to the point of cut-off. The initial pressure is often considerably above the pressure at the point of cut-off, and the indicator diagrams often give a curve in which the point of cut-off is not apparent; the expansion curve after cut-off appearing to be part of a regular curve commencing with the initial pressure.

This has given rise to misconception with casual observers. Cornish engines have been said to work with ten to fifteen expansions, judged from an imperfect inspection of the indicator diagrams. We have in another chapter discussed the principles which govern expansive working in the Cornish engines. As a matter of fact the highest rate of expansion obtained in the old or single Cornish engine is about four to one.

The cataract consists of a pump (fig. 163), placed in a tank so arranged

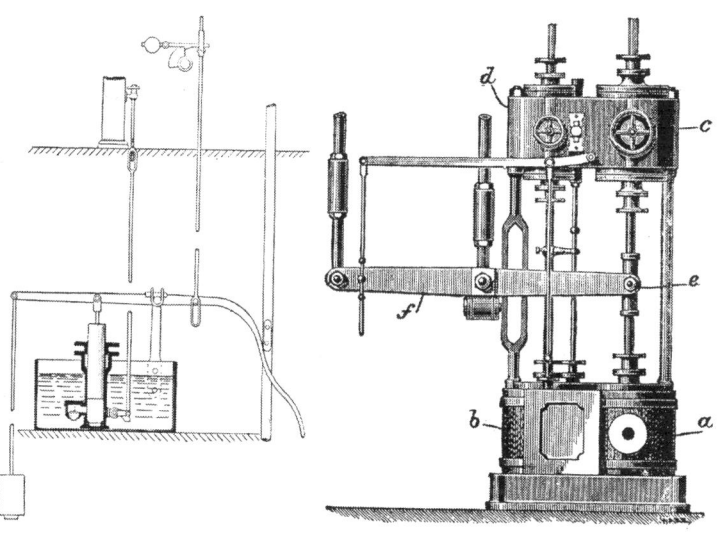

FIG. 163.—Cataract Gear of Cornish Pumping-Engine.

FIG. 164.—Davey's Differential Valve Gear.

that when the engine makes its indoor or steam stroke the pump plunger is lifted by the operation of a tappet on the side of the plug rod as shown. The water taken in under the plunger on the up stroke is discharged back into the tank through an adjustable valve on the down stroke by means of the weight acting on the plunger. The adjustment of the discharge valve determines the speed at which the plunger descends, and thus regulates the time at which the cataract releases the catch on the valve arbor. This determines the time when the valve is opened. With two cataracts there may be a pause at each end of the stroke. Generally only one cataract is used, and the engine only pauses at the completion of the equilibrium stroke.

The Differential Valve Gear.—This gear dispenses with the tappet

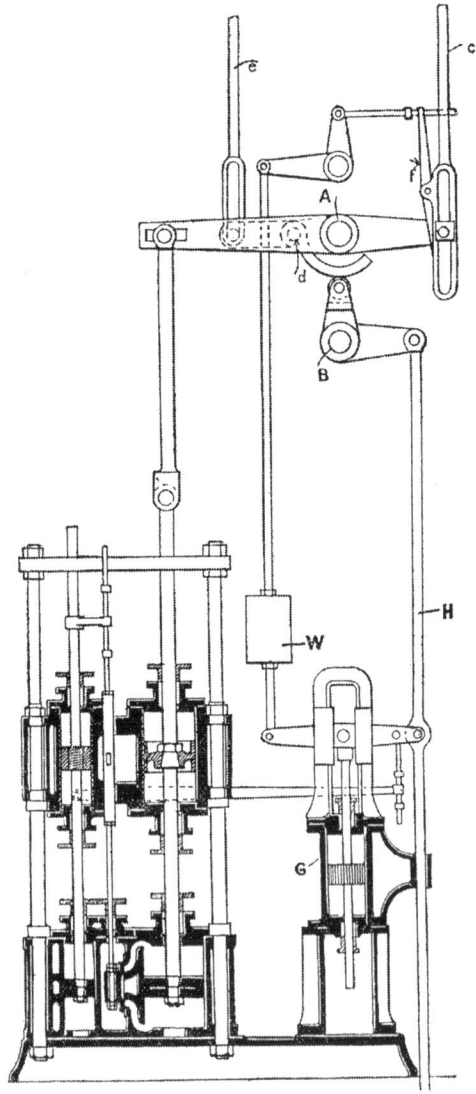

FIG. 165.—Davey's Differential Cut-off Motion and Trip Gear.

motion of the Cornish gear, and introduces a governing element by which
the valves are all closed or partially so when any erratic motion of the

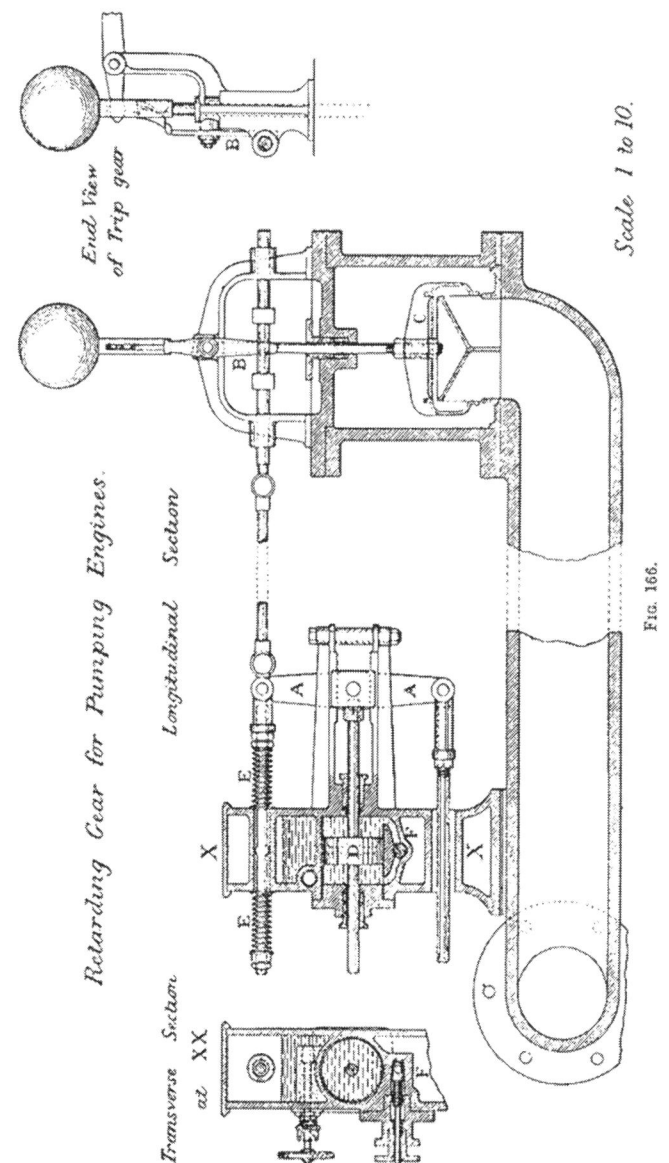

End View of Trip gear

Retarding Gear for Pumping Engines.

Longitudinal Section

Transverse Section at XX

Scale 1 to 10.

Fig. 166.

engine takes place owing to a loss of load.   The general principle is that of
causing the valves to be opened by means of a subsidiary steam cylinder at
a fixed ratio of speed determined by the adjustment of a cataract.   At the
same time the engine by its motion tends to close the valves.

As long as the motion of opening is faster than that of closing, the

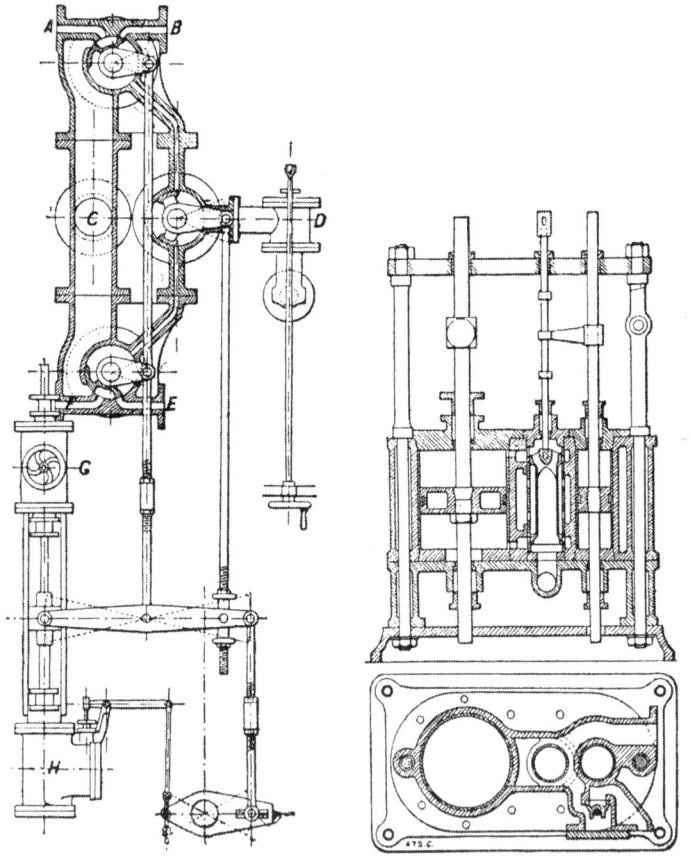

Fig. 167.—The Differential Gear as
applied to cylindrical slide valve
of compound engine.

Fig. 168.—Cataract Gear worked by
means of water pressure from the
pumping main.

valves remain open, but as soon as the engine motion becomes quicker
than that of the cataract, then the valves begin to close.

Whilst the engine is working under normal conditions, then the
opening and closing of the valves remains the same on each succeeding

stroke ; but when anything happens to
cause the engine to race, then the
valves close earlier.

The standard form of gear is shown
in fig. 164.

a and *b* are subsidiary steam
cylinders, and *c* and *d* cataracts with
adjusting plugs to enable the speed of
the subsidiary pistons to be regulated.

To the main cataract rod *e* a lever *f*
is attached ; at the centre of this lever
is a rod for giving motion to the
engine valves through a rocking shaft
or other convenient means. To the
outer end of the differential lever *f* is
a rod attached by any convenient means
to the engine so that it gets the motion
of the engine piston on a reduced scale.
The motion of the opposite ends of the
differential lever are in opposite directions. The cataract *d* is for the purpose of determining the pause at the
end of the stroke. It simply operates
the steam valve of the main subsidiary
cylinder by tappets on its piston rod,
and in that way the time occupied in
opening the valve of the steam cylinder
a determines the pause between the
strokes. It will be seen that in applying this gear to engines with drop valves
it is only necessary to have a rocking
shaft between the gear and the valve rods,
with suitable connections to the valves.

There are several additions to this
gear, such as separate cut-off motion
to the expansion valves, and a trip
motion to cause a sudden closing of
the valves in the case of breaking of the
pump rods, etc.

Both those motions are illustrated
in fig. 165.

The steam valve is opened by the
gear through the curved lever *d*, and
the cut-off takes place by the engine
motion transmitted through the rod H.

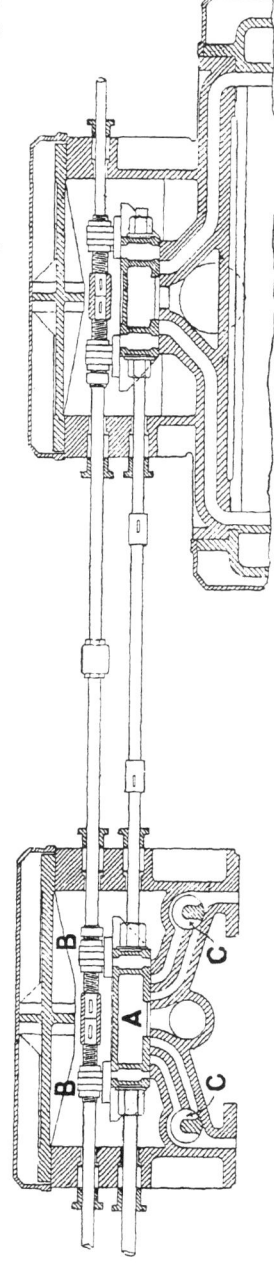

Fig. 169.—Expansion and Cushioning Valves of Compound Non-rotative Engine.

The roller on the bent lever B forms an abutment on which the lever *d* is kept in position during the opening of the steam valve, but the engine motion produces an upward motion of the rod H, and removes the abutment, causing the valve to drop. A trip motion is applied to the valve rod *c*.

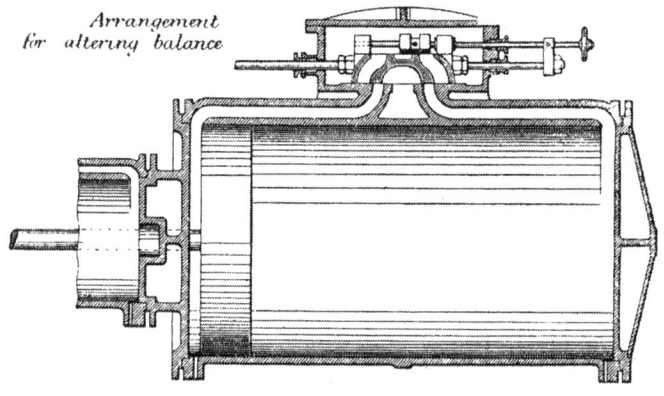

Arrangement for altering balance

Scale 1 to 32

FIG. 170.—Shutter Valve on the back of main valve of low-pressure cylinder of non-rotative engine, for the purpose of enabling the engine to work out of balance when used in working sinking pumps.

The weight W keeps the bent lever A from releasing the trip *f*. The adjustment of the cataract G is such as to cause the weight W to be just raised, but not sufficient to release the trip *f*. Should the load suddenly

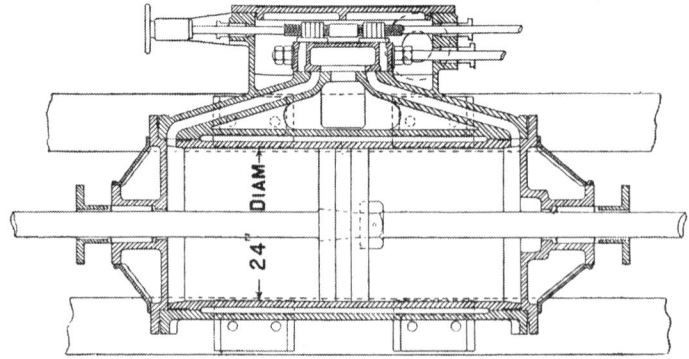

FIG. 171.—Meyer Cut-off Valves on the back of an ordinary D-slide valve.

fall off the engine, then the rod H would move relatively faster than the cataract piston G, with the result that the weight W would be raised, releasing the trip *f*.

This trip gear is applied in a variety of ways—sometimes in compound engines to a valve between the cylinders, as in fig. 166.

In this instance the weight W of the last illustration is substituted by the springs E E, the engine being double-acting, and the trip taking place on either stroke, closing a valve between the cylinders of the engine.

Fig. 167 shows the gear applied to the working of cylindrical valves with a definite cut-off valve in the middle position. In this illustration

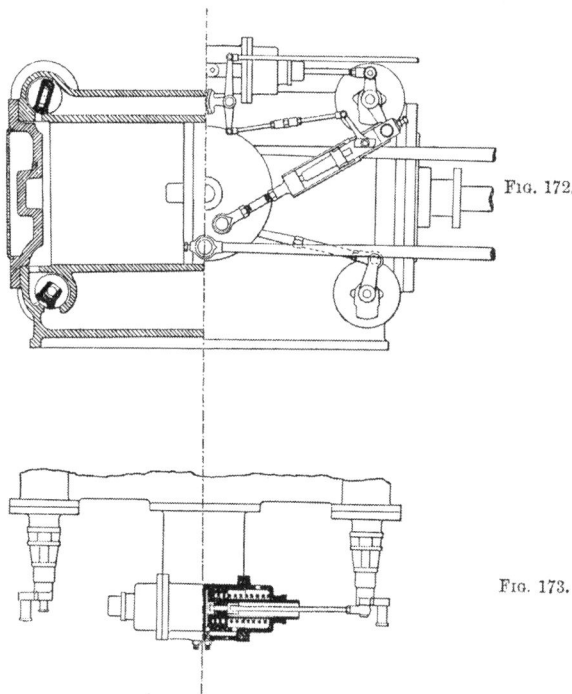

Fig. 172.

Fig. 173.

Figs. 172 and 173.—Corliss Gear.

the valves are those of a compound engine, having pistons moving in opposite directions.

Sometimes the gear is worked by water taken from the pumping main, as in fig. 168. In this case the subsidiary cylinders form cataracts in themselves, the regulation being effected by means of the water stop valves.

Sometimes the differential action is produced without the differential lever, as in fig. 169, where the main valves A are actuated by the direct cataract motion and the cut-off valves B B by the engine motion. In this illustration cushioning valves C C are shown on the high-pressure cylinder.

Balancing Valve.—With double-acting mining engines working the
double system of pump rods it is useful to be able to work the engine
when it is out of balance, during sinking operations, or when one pump is

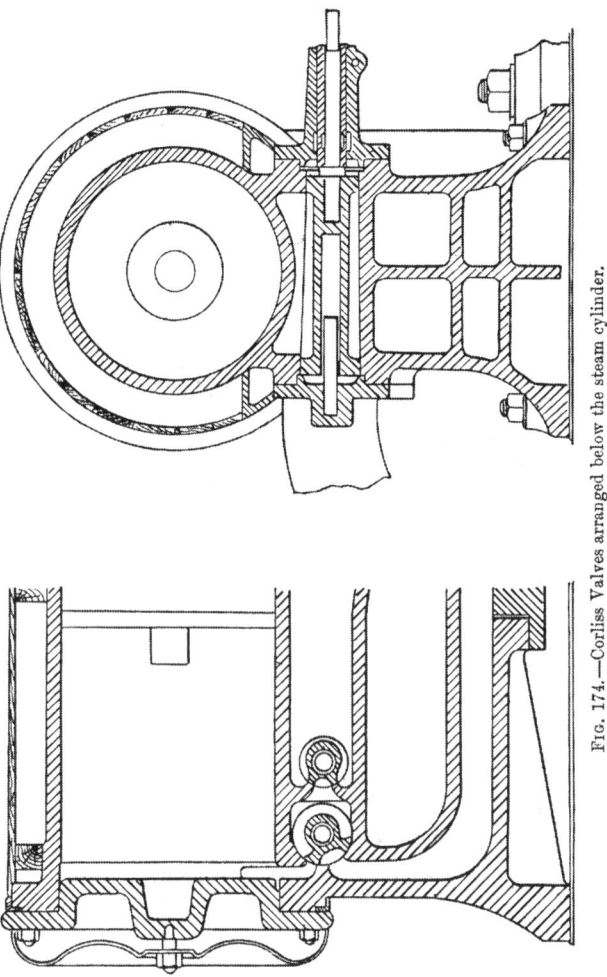

FIG. 174.—Corliss Valves arranged below the steam cylinder.

completely out of use, in the case of a breakdown of one pump, or from
other causes. With the differential gear working drop valves, it is quite easy
to work the engine under such conditions by throttling on the idle stroke,
and in slide-valve engines a shutter valve has been applied, as in fig. 170.

On the back of the low-pressure valve is placed a plate or shutter carried with the valve and held in position by a screw and nut on the valve itself. This shutter is adjustable from the outside whilst the engine is running, and it is only necessary to draw the shutter over the port to throttle the steam on the side where the load is lightest.

Rotative Engine Valve Gears.—Almost all kinds of valve gears in use on ordinary rotative engines may be applied to pumping-engines, but as pumping-engines generally work slowly and have a constant load, automatic cut-off gears are not often required. Slide-valve engines have

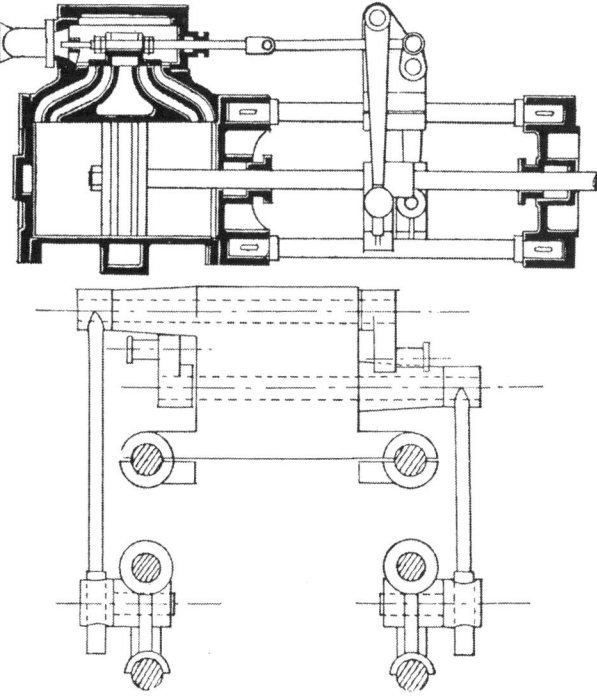

FIG. 175.—Duplex Steam Pump Valve Motion.

usually the Meyer cut-off valve on the back of the D-slide valve with a hand adjustment, as in fig. 171.

Corliss valves are largely used, and are sometimes placed in the cylinder covers to reduce the clearance spaces, as in fig. 8, Chap. II.

The ordinary Corliss gear is shown in figs. 172 and 173. An oscillating disc is placed on the side of the cylinder, and from this disc is taken the motion for the opening and closing of the exhaust valves and the opening of the steam valves. The opening of the steam valves is effected against the resistance of a spring, as shown in fig. 173, and the valve is released by the opening of a trip or catch. On the release of the valve the spring closes it, and a dash pot introduced under the spring cushions the sudden motion.

The release of the valve may be determined by the governor, as in fig.

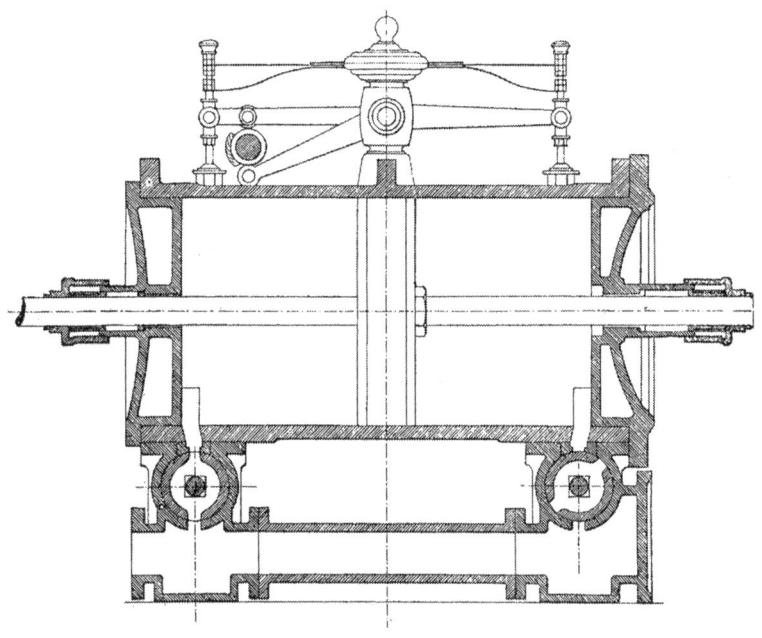

Fig. 176.—Example of Valve Gear in which the steam valves are worked by cams, and the exhaust valves from a crank action.—Shown in cross section in fig. 177.

172, or there may be a hand adjustment, which is generally sufficient in pumping-engines.

Fig. 174 shows another form, the Corliss valves being arranged under the cylinder.

Steam Pumps.—Steam pumps with steam-moved valves are numerous and varied. The steam piston is made to cause the main steam valve to be reversed at the end of the stroke by means of a tappet, or in other ways causing steam to press on a subsidiary piston connected with the main valve so as to reverse its position.

The tappet and lever are often dispensed with by arrangements of steam passages, covered and uncovered by the main piston, the main piston really performing the function of a subsidiary valve.

Duplex Valve Gear.—The duplex steam pump consists of two steam pumps arranged side by side, the piston of one pump being made to give

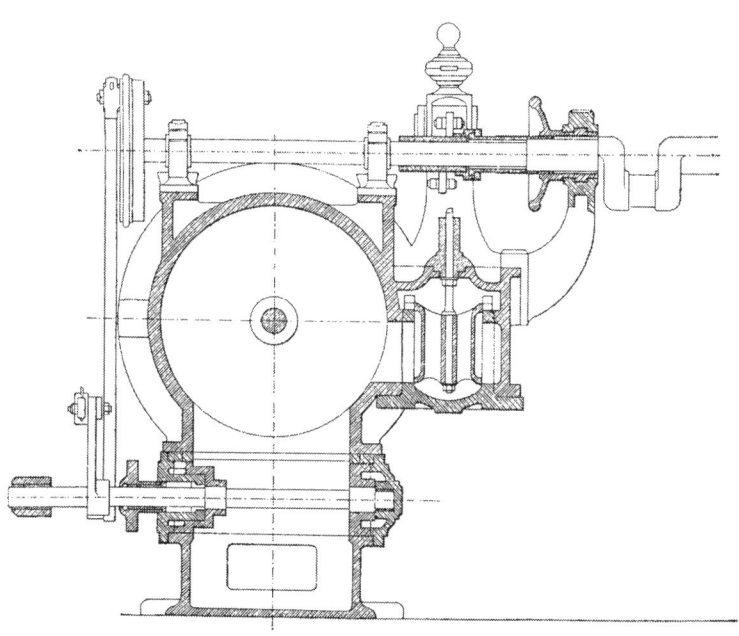

FIG. 177.—Example of Valve Gear in which the steam valves are worked by cams, and the exhaust valves from a crank action.—Shown in longitudinal section in fig. 176.

motion to the steam valve of the other. Its simplest form is given in fig. 175.

It will be seen that cushioning is necessary at the end of the stroke to prevent the pistons striking the covers. That is effected by the double ports shown in the engraving.

Steam is admitted to the cylinder through the outer ports, whilst the exhaust takes place through the inner ones. The exhaust port is closed by the piston before the end of the stroke, and in that way the piston is prevented from striking the cover.

The device of making the working of one engine dependent on that of the other is old. Cornish engines in waterworks have been made to work alternately, when two engines were pumping into the same main, by coupling the valve gears of the engines by suitable connections, and as the steam stroke is made quicker than the water stroke, one water stroke commenced before the other was completed, thus keeping up a nearly constant flow in the mains.

The author observed many years ago in using two independent steam pumps of precisely the same dimensions, and employing them to pump into the same rising main in a mining shaft, the rising main having no air vessel, that when the engines were working together they worked approximately, as if their pistons had been coupled together by means of cranks at right angles. The reason was obvious. The pump plungers simply followed the line of least resistance.

In almost all non-rotative engines having the main steam valves worked by a subsidiary cylinder, it is quite easy to arrange a pair of engines side by side and so connect the valve gears that the duplex action may be

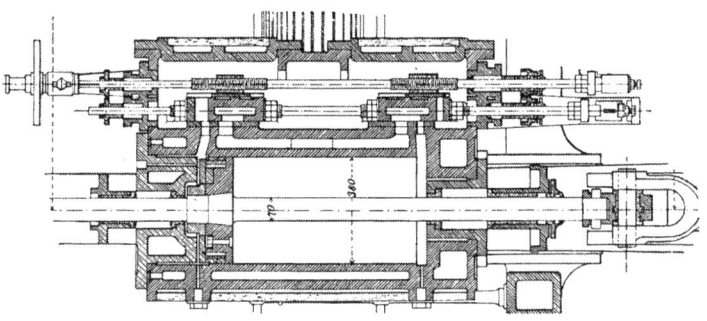

FIG. 178.—Example of Meyer Expansion Valves. Main and expansion valves worked from separate eccentrics.

secured. By the duplex action we mean causing one piston to commence its stroke before the other finishes its stroke, and that can be done by means of two engines, each capable of independent action. An example of such an arrangement is given in connection with a pair of engines in the chapter on Hydraulic Engines.

Valve gears of pumping-engines are so varied that we can only indicate the general features of different valve motions. The method of application is quite simple when the motion is understood.

Cam Motion is applied to drop, piston, and Corliss valves. It may also be applied to ordinary slide valves. With cams, quick opening and closing of the valve with an interval of rest can be secured; the valve has therefore little useless motion. The valves most suitable for cam motion are those moved with little resistance.

Eccentric or Crank Motion.—This is the motion most commonly used. It is combined with a trip motion in the Corliss gear, and it is applied with modifications to actuating almost all kinds of valves. In figs. 176 and 177 there is a cam motion to the steam valve, and a crank motion to the exhaust. In fig. 178 we have eccentric motion to the main and expansion valves. It may here be observed that an eccentric is a crank with the crank pin larger than the shaft. Two eccentrics at right angles to each other on one shaft will drive another shaft if coupled to the driven shaft by connecting rods attached to cranks, the cranks of course being at right angles to one another.

CHAPTER XI.

WATER-PRESSURE engines are employed in pumping water, in winding materials from mines, and in driving machinery of various kinds.

For pumping water, the engine is usually applied direct to the pump, and in its simplest form is illustrated in fig. 179, which represents the water-pressure engine erected by Trevithick at the Druid Copper Mines, Illogan, near Redruth. The engine consisted of a 10-in. double-acting hydraulic cylinder A, provided with a piston having a stroke of 9 ft. From the cross-head of the engine, spear-rods descended to work the pumps. The valves consisted of two lead 'plugs,' or pistons, B and C. Rods attached to these plugs passed through stuffing-boxes in the valve-chest cover, and were connected by means of a chain passing over a chain-wheel D. A lever on the axis of the chain-wheel was connected to the tumbling weight W. The movement of the main piston towards the end of its stroke, acting by tappets fixed on the piston-rod, caused the tumbling weight to be pushed over the centre, as in fig. 180; and in that way the valves or plugs B and C were reversed, and the return stroke of the engine was effected.

If the plugs B and C were made to cover the ports, and thereby to prevent all slip of water past them, then the driving column would be absolutely stopped at the end of each stroke, as would be also the delivery column of the pumps, with consequent loss of energy. Severe shocks would also be produced at the end of each stroke. To reduce the severity of these shocks, Trevithick made the plugs B and C less deep than the width of the ports, as seen in fig. 179, and thereby allowed water to pass through from the driving column to the exhaust during the reversal of the stroke, thus preventing the absolute stoppage of the driving column. This device reduced the shock at the expense of water; and the efficiency of the engine must have been small.

The loss of energy arising from fluctuations of velocity in the driving column is directly proportional to the weight of the driving column, and therefore to its length; consequently, when the height of the column is inconsiderable compared with its length, and therefore its power small compared with its weight, the percentage of loss is very greatly increased.

In applying hydraulic power for draining the dip and distant workings in mines, the author has practically experienced the difficulty of having to deal with a column very long compared with its height. A practical example is given in fig. 181, which represents two engines at Griff Colliery, near Nuneaton: these are now employed to pump out a very long dip

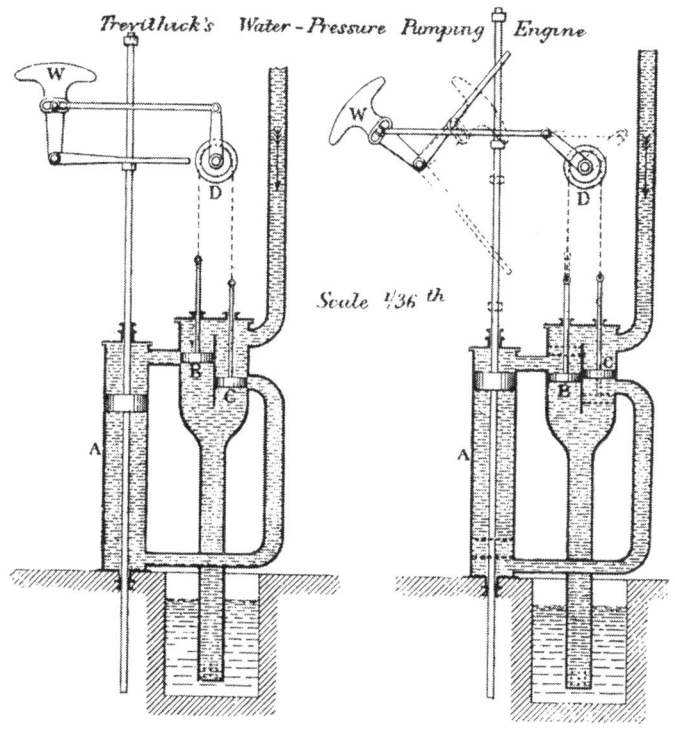

FIG. 179.   FIG. 180.

Reference Letters.—A, power cylinder ; B C, piston valves ; D, chain pulley ; W, tumbling weight.

working, and will be kept for draining it permanently. There are two hydraulic engines A and B, each capable of raising 150 gallons per minute to 150 ft. height, through 800 ft. of 7½-in. delivery pipe C C, under an effective head in the driving column of 450 ft., this head being supplied through 1900 ft. of 5-in. supply pipe D D. In commencing to get out the water from the dip, the first engine was placed as near as possible to the surface of the water. Suction pipes E were then added, as the

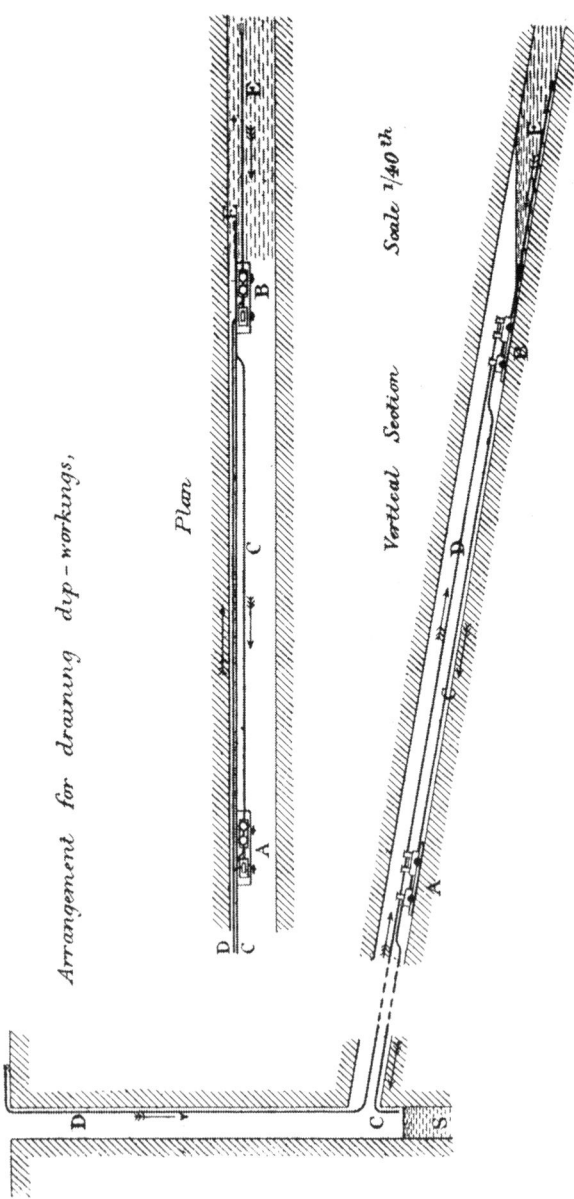

Arrangement for draining dip-workings.

Plan

Vertical Section

Scale 1/40th.

FIG. 181.

water was lowered, until a vertical depth of 25 ft. was reached; the second engine was then placed close down to the water-level, and the operation continued. In that way the pumping was uninterrupted, and a distance of 150 ft. down the dip could be reached before it became necessary to move either engine, the inclination of the dip being 1 in 6. The water delivered at the top of the dip working is conducted to the sump S of the main pumping-engines, and the water for driving the hydraulic engines is conveyed down the shaft from the surface of the mine; so that the main pumping-engines have to pump the water delivered at the sump, which includes the water used in working the hydraulic engines. The exhaust water from the power cylinder is delivered direct into the rising main of the hydraulic pump, so that the hydraulic engines work with an effective head equal to the height of the supply cistern above the main engine sump S.

A longitudinal section of the hydraulic engines is given in fig. 182. The stroke is 2 ft. 6 in., diameter of power cylinder $6\frac{1}{4}$ in., of pump $8\frac{3}{4}$ in. Thus the useful effect is represented by $\left(\dfrac{8\cdot75}{6\cdot25}\right)^2 \times \dfrac{150}{450} = 65$ per cent. The speed of working is twelve double strokes per minute. It will be seen that the pump is provided with a loose liner M; by withdrawing this liner, and putting in a larger piston of $12\frac{1}{2}$ in. diameter, the capacity of the pump is doubled, enabling the engine to pump double the quantity to half the height. The larger piston is used until half the total depth is reached, when it becomes necessary to insert the liner, and use the smaller piston for the remaining depth.

The valves of the power cylinder are of a very peculiar construction, and were designed by the author with a view to get rid of the difficulties encountered in applying slides, and other ordinary valves, to water-pressure engines. They are shown to a larger scale in figs. 183 and 184. Referring to fig. 183, the top orifice F is the inlet, and the bottom one G the outlet or eduction pipe; and the pipes J and K form communications to the two ends of the power cylinder. The eduction valves are annular gun-metal pistons H H, working vertically, and each having two valve-beats, one on the inner edge I and one on the outer edge O of its bottom face. As the annular valve descends, the outside beat closes the communication to the eduction pipe; and the inlet valve L, rising against the inner beat, closes the supply. This inlet valve is an ordinary single-beat mushroom valve, with its stalk projecting upwards and attached at the top to a piston N; the bottom face of this piston is constantly under the pressure of the driving column, while the top face is exposed alternately to the pressure of the driving column and to the pressure in the eduction pipe by means of a small gun-metal slide-valve P (fig. 184), actuated by a lever Q and tappet-rod R (fig. 182). The action of the two valves in combination will be readily seen by supposing that the exhaust valve is closed and the pressure valve is open, as on the left-

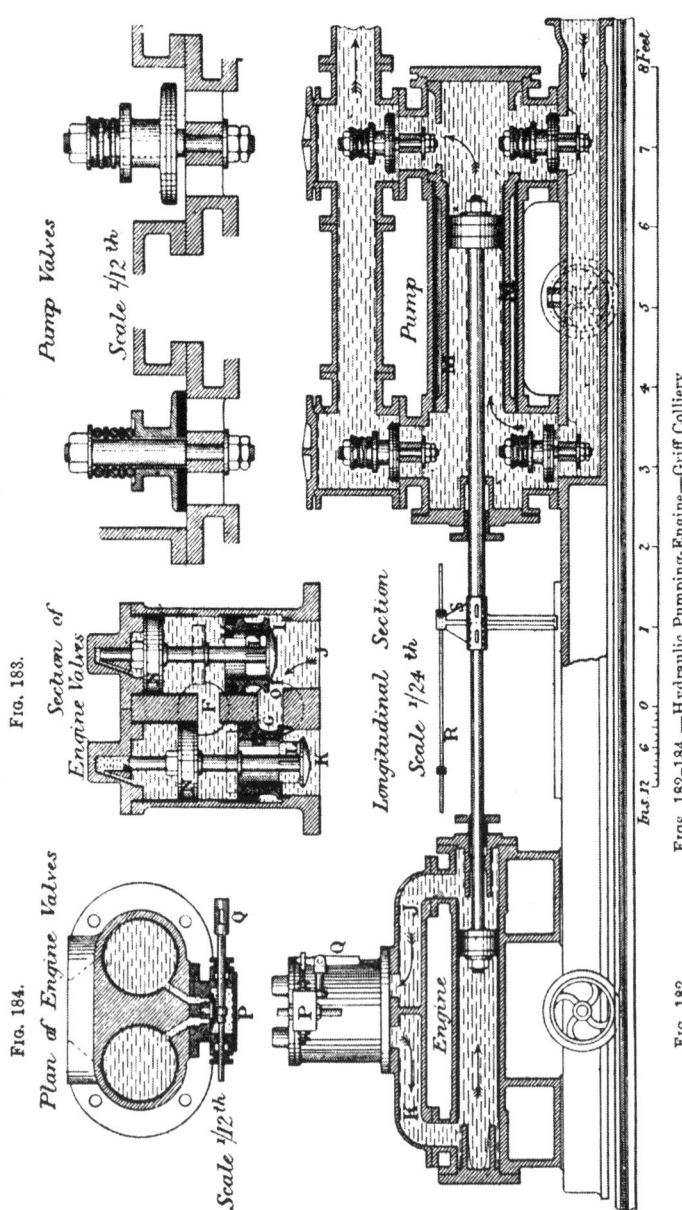

FIG. 183.

Pump Valves

Scale 4/12 th

Section of
Engine Valves

FIG. 184.

Plan of Engine Valves

Scale 1/12 th

Longitudinal Section
Scale 1/24 th

Pump

Engine

Ins. 12   6   0   1   2   3   4   5   6   7   8 Feet

FIG. 182.

FIGS. 182-184.—Hydraulic Pumping-Engine.—Griff Colliery.

hand side of fig. 183; then the pressure valve L, in closing, rises up against the annular exhaust valve H, and lifts it, opening the exhaust orifice G. The valves are now in the position shown on the right-hand side of fig. 183. Towards the end of the stroke, an arm S attached to the cross-head of the engine strikes the tappet and so pushes the slide-valve over, into such a position that the top of the right-hand valve-piston is exposed to the pressure column and that of the other to the eduction column. The main valves are thereby reversed: the right-hand piston being under equal pressure top and bottom, the pressure on the top of the annular valve H forces it downwards, carrying the pressure valve L with it. When the valve H has come down on its seat O, closing the exhaust orifice, the valve L continues to descend under the pressure above it, and opens the pipe K to the pressure water, as on the left-hand side of fig. 183. On the other side the ascending piston causes its inlet valve to close, and its eduction valve to open. With this successive and alternate action of the valves they cannot in working be placed in a position which would allow any water to slip through uselessly; that is to say, it is impossible for the exhaust valve and the pressure valve ever to be both of them open at the same time. The opening of the exhaust valve depends on the closing of the pressure valve, and the pressure valve cannot open until after the exhaust valve has closed, so that it is impossible for either valve to be open except while the other valve is closed.

The double-acting engines just described were specially designed for the purpose they are applied to, which necessitated their occupying a very small space, and being very portable. But in the majority of cases, especially where the engines are applied under considerable pressure, and where the water is at all dirty, the author uses plungers instead of pistons.

An example of a pair of engines with plungers, and applied under very heavy pressure, is given in fig. 185. These engines were made for pumping brine of 1·3 specific gravity.

Each hydraulic engine is capable of raising per minute 66 gallons of brine 1000 ft. high, with an effective pressure of 560 lbs. per sq. in. in the accumulator A, equivalent to 1000 ft. head of brine. In the larger view (fig. 186), B B are the power plungers of 8-in. diameter, reduced to 5⅛-in. diameter in the pumping cylinders C C. The rising and falling columns of the power water balance each other, and the head of the accumulator pressure exactly balances that of the brine in the rising main, so that the useful effect is simply given by the ratio between the total area of the plunger at B and the reduced annular area at C. Hence useful effect $= \dfrac{8^2 - (5\cdot125)^2}{8^2} = 59$ per cent. The power valves D D are precisely similar to those already described, but are placed in separate boxes, because the power cylinders are separated; they are shown in section to a larger scale in fig. 187, and one of the pump valves E E in

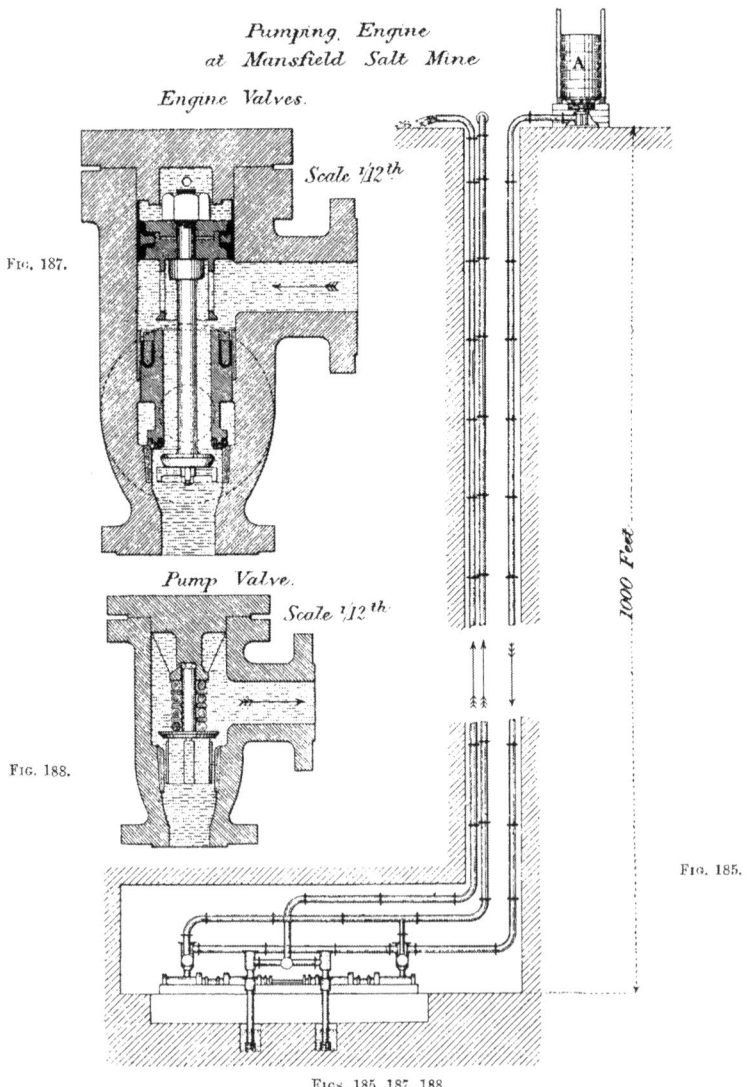

Pumping Engine
at Mansfield Salt Mine

Engine Valves.

Scale 1/12th

Fig. 187.

Pump Valve.

Scale 1/12th

Fig. 188.

1000 Feet

Fig. 185.

Figs. 185, 187, 188.

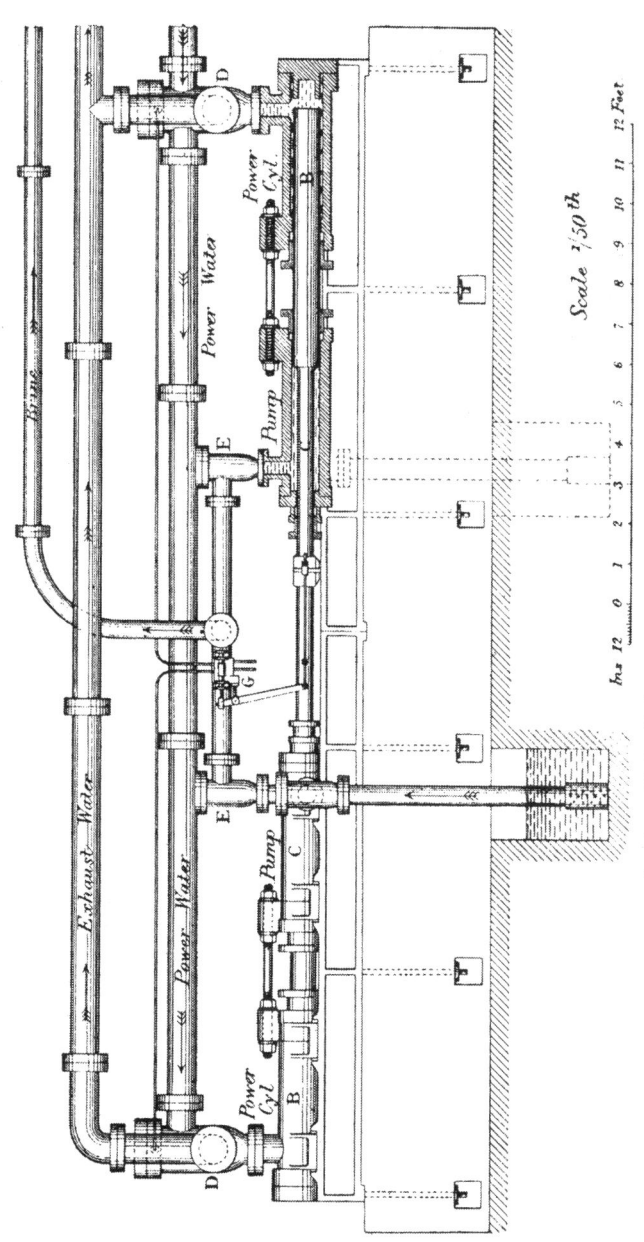

Fig. 186.—Hydraulic Pumping-Engine—Mansfield Salt Mine.

fig. 188. The change valve G (fig. 186) is worked by a tappet arrangement similar to that already described for the dip-engines.

The effect of fluctuations of velocity in the driving column from an accumulator is similar to that in a column with a natural head ; and the effect of enlarging the ram of the accumulator is to reduce the velocity of the moving weight, in the same way as the velocity of the water is reduced by enlarging the pipe in the other case. Hence it is highly important that when an accumulator is used it should be very large, thereby making its pulsations very slow.

As an example of the practical application of water-pressure machinery to the working of a metalliferous mine in a hilly district, the author has prepared figs. 189 to 195, which represent the machinery designed by him for the A. D. Lead Mine, near Richmond, Yorkshire.

It will be seen from fig. 189 that there is a reservoir situated on the hillside above the mine, at an elevation of about 500 ft. above the adit level. Pipes are led from this reservoir down the hillside for a distance of 1800 ft., and are then taken 240 ft. down a vertical shaft to the interior of the mine, at the inner end of the adit level. At this point a large chamber is excavated, to contain the pumping and winding engines.

The pumping-engine (figs. 190, 191, and 192) consists of two vertical hydraulic cylinders A A, each having a power ram 12 in. in diameter with a 7-ft. stroke. The rams are connected together by a chain passing over an overhead chain-pulley P (fig. 191), so that one ram makes its up stroke whilst the other is descending. A rod 3 in. in diameter, fixed to the ram, passes down through a stuffing-box in the bottom of the hydraulic cylinder, and is attached to the pump-rod. The pumps are each 13 in. in diameter, and of course have the same stroke as the hydraulic rams. On each hydraulic cylinder is placed a valve-box B, shown in section in fig. 192, with valves similar to those already described in connection with the other hydraulic engines ; both valve-boxes are connected with a single change-valve C. The pumps (fig. 190) are of the ordinary bucket type, provided with clack pieces, door pieces, and wind bores, such as are generally used for sinking purposes. The pumps will be used in deepening the shaft, and the hydraulic engines are proportioned for raising, at $6\frac{1}{2}$ strokes per minute, 500 gallons of water from a depth of 360 ft. to the adit level, the present depth being about 120 ft. At the full depth, with 534 ft. head on the rams, the useful effect will be represented by $\dfrac{13^2}{12^2-3^2} \times \dfrac{360}{534} =$ 84 per cent.

The winding-engine (figs. 193 and 194) consists of a pair of double-acting hydraulic cylinders, coupled to right-angled cranks on the driving shaft, which latter is geared to the winding drum by a spur pinion, the general arrangement of the engine being very similar to that of a steam winding-engine. The gearing has a proportion of 1 to 6, and the winding drum is 6 ft. in diameter. The weight to be raised is 2 tons of ore at a

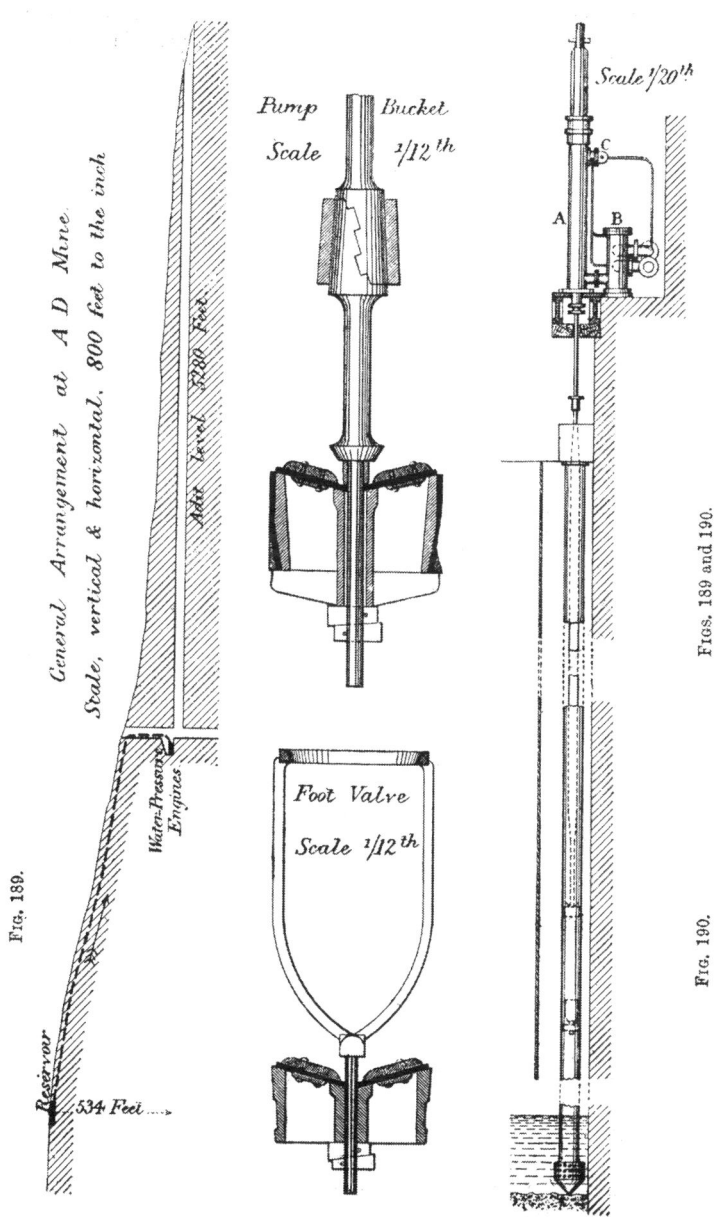

Fig. 189.

General Arrangement at A D Mine.

Scale, vertical & horizontal. 800 feet to the inch

Reservoir

534 Feet .....

Water-Pressure Engines

Adit level 5280 Feet

Pump Bucket

Scale ¹/₁₂ᵗʰ

Foot Valve

Scale ¹/₁₂ᵗʰ

Scale ¹/₂₀ᵗʰ

A   B   C

Figs. 189 and 190.

Fig. 190.

time. The cylinders are 5½ in. in diameter, have a 16-in. stroke, and run at 19½ rev. per minute, giving a speed of 60 ft. per minute to the

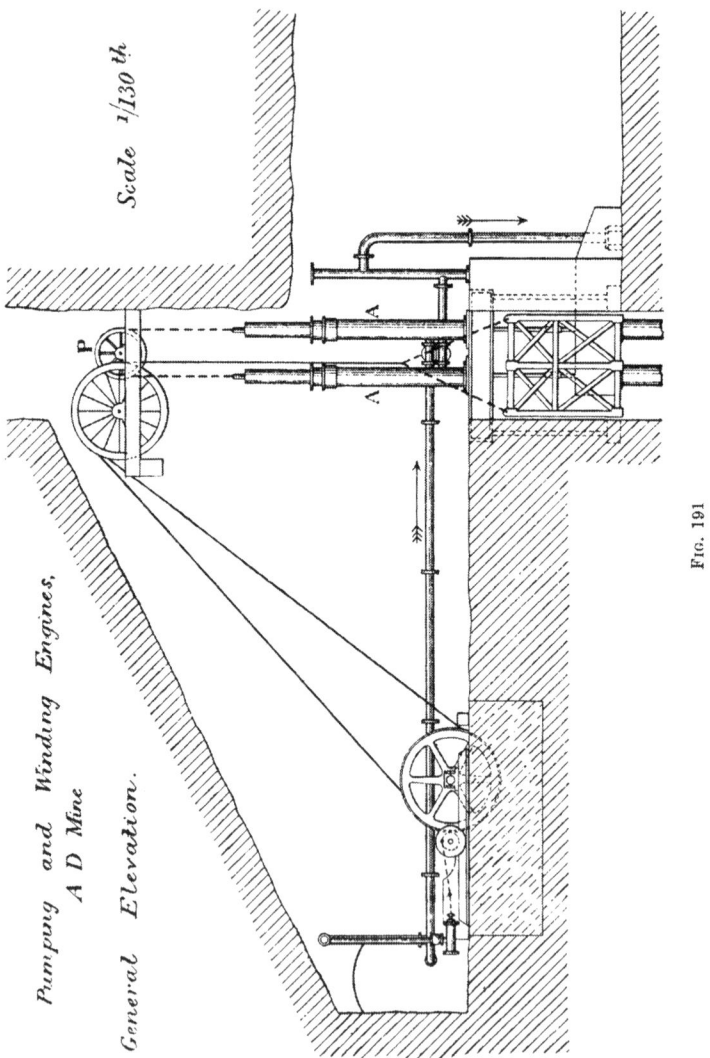

Scale ¹/₁₃₀ th.

Pumping and Winding Engines,
A D Mine
General Elevation.

Fig. 191

rope. The admission and eduction valves are somewhat similar in construction to those already described, but are driven by means of eccentrics, having a link-reversing motion, and are put in equilibrium by the some-

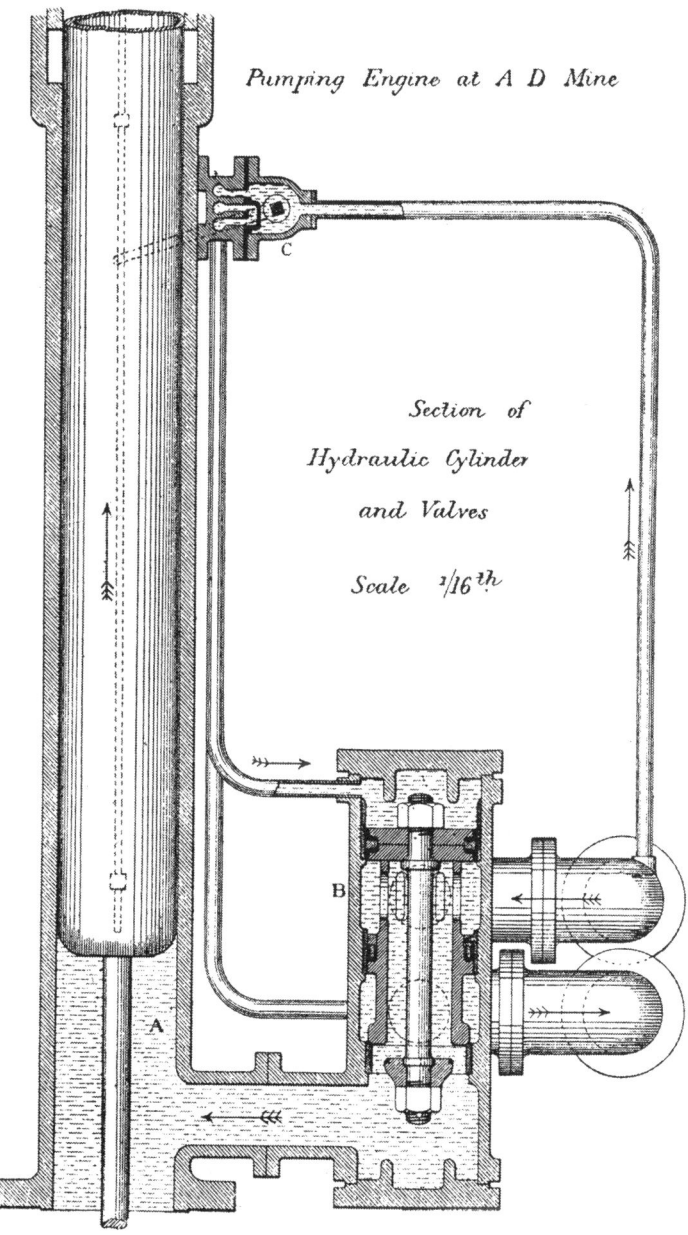

Pumping Engine at A D Mine

Section of Hydraulic Cylinder and Valves

Scale ¹/16ᵗʰ

Fig. 192.

Winding Engine at A D Mine.

Side Elevation.   Scale ¹⁄48th

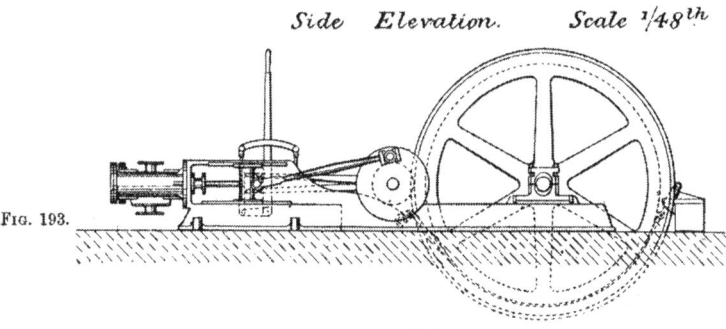

FIG. 193.

Plan

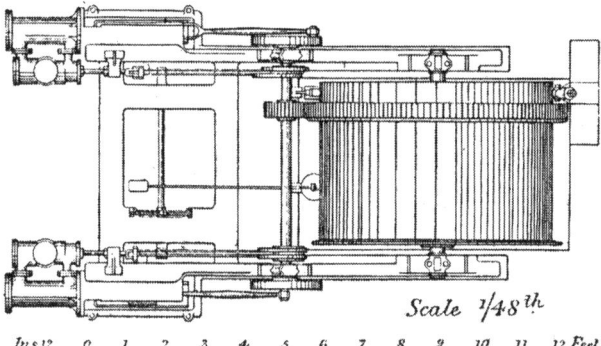

FIG. 194.

Scale ¹⁄48th

Ins.12   0   1   2   3   4   5   6   7   8   9   10   11   12 Feet

Longitudinal Section of Cylinder and Valve.

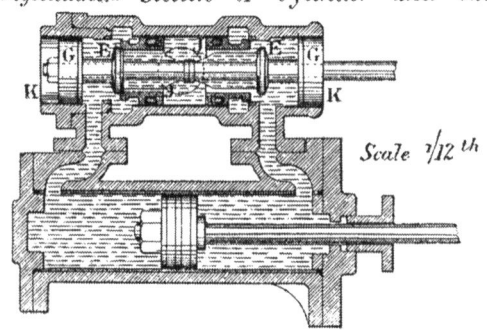

FIG. 195.

Scale ¹⁄12th

FIGS. 193, 194, and 195.

what novel arrangement shown in fig. 195. The two mushroom admission valves E E are on the same spindle, on which are also fixed two pistons G G, each equal in diameter to the annular eduction valves J J. These pistons work in cylinders K K, placed beyond the engine ports, and forming continuations of the valve-box. By following out the motion of the valves, as given by the eccentric, it will be seen that the eccentric has only to overcome the resistance due to friction, because in all positions the valves are balanced as regards pressures. This the author considers an important improvement in this type of hydraulic valve, as it enables very large valves to be used, and thus prevents any loss from throttling. It also enables the reversing to be done by means of a link motion, and gives easy and complete control over the engine.

Having described some examples of water-pressure engines of general application, the author proceeds to describe two engines specially designed for particular cases.

In fig. 196 is shown a peculiar application of a hydraulic pumping-engine, and one which the author has had occasion to adopt in several instances for mining operations. At the Hutton Henry Colliery, near Wingate, Durham, a certain quantity of water comes out of the strata at an intermediate point A, in a shaft where it is not convenient to place a pump except at the bottom. The water issuing at A has therefore to be taken down to the bottom before being forced to the surface. It is taken down from the point A in a down suction-pipe C C to the hydraulic pump B, and is delivered through the delivery pipe D D to the surface; so that the work done by the pump is that due to the difference between 866 ft. head in the delivery pipe and 502 ft. in the down suction-pipe, or to 364 ft. only. The power cylinder, $6\frac{1}{8}$ in. in diameter, with a 1 ft. 3 in. stroke, is actuated by means of a driving column from the point E, having an effective pressure of 260 ft. head. The power cylinder and pump, fig. 197, are both single-acting; but the pump is a piston pump of a peculiar construction. The pressure of the down suction column is constantly in the annular space between the piston-rod, $1\frac{3}{8}$ in. in diameter, and the inside of the pump barrel, 4 in. in diameter. During the forward or delivery stroke of the pump, the pressure behind the annular area of the pump piston assists the plunger of the power cylinder; and the return stroke is produced entirely by the pressure from the down suction column being brought to bear on the full front area of the pump piston, the effective pressure for the return stroke being therefore that due to the difference between the full front area of the pump piston and the annular area of its back face, or, in other words, to the area of the piston-rod. The useful effect is $\left(\frac{4}{6 \cdot 125}\right)^2 \times \frac{364}{260} = 60$ per cent. The engine is designed to work at $10\frac{1}{2}$ double strokes, and to raise 7 gallons per minute. It should be added that the driving water is water which would run down to the bottom of this shaft in any case, and is simply utilised for pumping.

FIG. 196.

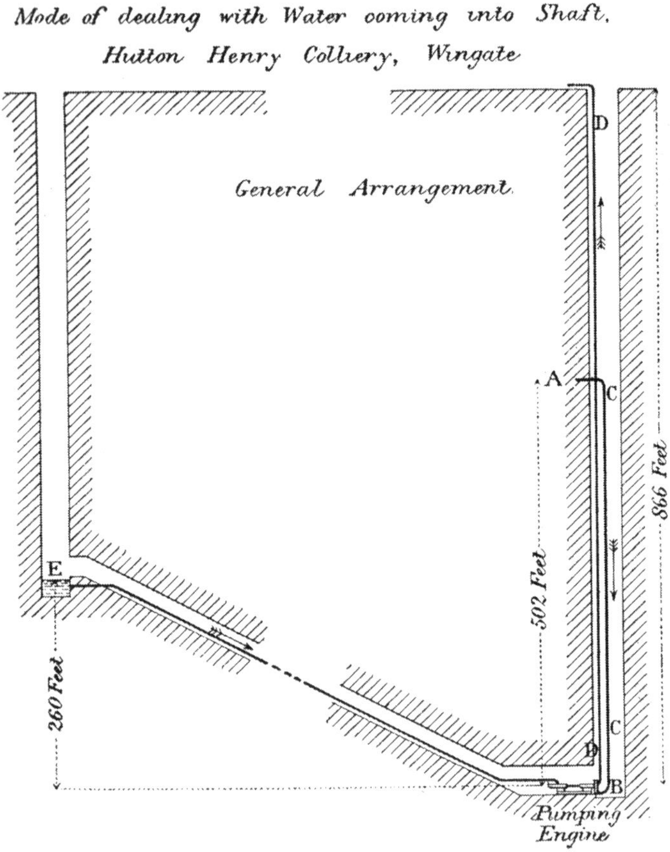

Mode of dealing with Water coming into Shaft,
Hutton Henry Colliery, Wingate

General Arrangement.

866 Feet

502 Feet

260 Feet

D

A
C

E

C
D
B

Pumping
Engine

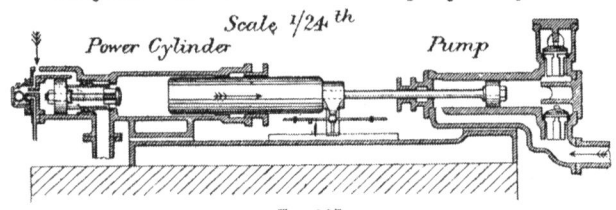

Longitudinal Section of Pumping Engine
Scale 1/24th

Power Cylinder　　　　　　　Pump

FIG. 197.

FIGS. 196 and 197.

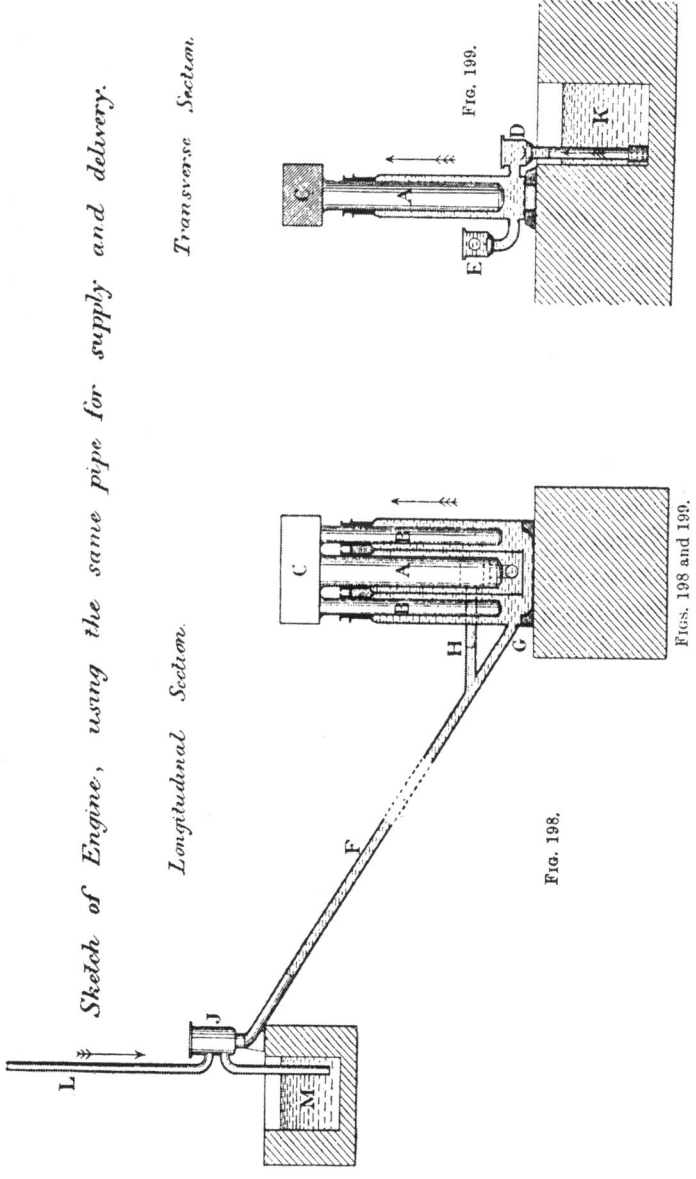

Sketch of Engine, using the same pipe for supply and delivery.

Transverse Section.

Longitudinal Section.

FIG. 199.

FIGS. 198 and 199.

FIG. 198.

In figs. 198 and 199 is shown a pumping-engine designed by the author to work under a most peculiar condition, namely that of making a single pipe serve both for the driving column of the engine and as the delivery pipe of the pump. A is the vertical pump plunger, and B B a pair of power plungers, all three coupled side by side to the same cross-head C. D is the suction valve of the pump, and E the delivery valve. F is the pipe which serves both for the delivery from the pump and for the supply to the power cylinder. It is in connection with the power cylinders at G, and has a branch H connected to the delivery valve-box above the delivery valve E. The cross-head C is of sufficient weight to cause the descent of the three plungers against the head of water in the pipe F. Water has to be pumped from the point K to the point M. The power water is obtained from the column L. J is a single-acting valve-box, similar to those already described in connection with the other hydraulic engines, and has a change valve actuated through a tappet-rod and wire by the rise and fall of the plungers.

The *modus operandi* may be thus described. During the up stroke the plungers B B are being raised by means of the pressure in the driving column L, and water is being drawn up into the pump cylinder A through the suction valve D. On the completion of the up stroke, the change valve in the valve-box J is reversed by the tappet-rod and wire, and the other valves are also reversed by the water itself, so that the communication to the plungers B B from the driving column L is closed, and also the suction valve D belonging to the lower sump K; while the delivery pipe to the upper sump M is opened. The weight of the cross-head C then causes the plungers to descend, forcing the water out of the pump and the power cylinders, through the pipe F, up into the sump M; and the operation is repeated in the same way. It will thus be seen that the pipe F serves as a supply pipe to the engine during the up stroke, and as a delivery pipe from the pump during the return stroke. This engine is specially applicable to the draining of dip workings, but can only be used to advantage in fixed positions, since it is not so portable as the other dip-working engines already described.

If a pump situated at J be substituted for the driving column L (fig. 198) the valve-box J can be done away with, and the pipe F connected directly to the pump barrel. The dip-engine would then derive its motion from this pump, and would work simultaneously with it.

The two engines last described are of very limited application; but they present several points of interest, particularly as, in the application of water power, engineers are often called upon to devise special means in order to meet special contingencies.

The question of applying hydraulic power economically by means of rotative engines to varying resistances has received considerable attention. A very clever and ingenious device—that of automatically altering the stroke by means of a resistance governor—has been used. There are other methods which readily suggest themselves, such as levers having shifting fulcrums, and similar devices; but the author has never found any such mechanism sufficiently practical for general application.

The dip engine already described (fig. 182) might be provided with a lever having a shifting fulcrum, between the power-cylinder and the pump : but the economy so gained, beyond that effected by the loose liner and change of piston, would not compensate for the extra complication.

In rotative engines having fly-wheels the admission valves of the power cylinders may be made to close at varying points in the stroke, by means of known mechanisms; and if the ends of the cylinders are provided with vacuum valves opening inwards, from pipes dipping into a waste-water cistern, the speed of the engine may be kept constant, whilst the supply of pressure water is varied to meet the varying resistances to which the engine may be applied. The cylinder would be partly filled from the pressure pipe, and then the remaining space would be automatically filled from the waste tank. By the use of three double-acting cylinders working on one crank-shaft, the portion of the stroke during which the pressure water would be admitted might be considerably varied, without causing a great fluctuation in the velocity of the driving column.

The foregoing is from a paper by the author in the *Proceedings of the Institution of Mechanical Engineers.*

Inertia of Water Column and Pump Rods.—In applying water-pressure engines to the working of mining pumps through the medium of pump rods, the author met with the difficulty arising from the inertia of the pump rods and attachments in addition to the inertia of the motive power column. The effect of inertia of pump rods has been dealt with in Chap. IV. Assuming that by the use of an air vessel, or by other means, the water pressure may be preserved nearly constant at the engine, how is the inertia of the pump rods to be compensated for?

Fig. 200 represents the application of a water-pressure engine to the working of mine pumps by means of the quadrants $g\,g'$ and the rods $h\,h'$. Assuming the rods to possess considerable weight, then a pressure above that required by the resistance of the pumps is necessary at the beginning, and a less pressure towards the end of the stroke.

This engine is employed to actuate pumps having heavy reciprocating parts, such as well or shaft pumps actuated by means of pump rods. a and b are motive power cylinders provided with admission and exhaust valves; c is a plunger or piston connected to the pump by means of the rod d, connecting rods e and f, quadrants g and g', and pump rods h and h'. The said plunger or piston is connected directly or indirectly by means of a rod i to the piston or plunger j, working in a water cylinder k, having at the ends air vessels $l\,l$; these air vessels $l\,l$ are partly filled with water, so that when the piston or plunger j is moved in one direction or the other, the air above the water in one of the air vessels is compressed, whilst the air in the other is rarefied.

In this way when the piston or plunger is moving from one end of its

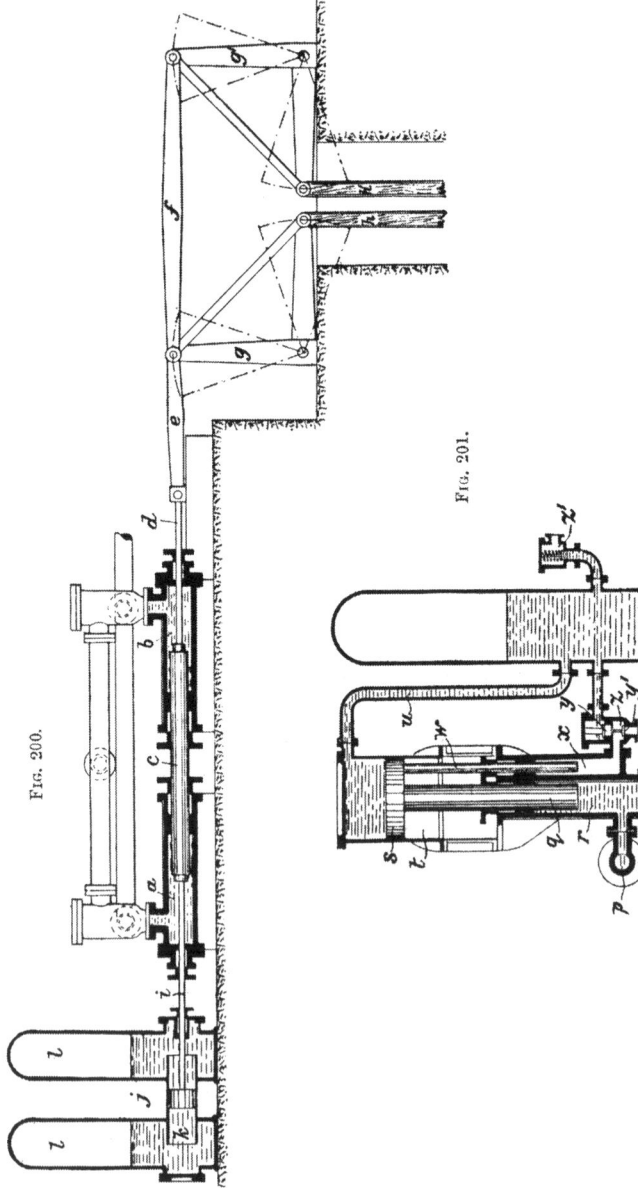

Fig. 200.

Fig. 201.

Figs. 200 and 201.—Davey's System of compensating for inertia in water-pressure engines having a heavy mass to put in motion, and in hydraulic mains.

stroke, it is assisted by the pressure of the air compressed in the air vessel at that end, and in moving from mid position towards the other end its motion is retarded by the compression of the air in the air vessel at the end towards which the piston is moving. Thus the motion of the motive power piston or plunger c is assisted at the beginning and resisted towards the end of the stroke, in this way compensating for the inertia of the pump

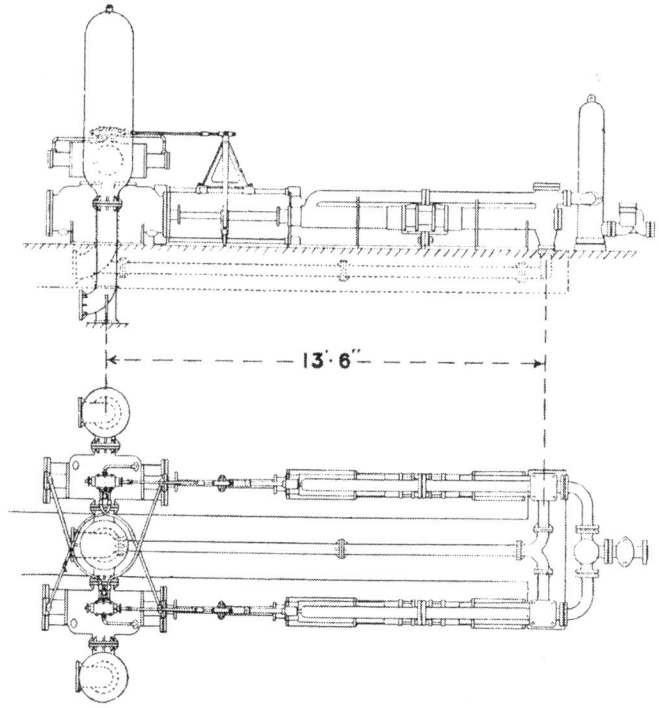

Fig. 202.—Duplex Hydraulic Pumps.

rods and connections, and the inertia of the water column actuating the motive power piston or plunger c.

To minimise the loss of energy arising from shock in the motive power water column caused by the fluctuations in velocity of the column, the author devised the apparatus shown in fig. 201. p is the motive power water pipe, q a plunger or piston working in a cylinder r, connected to the pipe p. Attached to the plunger q is a piston or plunger s working in a cylinder t; from the cylinder t passes a pipe u connected to the air vessel v, partly filled with water. w is a plunger attached to the piston s, and working in the pump cylinder x, $y\,y'$ are suction and delivery valves in the chamber z. Any leakage past the piston s passes into the cylinder x, and on the descent of the plunger w, the air and water present is forced out and into the air vessel v, so that the latter is by this means kept charged.

The areas of the pistons q and s respectively are so proportioned, and the pressure of the air in the air vessel v is so regulated that the pistons or plungers shall be towards the bottom of their strokes when the pressure in the motive power pipe is normal or below the normal, and the water in the pipe p is in motion. Should the motion of the water in the pipe p be suddenly stopped, then the increased pressure in the pipe p, due to the inertia of the water, would push up the pistons or plungers q and s, and

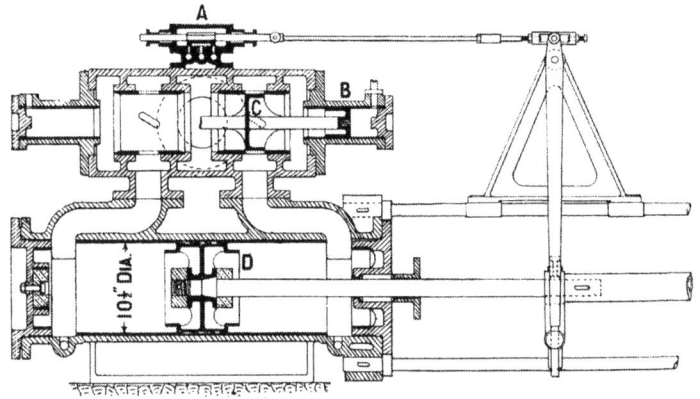

Fig. 203.—Power Cylinder and Valves of duplex hydraulic pump.

thus force the water above the piston s into the air vessel v, thereby compressing the air.

The energy thus stored in the air vessel would be given out again to the water in the pipe p by the return of the plunger when the pressure in the pipe p fell below the normal pressure.

A loaded relief valve of any suitable form is placed at z', to prevent any excess of pressure in the air vessel v.

We have already described very fully the principles and construction of water-pressure engines. Fig. 202 illustrates a pair of low-pressure engines employed to pump water to a higher elevation than the source of power. Water flowing from one reservoir to another at a lower level is made to pump water to supply a district above the higher reservoir. The engines are arranged to work on the duplex plan, but can easily be altered to enable them to be worked singly.

The hydraulic piston D (fig. 203) is cushioned by causing it to enter a recess at each end of the cylinder, and so confine the water being exhausted.

The power valves C (fig. 203) are of the piston type, with a subsidiary valve A and subsidiary piston B, by means of which they are actuated. The duplex action is secured by means of connecting pipes between the subsidiary valve of one engine and the subsidiary cylinder of the other. To enable the engines to work separately it is only necessary to connect the pipes from the subsidiary valve box of each engine to its own subsidiary cylinder.

CHAPTER XII

Town Water Supply.—Machinery for this purpose is very varied in character and design, partly owing to the varied nature of the application and partly owing to the changes which have taken place arising from the introduction of high-pressure steam and the progress which has been made in design, in the simplification of parts, and in improved details. With the older form of engine, the house and engine were combined into one structure, as is the case with beam engines generally.

The modern engine is self-contained and only requires a house as shelter.

Pumping from wells requires a special design of engine and pump to suit the local circumstances. It may be advisable to use the permanent engine to drain the well during sinking operations, or it may be advisable to use bore-holes as wells. It may be necessary to pump the water raised from the well into an elevated reservoir or into the distributing mains by the same engine; or it may be better to have a separate engine. These are questions influenced by local circumstances.

Water supply from wells is, generally speaking, obtained from a depth not exceeding 200 or 300 ft.

At that depth pumping is not so expensive as to become of great import-ance, but as the expense of fuel, upkeep and some of the working charges increase somewhat in the proportion to the increase in depth from which the water is pumped, it is worth while to expend considerable capital in order to obviate having to pump water from excessive depths. When water supply is taken from rivers it often becomes necessary to have pumps to lift the unfiltered water to filter beds, and others to pump this filtered water into the distributing mains. The two sets of pumps are sometimes actuated from the same engine, but the more modern and approved plan is to have a separate engine for each purpose.

The filter pumps are generally of small power and large capacity. The simplest forms are the centrifugal and the ordinary steam pump, but both are large steam consumers. Such pumps are, however, simple, and occupy but little space. The author has used the ordinary non-condensing steam

PLATE V.]

[*To face page* 216.

Triple-expansion Pumping-Engine, City of Leeds Waterworks.

pump for the purpose of pumping water to filters on a plan which secures economy. The system is illustrated in fig. 12, Chap. II., and described on p. 39. A separate boiler having a working pressure of 30 or 40 lbs. per sq. in. above that of the boilers of the main engines may be used for the filter pumps, and the filter pump engines may be made to exhaust into the boilers of the main engines. In this way a simple single-cylinder non-condensing steam pump may be made use of without the very great waste of steam usually resulting from the use of such pumps.

Water-pressure engines taking their power from the water mains may under certain circumstances, be also used for the purpose.

Where the lift is only a few feet, a very convenient method is the use of a centrifugal pump with a 'Pelton' wheel fixed on the same spindle (fig. 235, Chap. XIV.), the Pelton wheel being driven by means of water taken from the mains.

Generally speaking, the filter pumps require more power than it would be advisable to apply by the methods just named. Assuming that the main engines are required to be of 300 H.P. with a lift of 200 ft., then the filter pumps would be required to be considerably over 30 H.P. for a lift of 20 ft.

Engines are sometimes used to pump from a water main at one pressure into another at a higher pressure. In Birmingham the distribution is divided into four zones with a difference in level between the one zone and the next of from 100 ft. to 200 ft. Two compound pumping-engines each of two million gallons per day capacity are used to pump water from the low-level zone main into the mid-level zone main, the pressures being 110 ft. in the low-level and 250 ft. in the mid-level main.

Water-pressure pumps are sometimes used for water supply, and where a natural head of water is available the economy is obvious. Such pumps have also been used with the power derived from water falling from one reservoir level to another, to pump a supply to a district above the higher reservoir. The water-pressure pump is a simple machine requiring little attention, and when a counter is attached it becomes a fairly good water meter.

Sometimes waterworks engines are employed to pump direct into the distributing mains without the use of a reservoir for the purpose of maintaining a constant pressure, the pressure being maintained during the fluctuations in demand by alterations in the speed of the engines.

Balanced valves have also been used to maintain a constant pressure against the pump.

With waterworks plant reserve power is necessary, and that reserve must be based on the maximum supply which takes place either in dry summer or severe winter.

The capacity of each engine must also be above the maximum requirements to allow for reduced delivery of the pump when out of order.

In addition to reserve engines, each engine should have a reserve capacity of say 30 per cent. The question as to what reserve power engines should have depends on the extent of the water supply. With small supplies the unit of power becomes small, and 100 per cent. reserve power is necessary to avoid having very small units.

In large establishments, where the units of power become numerous, then 33 per cent. may suffice, say one spare engine in three.

A town having a mean supply of 20,000,000 gallons per day might make a demand on the pumping machinery to the extent of 28,000,000 or 30,000,000 gallons per day during a dry summer or a severe frost. The following diagram (fig. 204) shows the fluctuation in water supply in the city of Birmingham for the years 1896–7.

Where engines pump direct into the distributing mains without a service reservoir, the reserve power of the engines must be increased, because the variation in demand during twenty-four hours largely exceeds the mean supply. Service reservoirs should contain at least twenty-four hours' supply.

It is not desirable to pump into distributing mains without connection to a service reservoir, although sometimes it becomes necessary.

Types of Waterworks Engines.—The single cylinder Cornish engine (fig. 205) was formerly the most favoured, and very justly so, because, with the pressure of steam then available—30 to 40 lbs.—it was the most economical engine made.

Under proper conditions, large engines required only 24 to 26 lbs. of steam per pump H.P., a performance which was only equalled by compound rotative Woolf engines working with much higher boiler pressure. The Cornish engine in the single cylinder form is only suited to low pressures, and for that reason it has given place to compound engines of the double or triple class.

The general form of compound rotative beam engines is shown in figs. 206 and 207.

A double-acting pump is actuated from the crank end of the beam, as in fig. 206, or two single-acting pumps are placed one on each side of the beam centre, as in fig. 207.

This type of engine was generally worked with the Woolf distribution of steam—that is, with a cut-off valve on the high-pressure cylinder only.

To work such engines on the receiver plan with a cut-off on each cylinder a very heavy fly-wheel is required ; but if two engines are combined to form a compound engine coupled to the same shaft with cranks at right angles, then a lighter fly-wheel suffices. Since it has become practicable to make boilers which are durable and safe, carrying very high pressures, new types of pumping-engines are taking the place of the old beam engines.

It had long been recognised that there was a distinct advantage in making the engine what is termed 'direct acting'—that is, in attaching the

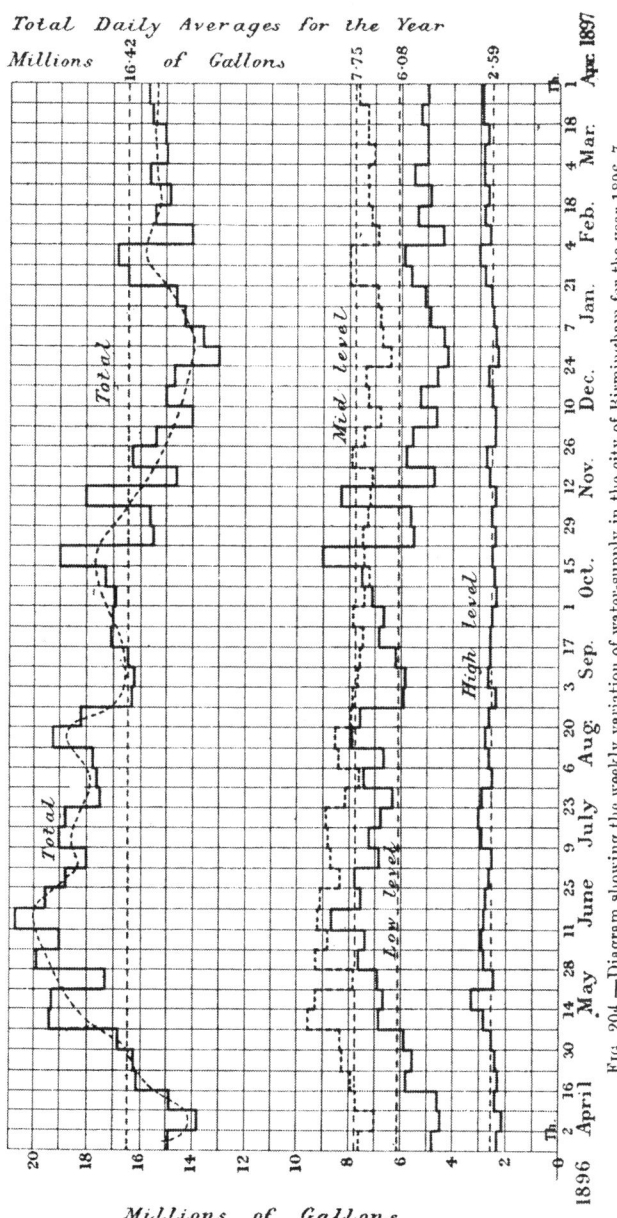

Total Daily Averages for the Year
Millions of Gallons

Fig. 204.—Diagram showing the weekly variation of water-supply in the city of Birmingham for the year 1896–7.

Millions of Gallons

pump direct to the piston rod instead of by an indirect connection through a beam and connecting rod. When higher boiler pressures became usual,

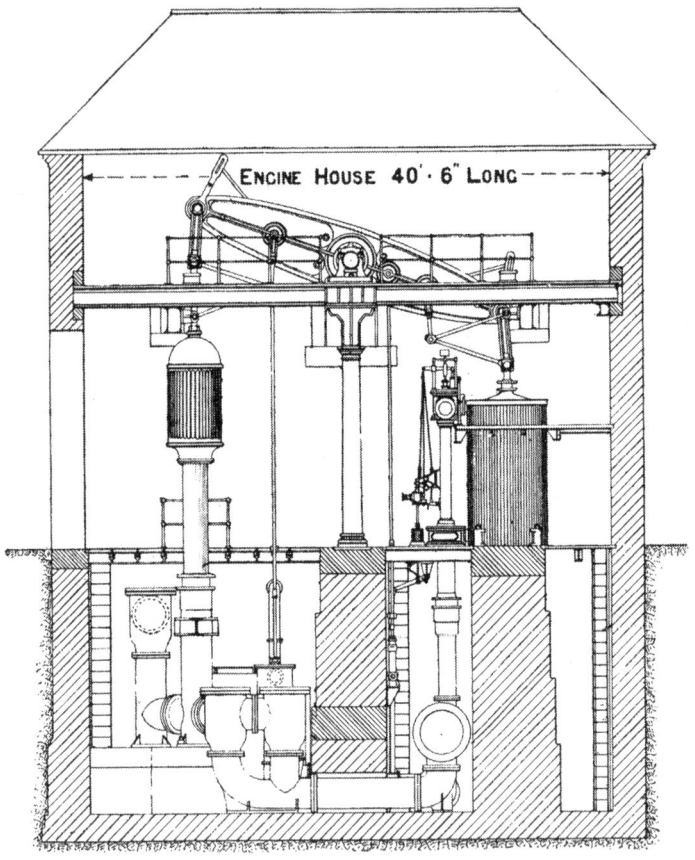

ENGINE HOUSE 40' · 6" LONG

FIG. 205.—Cornish Pumping-Engine—Wolverhampton Waterworks.

Diameter of cylinder, . . . 56 in.	Height of lift, ... 200 ft.
Length of stroke, . . . 8 ft.	Valve gear—Davey's differential.
Diameter of plunger, . . . 23 in.	Surface condenser, 380 sq. ft. of surface.
Length of stroke, . . . 8 ft.	Duty = 63 millions on an evaporation of
Boiler pressure, ... 35 lbs.	8 lbs. of water per 1 lb. of coal.
Number of strokes per minute, . 12	

new types of engines were produced. An example of a triple compound engine is shown in figs. 208 and 209.

In this engine there are three plungers connected directly to the piston rods, and coupled by connecting rods to a three-throw crank shaft.

The pumps are single-acting, whilst the steam cylinders are double-acting. To balance the engine the plungers are sometimes weighted to half the resistance. This engine has Corliss valves placed in the covers for the purpose of reducing the capacity of the steam ports. Receivers are placed

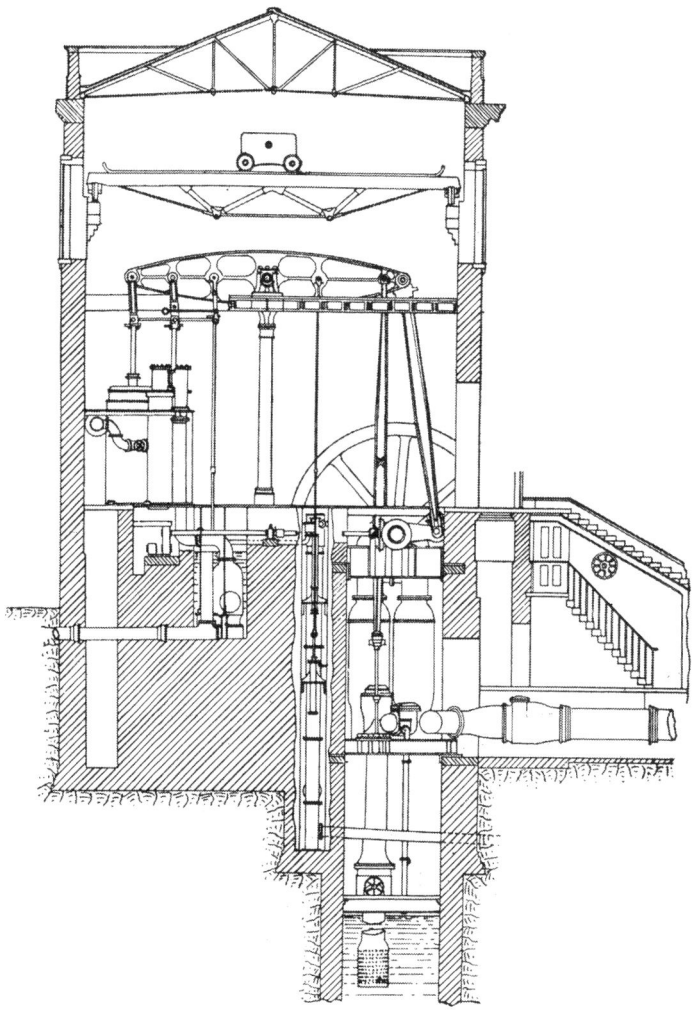

Fig. 206.—Compound Beam Engine—London Waterworks.

between the cylinders, in which are superheaters kept charged with steam at the full boiler pressure, 150 lbs. per sq. in.

Fig. 210 is that of a compound inverted direct-acting engine working with a boiler pressure of 70 lbs. per sq. in. It has cylinders 33 and 60 in. in diameter, a 10-ft. stroke and 26-in. pumps; its capacity is 6,000,000 gallons of water per day to a height of 260 ft.

The pump valves are of the hat-band type. The pumps have plungers weighted to the full resistance, the weight performing the function of a

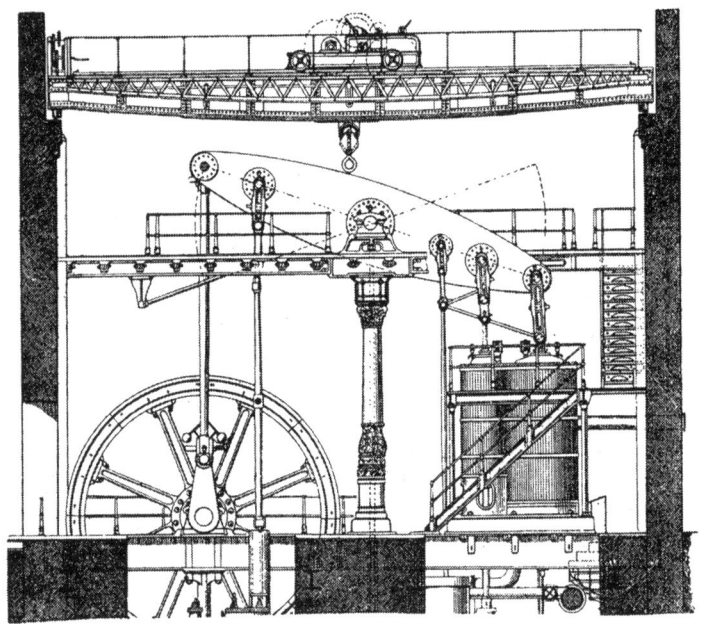

FIG. 207.—Leicester Sewage Works Compound Beam Pumping-Engine.
High-pressure cylinder, 30 in. diameter × 5 ft. 9¼ in. stroke.
Low-pressure cylinder, 48 in. diameter × 8 ft. 6 in. stroke.
Boiler pressure, 80 lbs.
Pumps, piston and plunger type, piston 27¼ in. diameter × 5 ft. 9¼ in. stroke.
Fly-wheel, 21 ft. diameter, weight 21 tons.
Revolutions per minute, 12.
Feed water per I.H.P. per hour reported to be 15·4 lbs. to 18·3 lbs.
Indicated horse power, 200.

fly-wheel in admitting of expansive working, as explained in the chapter on the general principles.

The engines and pumps are coupled together by means of a parallel motion, shown in detail in fig. 211. This motion is more perfect when the outer ends of the outer beams are guided in horizontal slides, instead of being attached to links (see fig. 97, Chap. VI.), but the link secures sufficient accuracy for practical purposes.

In fig. 212 is shown a triple compound horizontal engine actuating well pumps by means of rocking discs with varying stroke, the principle of which is explained in Chap. IV., fig. 72.

Fig. 213 is a horizontal cross compound pumping-engine such as is

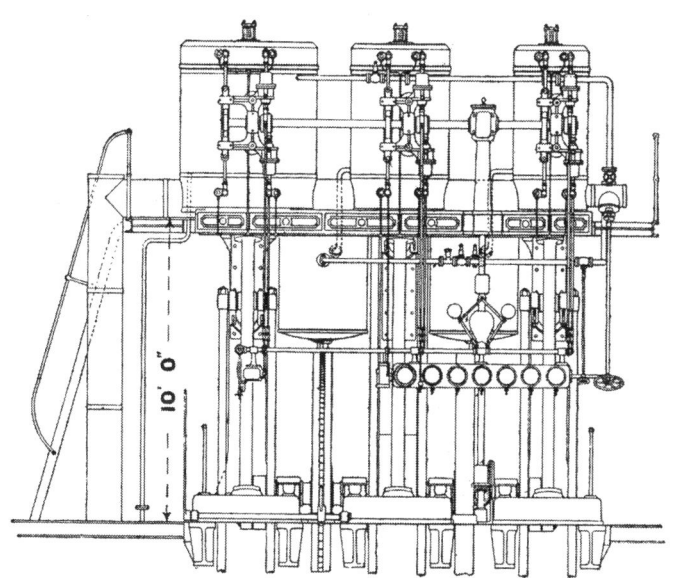

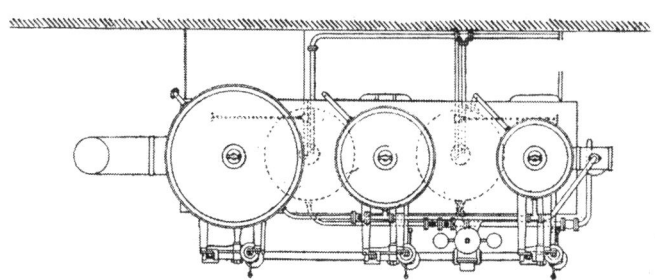

FIG. 208.—Triple-expansion Pumping-Engine—City of Leeds Waterworks.

Cylinders, 15 in., 24 and 40 in. diameter } × 3-ft. stroke.
Plunger pumps, 13½ in. diameter

To raise 1850 gallons per minute against a head of 300 ft.
Boiler pressure, 150 lbs.

often used for town water supply. It has cylinders 24 in. and 41 in. in diameter and a 5-ft. stroke, and actuates two double-acting piston pumps. To get the best economy from this type of engine it should be worked on the receiver system with a cut-off on the low-pressure cylinder. The

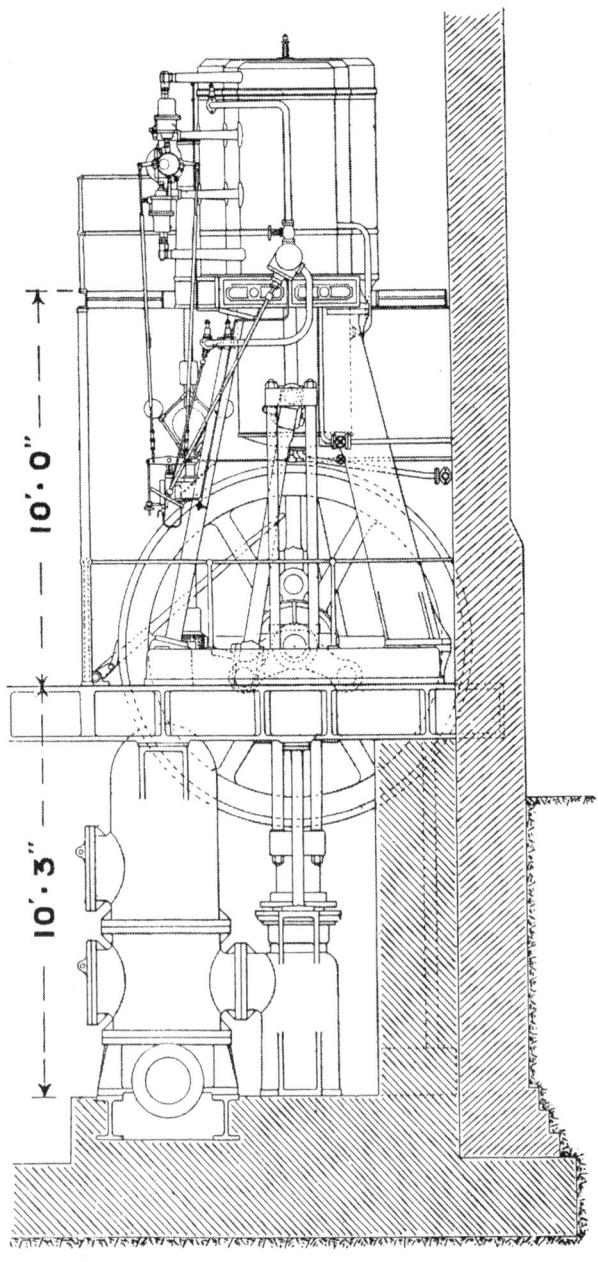

FIG. 209.—Triple-expansion Pumping-Engine—City of Leeds Waterworks.

engine is provided with D-slide valves, with a cut-off valve on the back of the main valve, the cut-off being adjustable by hand whilst the engine is

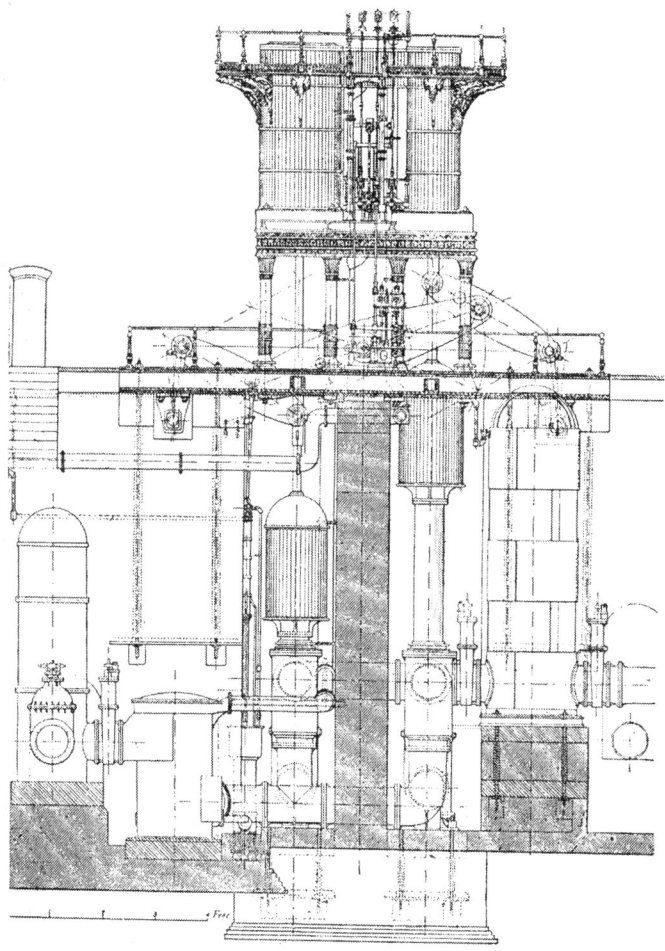

Fig. 210.—Compound Non-rotative Pumping-Engines—Birmingham Waterworks.
Cylinders, 33 and 60 in. in diameter ⎫
Pumps, 26 in. in diameter ⎬ 10-ft. stroke.
To raise 6 millions of gallons per day 260 ft. high.

running. The pumps are 16 in. in diameter, and pump to a total height of 150 ft. at 35 revolutions per minute with 80 lbs. boiler pressure. The valves of the pump are small hard rubber valves working on gun-metal

P

grid seatings and kept down by means of a light spring. There are forty-eight valves in each pump, each $7\frac{1}{2}$ in. in diameter.

The condenser is of the injection type, the air pump being worked from a small crank shaft driven from the crank of the low-pressure cylinder by means of a drag link.

Fig. 214 represents a compound steam pump having cylinders 26 in. and 37 in. in diameter and a 5-ft. stroke. It was designed by the author to work from a set of boilers at 130 lbs. pressure, and to exhaust into a battery of boilers at 20 lbs. pressure, supplying several Cornish engines. In this way economy was secured at a small capital outlay as the low-pressure engines and boilers already existed and worked constantly. The system is described in Chap. II., fig. 13.

The pump is 22 in. in diameter, and at fifteen double strokes per

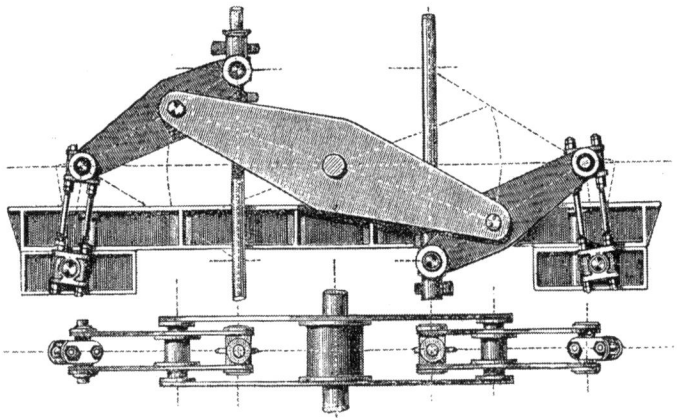

FIG. 211.—Parallel Motion Beams for Compound Pumping-Engines.

minute pumps 3,000,000 gallons per day against a head of 220 ft. The valves are of vulcanite, each $4\frac{1}{2}$ in. in diameter, working on a gun-metal seating, and kept in place by a light spring.

Fig. 215 is a drawing of a horizontal compound Cornish engine actuating a single-acting bucket pump. The pump is suspended by steel rods and screws in a well to a depth of 220 ft. The wooden gantry is for the purpose of supporting and lowering the pump.

When the water is not being pumped it rises to the top of the well. The pump was lowered through the water and worked while still suspended in the rods. By pumping, the water was lowered to the level of the pump barrel ; at that depth bearers were put in to take the weight of the pump, while the rods and screws were allowed to remain for the purpose of drawing the pump when necessary, as the pump itself was not fastened down to the bearers.

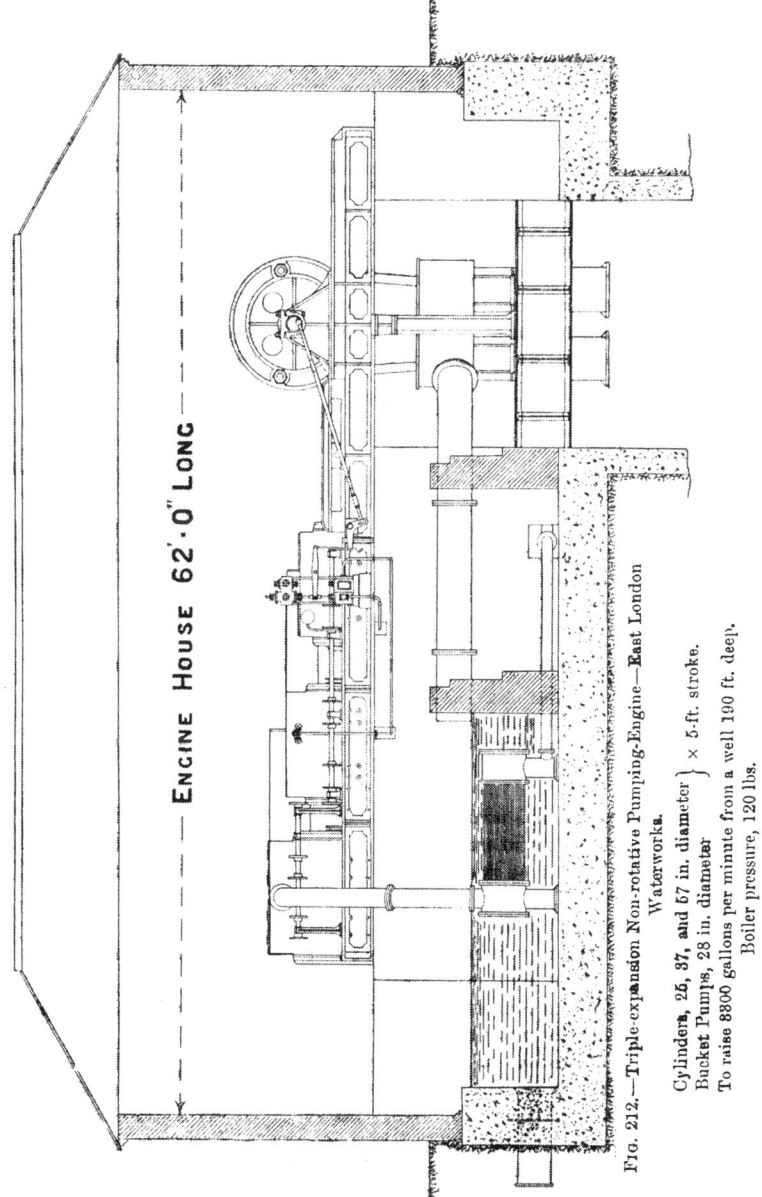

Fig. 212.—Triple-expansion Non-rotative Pumping-Engine—East London Waterworks.

Cylinders, 26, 37, and 67 in. diameter } × 5-ft. stroke.
Bucket Pumps, 28 in. diameter
To raise 8800 gallons per minute from a well 190 ft. deep.
Boiler pressure, 120 lbs.

ENGINE HOUSE 62' 0" LONG

The engine has cylinders 25 in. and 54 in. in diameter, and a 5-ft. stroke. The pump is 24 in. in diameter and has a 5-ft. stroke, and delivers its water to a reservoir above the top of the well two miles distant, the total lift on the pump being 290 ft. The boiler pressure is 130 lbs.

The pump valves are of the ordinary double-beat type.

Fig. 216 is a drawing of a triple compound American engine designed by Leavitt. This engine has cylinders 13·7 in., 24·375 in., and 39 in. in diameter, all having a stroke of 6 ft. On a trial it is reported to have raised 12,350 gallons per minute 137 ft. high at 59·59 revolutions per minute with a steam pressure of 157·7 lbs. The consumption of feed water was 12·25, and of dry steam 11·2 lbs. per I.H.P. per hour, which is equal to an efficiency ratio of about 68 per cent. The pump plungers are 17·5 in. in diameter, have a 4-ft. stroke, and are three in number.

Bore-hole Wells for Town Water Supply.—In procuring water for town

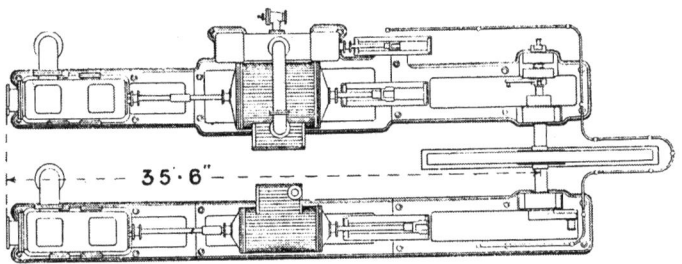

FIG. 213.—Cross Compound Rotative Pumping-Engine—Shanghai Waterworks.

Cylinders, 24 and 41 in. diameter } × 5-ft. stroke.
Pumps, 16 in. diameter
To pump a total height of 150 ft. at a speed of 35 revolutions per minute.
Boiler pressure, 80 lbs.

water supply, it is the usual and necessary practice to provide duplicate pumping-engines, and where two engines are made to pump from the same well, the well must be very large that it may accommodate two sets of pumps.

Such wells are usually 12 to 14 ft. in diameter.

To sink such a well in the ordinary way is a very long and costly undertaking, especially if soft strata are met with, where lining becomes necessary. On the completion of the well it may be necessary to drive adits to increase the water supply.

A simple bore-hole is made very cheaply, and very expeditiously. Four 30-in. bore-holes can be put down in a very small fraction of the time required to sink a 12-ft. well. Instead of making a large well, the author puts down four bore-holes to accommodate the pumps for duplicate pumping engines—a pair of pumps to each engine, as in figs. 217 and 218. The bore-holes being completed, the pumps are lowered into them, and

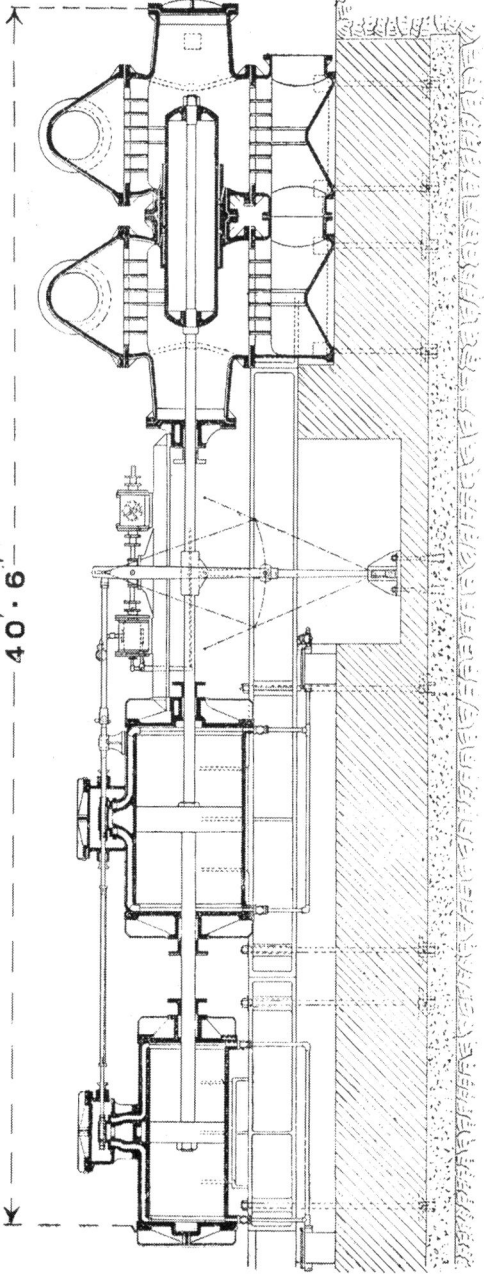

Fig. 214.—Compound Steam Pump—Birmingham Waterworks.

This steam pump takes steam from boilers at 130 lbs. pressure and exhausts into other boilers at 20 lbs. pressure, as described in Chap. II., fig. 13.

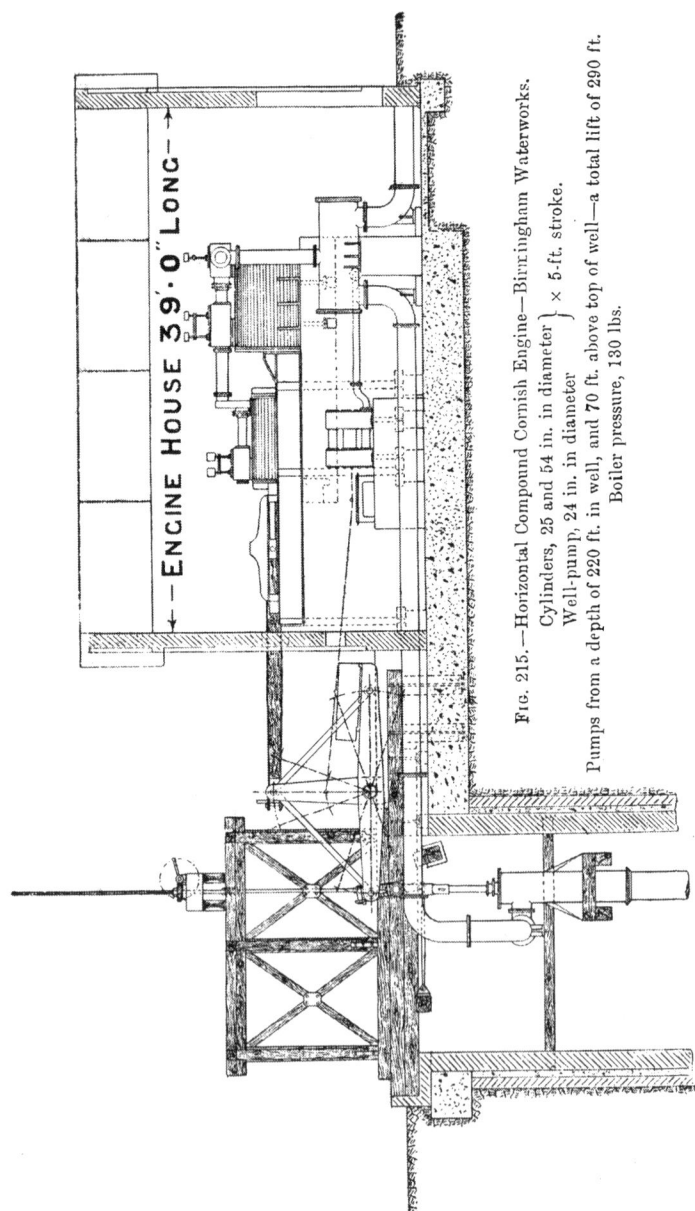

Fig. 215.—Horizontal Compound Cornish Engine—Birmingham Waterworks.
Cylinders, 25 and 54 in. in diameter } × 5-ft. stroke.
Well-pump, 24 in. in diameter
Pumps from a depth of 220 ft. in well, and 70 ft. above top of well—a total lift of 290 ft.
Boiler pressure, 130 lbs.

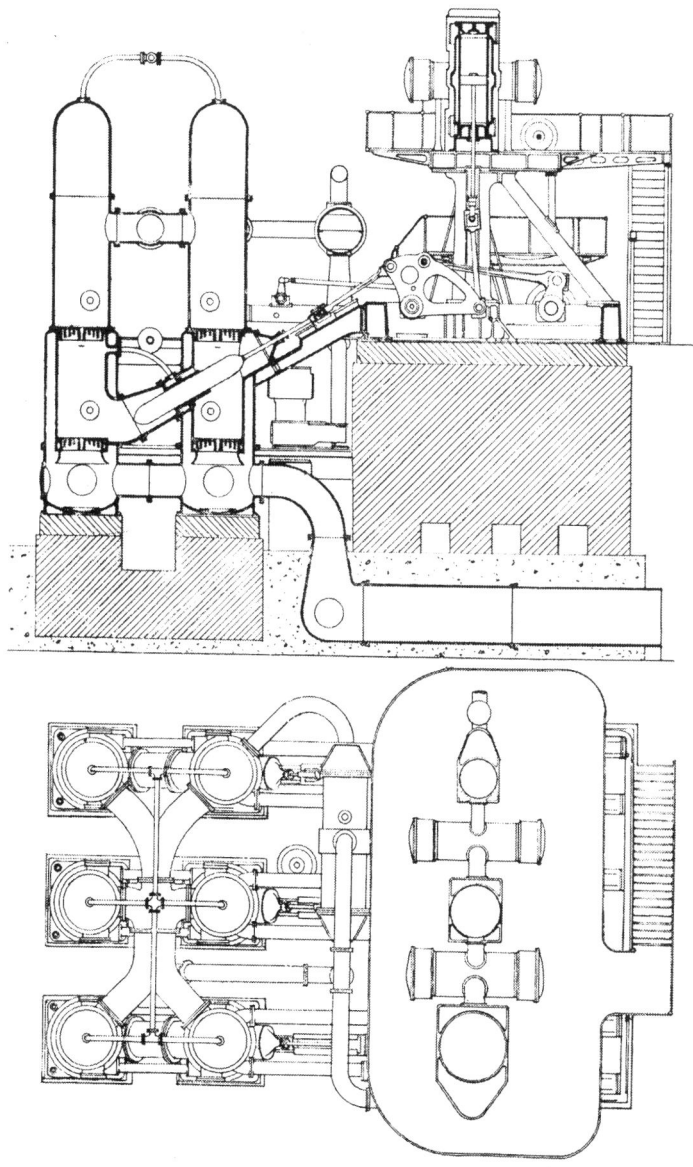

FIG. 216.—Leavitt's Triple-expansion Pumping-Engine.

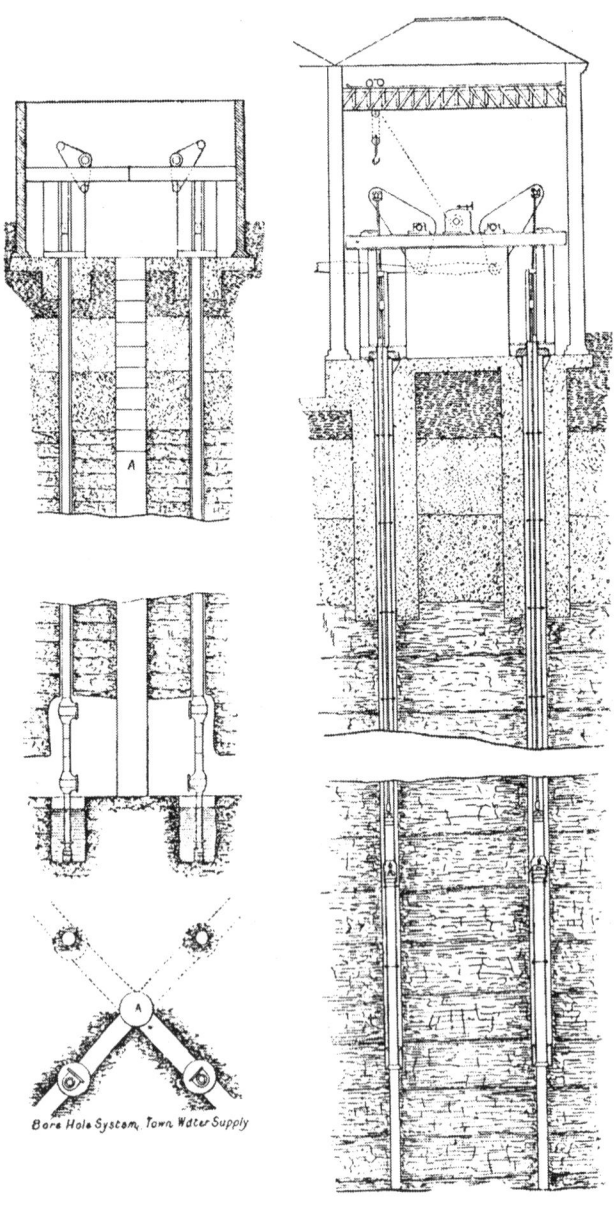

FIG. 217. FIG. 218.

FIGS. 217 and 218.—Davey's Bore-hole System of Wells for Town Water Supply.

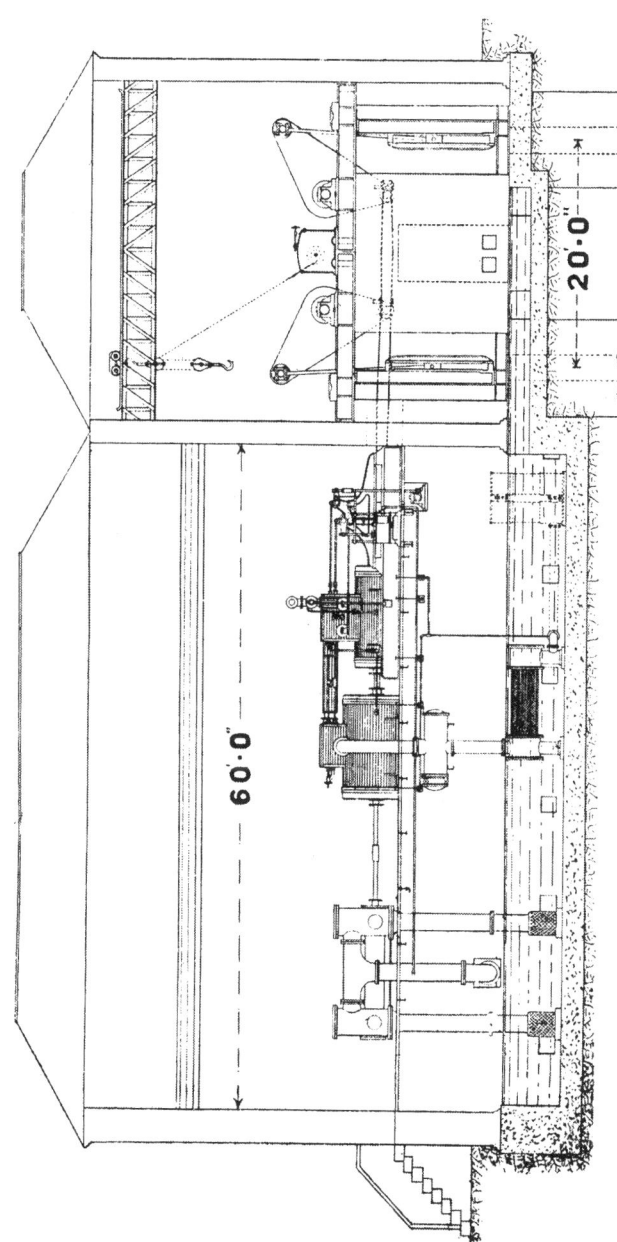

FIG. 219.—Compound Non-rotative Pumping-Engine—Widnes Waterworks.

coupled up to the permanent engines. Immediately that is done the water found in the bore-holes can be pumped and supplied to the town. Should it be insufficient, then a small well would be sunk in dry ground to the bottom of the bore-hole pumps. The water being kept down by the pumps, the bore-holes at the level of the pumps would be connected to the centre well, and adits driven to collect more water. Should the bore-holes yield sufficient water, it would not be necessary to sink the well. It would be absurd to advocate any particular system of well-sinking as being universally applicable; this system, however, of making wells offers advantages under favourable conditions, but the advisability of its

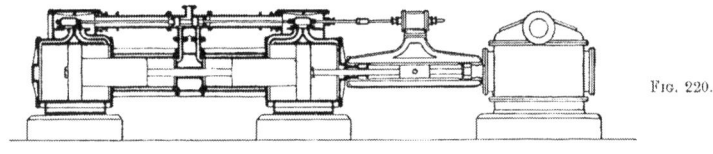

Fig. 220.

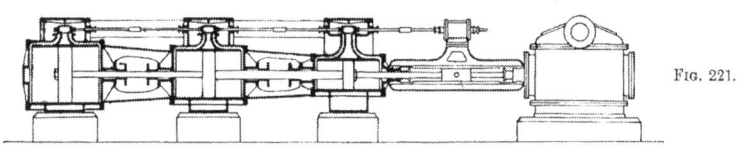

Fig. 221.

FIGS. 220 and 221.—Triple Compound Steam Pumps.

adoption in any particular case must be a matter of judgment with the engineer planning the work.

Having illustrated and described the system in outline, we will now describe a practical example of executed work.

For the water supply of Widnes, bore-holes two in number are sunk in red sandstone rock, and are placed 20 ft. apart, each bored to a diameter of 30 in. for a depth of 200 ft., and to a reduced diameter of 18 in. for a further depth of 200 ft. and 300 ft. respectively, thus making the first hole 400 ft. deep and the second one 500 ft. deep. On the completion of the boring the water stood 70 to 80 ft. from the surface of the ground, when the quantity pumped by an old engine on the same site was $1\frac{1}{4}$ million gallons per day. The main pumps were then lowered into the bore-holes, each pump extending to the bottom of the large part of the hole 200 ft. from the ground level. In that position the pumps were suspended from a cast-iron bed-plate supported on a concrete foundation formed round the top of the holes, a block of oak being inserted between the head of the pump and the bed-plate. In this suspended position the pumps work.

It must be observed that no temporary pumping plant being required, the main engines were erected whilst the bore-holes were being made, so that on the completion of erection of the engines and pumps, water could be at once supplied to the town, but it remained to be proved what was the quantity of water available.

The engines were made for the purpose of pumping $2\frac{1}{4}$ million gallons per day, but it was found that working up to their full capacity of $2\frac{3}{4}$ million gallons, the full yield of the bore-hole was not reached. On starting the new pumps it was found that when pumping $2\frac{3}{4}$ million

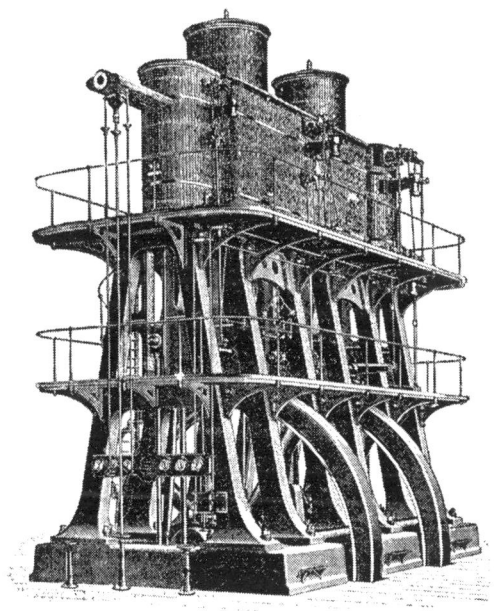

Fig. 222.—American Triple-expansion Pumping-Engine.

gallons per day, the water level was lowered to 100 ft. from the surface of the ground.

Had the yield of water been insufficient, then whatever water there was might have been sent continuously to the town, whilst the small well A shown in fig. 217 was sunk, and adits driven, say, to a depth of 180 or 200 ft. from the surface for the purpose of collecting more water. In chalk this would probably have been necessary, but in this and in other cases in sandstone it has not been necessary to drive adits. Similar bore-holes and pumps have been put down in red sandstone at Fleetwood, Kingswinford, and other places.

The motive power consists of a 230 H.P. compound surface-condensing engine employed to pump from the bore-holes into a masonry tank by the

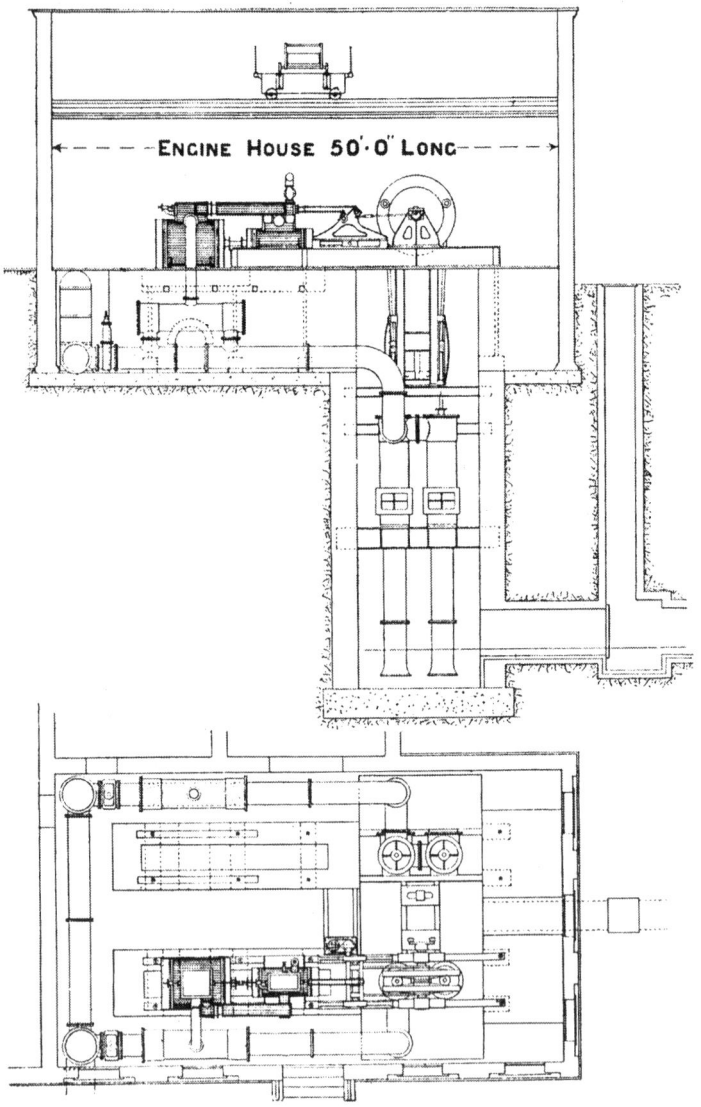

FIG. 223.—Compound Non-rotative Pumping-Engines—Cambridge Sewage Works.

engine foundations, from which tank┐the water is forced by the same engine to a reservoir at an elevation of 260 ft.

The engine is illustrated in fig. 219. It works the force pump by means of

FIG. 224.

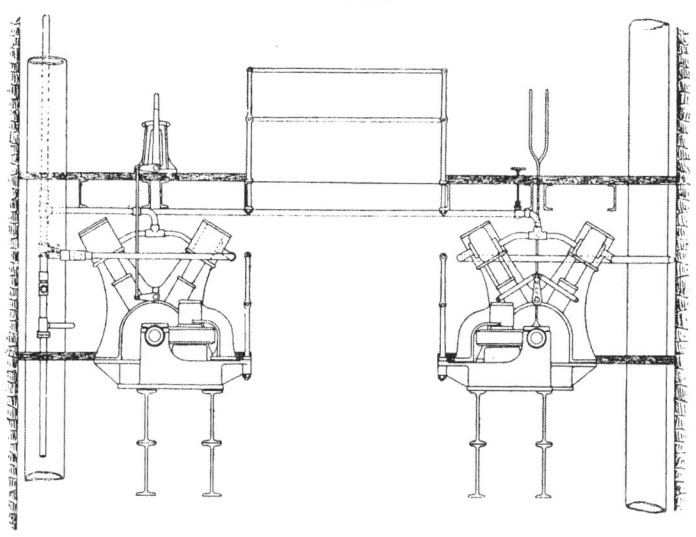

FIG. 225.

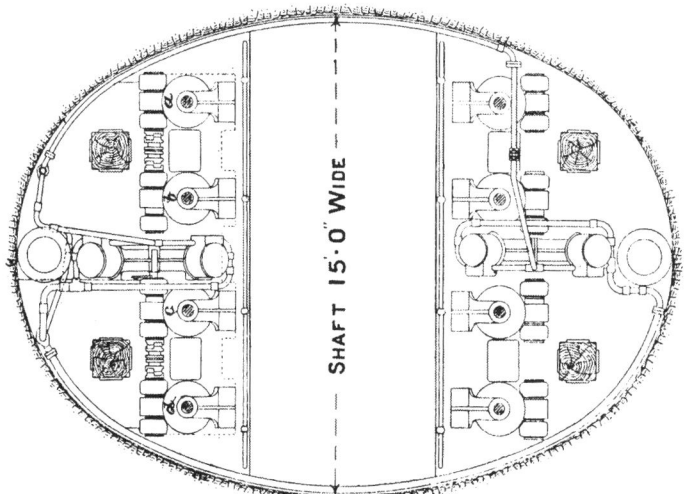

FIGS. 224 and 225.—Method of lowering pumps in deep wells for town water supply.

a tail rod from the low-pressure cylinder, and the bore-hole pumps by means of compensating rocking levers actuated by a connecting rod from the crosshead of the engine. The principle of the compensating lever is described in Chap. IV., p. 78.

Steam Distribution.—The engine is of the receiver type, having separate expansion valves on both high- and low-pressure cylinders, adjustable by hand whilst the engine is in motion.

Description of the Engine.—Both the steam cylinders are completely jacketed, bodies and ends, with steam at boiler pressure.

The valve gear consists of a D-slide valve on each cylinder actuated by

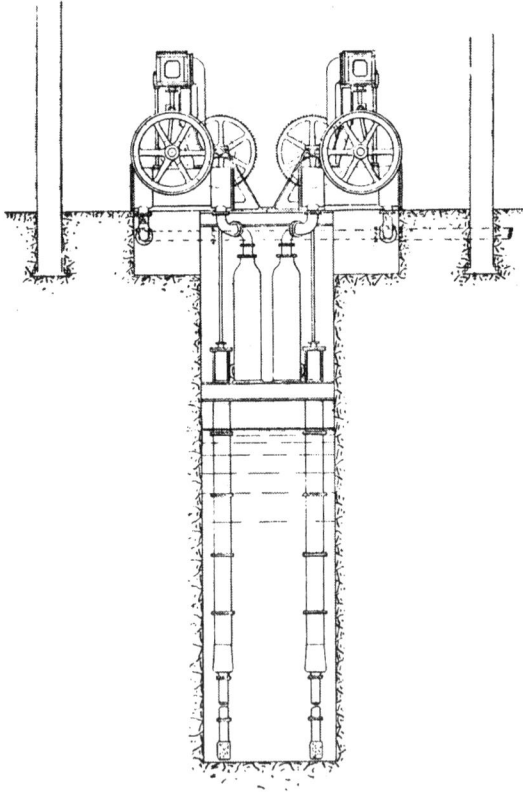

FIG. 226.—Pumping-Engines with deep well pumps for small water supply.

means of a rocking shaft deriving its motion from a water cylinder worked by pressure from the pumping main. This arrangement has several advantages, one of which is that, should the water main burst, the engine suddenly stops of itself. On the back of each of the D valves is a pair of Meyer expansion valves actuated by means of a rocking shaft, deriving its motion from the engine itself.

The main valves move uniformly and independent of the engine. The expansion valve rests on the back of the main valve, and is moved by the engine. Both travel in the same direction, and the quicker the engine goes, the quicker will the expansion valve cover the port; with no load on the engine it covers the port instantly the port begins to open, because the engine motion then responds very promptly to the opening of the port.

Between the engine and the surface condenser is placed a feed-water heater which raises the feed water to a temperature of 110° Fahr. before it passes into the boiler.

The bore-hole pumps are made of lap-welded steel tubes with wrought-iron flanges screwed on.

A steam winch is provided for drawing the pump rods, and for changing the working parts of the pumps.

For indicating the water level in the bore-holes, a pipe is put down outside the pumps. At the top of the pipe in the engine-room is fixed a sensitive pressure gauge ; below the gauge is a branch communicating with a small air pump worked by the engine. The pipe is thus kept filled with air, and the hydrostatic pressure on the outside of the tube is thus shown on the pressure gauge.

The following summary gives the general particulars of a trial of this engine and cost of the installation :—

Steam pressure,	70 lbs. per sq. in.
Diameter of cylinders,	32 and 60 in.
Length of stroke,	6 ft. 3 in.
Diameter of force pump,	18½ in.
Height of lift,	260 ft.
Length of stroke,	6 ft. 3 in.
Diameter of bore-hole pumps,	18½ in.
Height of lift,	100 ft.
Length of stroke,	6 ft. 6 in.
Number of strokes per minute,	12½.
Depth of bore-holes,	400 and 500 ft.
Diameter of bore-holes,	30 ins. for 200 ft. down, and then 18 in.
Tube surface of condenser,	420 sq. ft.
Tube surface of feed heater,	140 ft.
Water pumped in 24 hours,	2½ million gallons.
Duty on engine on an evaporation of 10 lbs. of water per 1 lb. of coal,	124 millions.
Lbs. of steam used per I.H.P. per hour,	15·6.
Lbs. of steam used per P.H.P. per hour,	18·0.
Indicated horse power,	230.
Pump horse power,	200.
Mechanical efficiency,	87 per cent.
Weight of engine and pumps,	170 tons.
Cost of engine and pumps,	£6000.
,, ,, per P.H.P.,	£30.
Total cost of engine, pumps, bore-holes, and buildings,	£9700.
,, ,, ,, ,, per P.H.P.,	£485.

Steam Pumps are of various forms, but all couple the steam piston to the pump piston or plunger direct without rotative mechanism. We have not space to notice the various details of such pumps.

Figs. 220 and 221 illustrate convenient forms of pumps with triple compound steam cylinders. Such pumps are constructed on the single or duplex plan, or a pair of pumps may be employed with the valve gear so

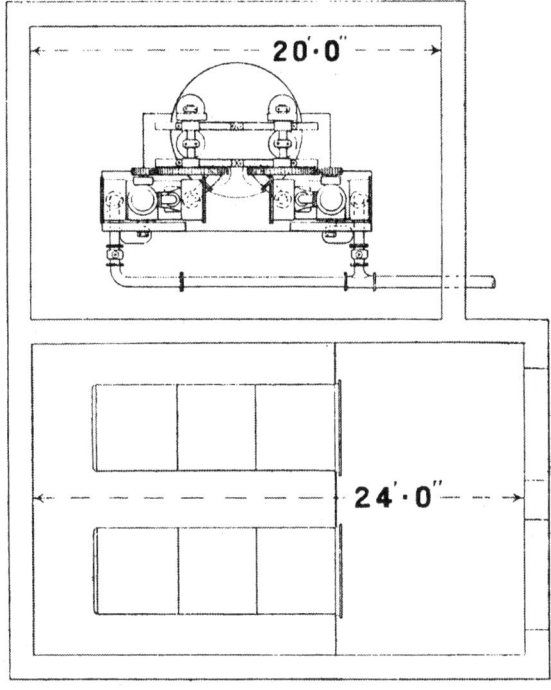

Plan.

Fig. 227.—Pumping-Engines with deep well pumps for small water supply.

arranged as to enable the two to work together on the duplex plan, or to work singly as desired.

When so designed, cushioning ports or cushioning valves necessary to the duplex action must be employed. See Chap. X., on Valve Gears.

Fig. 222 represents an American triple-expansion rotative pumping-engine with the Corliss type of valve gear.

Fig. 223 shows a pair of sewage pumping-engines at the Cambridge Sewage Pumping Station. The pumps are actuated by means of a rocking disc on the Davey compensating principle already described. The engines have cylinders 22 and 44 in. in diameter and a 4-ft. stroke. The pumps are of the bucket type, 34 in. in diameter, and have a 4-ft. stroke. The height of lift is 50 ft.

Figs. 224 and 225 illustrate the application of steam power to the working of lowering screws for suspending pumps during sinking operations.

Figs. 226 and 227 illustrate a convenient form of pumping-engine and pumps for small town or village water supply. The engines and pumps are in duplicate. The base of each engine forms a surface condenser through which all the water is pumped.

The pumps consist of two bucket pumps to each engine actuated by means of cranks set opposite to each other. The engine is geared to the pumps in the proportion of 4 to 1.

CHAPTER XIII.

Trials of Pumping-Engines.—In Chap. II. we have discussed the relative economics which should be derived from steam of different pressures, assuming that the engines were all equally efficient for the pressures employed.

All other things being the same, higher steam pressure secures greater economy per I.H.P. because there is more work obtainable from 1 lb. of high-pressure than from 1 lb. of low-pressure steam; therefore in drawing conclusions from steam-engine trials the steam pressures employed must be taken into account. In other words the efficiency ratio should be known.

In Chap. II. is given a summary of trials of pumping-engines, as also the efficiency ratio. The efficiency ratio is the relation between the performance of the engine tested and that of a perfect steam-engine working without loss, and having the same initial and exhaust pressures.

The tables give a summary of results of many reported trials, but it may be remarked that all who have had considerable experience in steam-engine trials know that small errors occur if great care is not taken. There may be undetected leakages in valves and piston, or the steam may not be what is called dry on admission, without its being observed, and the efficiency ratio of that particular engine may be low in consequence. The same type of engine better fitted and worked with drier steam would give better results.

On the other hand, if the efficiency ratio given is abnormally high, the accuracy of the trial may be questioned.

Engine trials are exceedingly useful, both scientifically and practically; but the engineer has to take into account the local conditions of the application of the engine, the capital expenditure, cost of upkeep, etc., in selecting the machinery to be employed. It must also be observed that the particular conditions under which the engine is to be worked may make it necessary to use a system of steam distribution which does not give a high efficiency ratio, as in the case of mining engines having heavy pit-work, where it is expedient to work compound engines on the Woolf distribution of steam, and not on the receiver system, for many practical reasons.

Waltham Abbey Pumping-Engine.—Fig. 228 illustrates a triple compound engine of the marine type, one of the first engines of the class used by Mr. W. Bryan at the Waltham Abbey Pumping Station of the East London Waterworks. Mr. Bryan and the author made a trial of this engine with the following results.*

The engine on which the experiment was made is a triple-expansion surface-condensing engine of the ordinary inverted direct-acting marine

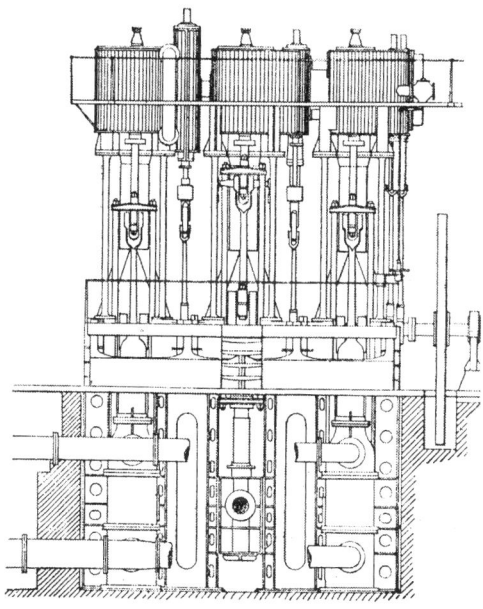

Fig. 228.—Triple-expansion Pumping-Engine, East London
Waterworks, Waltham Abbey.

type. It was fully described and illustrated in *Engineering*, 8th August 1890, pp. 158 and 162.

The diameters of the cylinders are 18, 30·5, and 51 in. respectively. The stroke of each of the three cylinders is 36 in. Each piston cross-head is connected to its crank by a single connecting rod, and to a plunger pump by a pair of pump rods.

There are thus three steam cylinders, three connecting rods, three pairs of pump rods, and three pump plungers. Each of the three cylinders is provided with an ordinary slide valve actuated by a separate eccentric on the crank shaft. The slide valve on the high-pressure cylinder is actuated direct by its eccentric and is provided with a Meyer expansion

* *Proc. Inst. Mechanical Engineers,* 1894.

valve, by means of which the speed of the engine was regulated during the experiment. The slide-valves on the intermediate and low-pressure cylinders are actuated through variable expansion links, the positions of which remain unchanged throughout the trial. The crank shaft is placed above the pumps, and the cranks rotate in the sequence—high, low, intermediate. The fly-wheel is placed towards one end of the crank shaft, and beyond it there is fixed a crank actuating a pair of well pumps by means of rocking quadrants. The bodies and both ends of all three cylinders are steam jacketed, the cylinders forming liners in the body-jackets.

Steam is supplied to these jackets through a pipe connected to the main steam-pipe on the boiler side of the stop valve. The steam to the jackets enters at the top on one side, and at the bottom on the opposite side; and in ordinary working the high-pressure jackets discharge directly into the boiler while the intermediate and low-pressure jackets discharge through steam-traps into the hot well. The jackets of the high-pressure cylinder receive steam at full boiler pressure, and by means of reducing valves the pressures in the jackets of the intermediate and low-pressure cylinders are maintained a little higher than the pressures in their respective valve-chests.

Each cylinder is therefore jacketed with steam a little above its own initial pressure.

Duration of experiment, ...	8 hours.	
Cylinder, diameter, high pressure, . . .	18 ins.	
,, ,, intermediate, . . .	30·5 ,,	
,, ,, low pressure, ...	51 ,,	
Stroke, length, ...	36 ,,	
Mean pressure in boiler above atmosphere, . .	130 lbs. per sq. in.	
Mean pressure in high-pressure jacket above atmosphere,	129 ,, ,,	
Mean pressure in intermediate jacket above atmosphere, .	28 ,, ,,	
Mean pressure in low-pressure jacket above atmosphere, .	5·5 ,, ,,	
Mean effective pressure, total reduced to low-pressure cylinder, ...	16·30 ,, ,,	
Number of expansions, ...	30	
Revolutions per minute, ...	22·9	
Piston speeds, feet per minute, mean, . . .	137·4	
Indicated horse-power, mean total, . . .	138	
Feed water, total used during trial, . . .	17,053 lbs.	
,, ,, per hour total, ...	2132 ,,	
,, ;, per I.H.P. per hour total, . .	15·45 ,,	
,, ,, percentage less, with steam in jackets, .	10·3 per cent.	
Jacket water, total during trial, ...	1900 lbs.	
,, ,, per hour, ...	237·5 lbs.	
,, ,, from high-pressure jacket per I.H.P. per hour,	0·81 lb.	
,, ,, from intermediate jacket per I.H.P. per hour,	0·53 ,,	
,, ,, from low-pressure jacket per I.H.P. per hour,	0·38 ,,	
,, ,, total from all jackets per I.H.P. per hour,	1·72 lbs.	
,, ,, total in percentage of feed water, . .	11·1 per cent.	
Coal used during trial, ...	1981 lbs.	
,, per I.H.P. per hour, ...	1·79 lbs.	

Lea Bridge Pumping-Engine.—Fig. 229 represents in outline a triple-expansion compound engine of the inverted marine type designed by Mr. W. Bryan, two of which have been erected at the Lea Bridge Pumping Station of the East London Waterworks. A trial of this engine was reported in the *Proceedings of the Institution of Mechanical Engineers*, 1894, from which report we extract the following results.

Description of Engines.—The engines tested were two in number, duplicates of each other. The cylinders are 20·03, 33·99, and 57·05 in. in

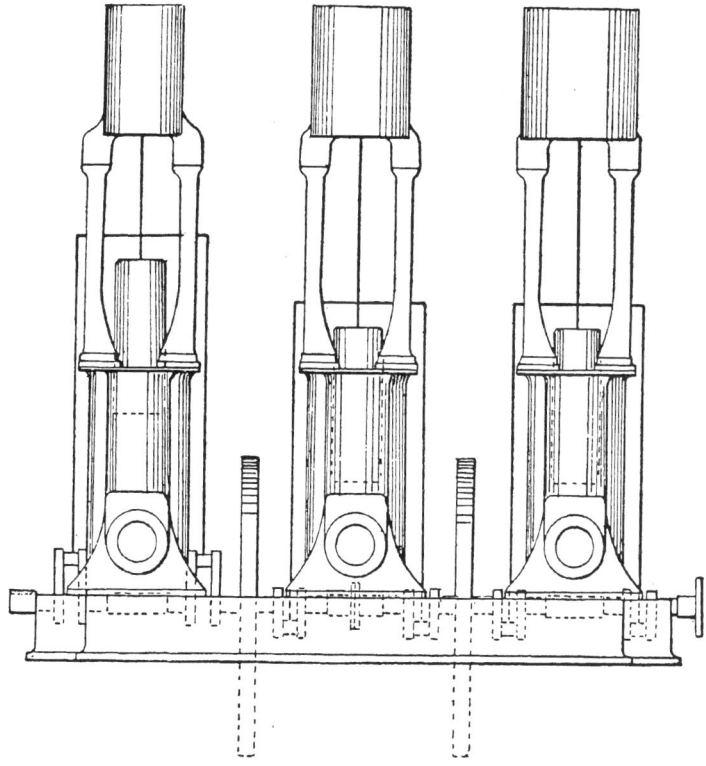

Fig. 229.—Triple-expansion Pumping-Engine, East London Waterworks, Lea Bridge.

diameter respectively, each having a stroke of 48 in. The piston rods are each 5 in. in diameter, and the pump plungers, which form prolongations of the piston rods, are 30 in. in diameter, each plunger having a displacement of 122·4 gallons per revolution.

The crank shafts are placed under the pumps and revolve in the sequence—high, low, intermediate. All the cylinders are provided with Corliss valves and the Corliss cut-off; the latter, however, is not under the control of a governor, but is regulated by hand whilst the engine is running. Each engine is provided with its own surface condenser having 1053 square ft. of cooling surface, and the steam passing through the engine during the trials was measured by the air-pump discharge. The water from the jackets of each cylinder was measured separately.

The two engines experimented with are named 'North' and 'Central,' from their positions in the engine house.

Jacketing.—Each cylinder has a jacket on the body, and one on each of the ends, the body jacket having a steam space of one inch. Drain pipes were provided, so that the water drained from the jackets of each cylinder could be measured separately. The pressure in any jacket could be varied by means of reducing valves.

Clothing.—The cylinders and steam pipes are clothed with about $2\frac{1}{2}$ in. of ordinary composition, with the exception of the bottom covers, which are not clothed.

Results and Comparisons.—In the following table the thirteen trials are divided into three groups, and are arranged in these groups in their order of merit, according to the total consumption of steam per indicated H.P. per hour; the steam pressures in the boilers and jackets are added.

Steam-pressures, lbs. per sq. in. above atmosphere.				Steam per I.H.P. per hour.
In Boiler.	In Jackets.			
	High-press. Jacket.	Inter. Jacket.	Low-press. Jacket.	
lbs.	lbs.	lbs.	lbs.	lbs.
Central Engine (check trials).				
118·3	113·0	42·0	29·6	12·50
119·7	...	...	...	12·99
Central Engine.				
118·3	...	...	112·5	12·89
117·4	112·7	41·6	9·8	13·16
117·1	...	113·1	111·3	13·19
117·5	...	...	...	13·47
117·9	...	113·2	...	13·66
111·1	− 13·7	− 18·7	− 13·7	13·94
North Engine.				
117·0	116·1	113·9	...	14·16
116·3	114·9	112·4	113·1	14·24
115·6	113·3	...	...	14·54
115·7	114·6	...	112·4	14·59
116·4	...	...	...	14·69

Pumping-Engine at Wapping.—Fig. 230 is an outline sketch of a triple compound engine at the Wapping station of the London Hydraulic

Power Co., and the following results of a trial are taken from a paper

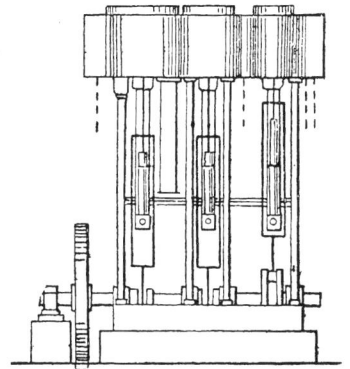

Fɪɢ. 230.—Triple-expansion Pumping-Engine, London Hydraulic
Power Co., Wapping Pumping Station.

by Mr. Ellington in the *Transactions of the Institution of Civil Engineers.*

TRIALS OF A TRIPLE-EXPANSION COMPOUND VERTICAL SURFACE-CONDENSING STEAM PUMPING-ENGINE AND FAIRBAIRN-BEELEY BOILER AT THE WAPPING PUMPING STATION OF THE LONDON HYDRAULIC POWER COMPANY.

	1 March 7, 1892.	2 March 11, 1892.	3 March 18, 1892.	4 March 25, 1892.
Duration of trial, ... hours	9	8	9	9
Engines.				
Diameter of high-pressure cylinder, in in.	15	...	...	...
,, ,, intermediate cylinder, ,,	22	...	...	...
,, ,, low-pressure ,, ,,	36	...	...	...
Stroke, ... ft.	2	...	...	...
Total revolutions, ...	31,972	28,235	32,815	49,006
Average revolutions per minute, .	59·76	58·82	60·77	90·70*
Barometer, . . . in. of mercury	29·8	29·4	29·9	Not taken.
Vacuum, . . . ,, ,,	28·0	27·74	28·43	
Total I.H.P., ...	179·80	185·63	206·55	
Thermal efficiency of engines, . per cent.	14·25	14·73	15·25	...
Boilers.				
Total heating surface, . . sq. ft.	800	...	...	...
,, grate area, . . . ,,	20	...	...	...
Pressure of steam in lbs., . per sq in.	118·3	142·8	143·0	141·3
Temperature of issuing gases, . ° Fahr.	...	...	557·5	...
Temperature of issuing gases before passing economiser, . . . ° Fahr.	355·7	345·3	323·1	436·2
Temperature of issuing gases after passing economiser, . . . ° Fahr.	247·3	253·8·	238·5	271·3
Amount of condensed steam from jackets, lbs.	2479	2620	2674	...

* Two engines running, one at 55·4 and the other at 35·3 revolutions per minute.

TRIALS OF A TRIPLE-EXPANSION COMPOUND VERTICAL SURFACE-CONDENSING STEAM PUMPING-ENGINE AND FAIRBAIRN-BEELEY BOILER AT THE WAPPING PUMPING STATION OF THE LONDON HYDRAULIC POWER COMPANY—*continued.*

	1 March 7, 1892.	2 March 11, 1892.	3 March 18, 1892.	4 March 25, 1892.
Coal.				
Quality, ...	(Inland)	(Nixon's Navigation.)		
Total weight fed on to grates, . lbs.	3287	3200	2774	3744
,, ,, of clinker, etc., . . ,,	114	196	237	162
Fuel burnt per hour (including clinker, etc.), ... lbs.	365·2	400·0	308·2	416·0
Fuel burnt per square foot of grate, ,,	18·26	20·0	15·41	20·8
,, consumed per I.H.P. per hour, ,,	2·03	2·155	1·49	...
Calorific value of dry coal, . . . } lbs. of water per lb. of coal }	13·75	11·76	15·65	15·5
Feed-water.				
Feed temperature on entering economiser, } ° Fahr. }	67·4	66·3	72·2	63·5
Feed temperature on entering boiler, } ° Fahr. }	204·7	203·2	194·1	215·8
Heating surface of economiser, sq. ft.	1375	...	...	...
Total water evaporated, . . . lbs.	26,199	23,005	28,365	41,584
Water evaporated per lb. of fuel (wet), ,,	7·97	7·19	10·22	11·11
,, ,, ,, ,, (dry), } from and at 212° (calculated) lbs., . }	9·96	9·27	12·25	13·44
Thermal efficiency of boiler, including } economiser, . . . per cent. }	72·5	78·9	78·2	86·7
Amount of feed-water per I.H.P. per } hour (engine), ... lbs. }	15·14	14·59	14·10	...
Steam-pipe condensation and loss * per } I.H.P. per hour, . . . lbs. }	1·05	0·9	1·16	...
Total feed-water, ...	16·19	15·49	15·26	...
Total quantity of water pumped, gallons.	160,800	140,600	160,880	235,228+
Accumulator pressure, . lbs. per sq. in.	727·3	728·5	795·0	800·0‡
Water pumped per cwt. of coal consumed, } gallons }	5530	4920	6495	7036

Pumping-Engines at Liverpool Waterworks.

TRIAL OF A TRIPLE-EXPANSION PUMPING-ENGINE AT THE LIVERPOOL WATERWORKS, by Mr. J. PARRY.

Diameter of cylinder in in., . . .	.	15, 23, and 38
Length of stroke, ...	.	3 ft.
Net area of cylinders, each end, . . .	. {	167 sq. in. 406 ,, 1125 ,,
Clearance, per cent. high pressure, . .	.	2·3
,, ,, intermediate, . .	.	2·1

---

* Long range of steam-pipes for several engines. The condensation alone was about 0·5 lb. per I.H.P.

† Calculated at 4·8 gallons per revolution.   ‡ Approximately.

Clearance, per cent., low pressure, ... 2
First receiver, capacity in cubic ft., ... 7·8
Second ,, ,, ,, ... 14·1
Total steam re-heating surface, in sq. ft., . . . 183·5
Number of single-acting pump rams, ... 3
Diameter of single-acting pump rams in in., . . . 22
Diameter of single-acting air pump ,, . . . 10
,, ,, feed pump ,, . . . 1½
Stroke of pumps in ft., ... 3
Number of gallons delivered per revolution, . . . 148

ENGINE TEST.

Average Temperatures—

Engine room, ... 71·5 deg. Fahr.
Water in suction tank, ... 58·28 ,,
,, hot well, ... 101·6 ,,
,, feed tank, ... 173·38 ,,
Exhaust steam from low pressure, . . . 126 ,,
Discharge from heater, ... 125 ,,

Average Pressures—

Steam gauge at separator, in lbs., . . . 132·6
Vacuum gauge, in in., ... 26·58
Water-pressure gauge, in ft., ... 85·85
Height of gauge above water level in suction tank, in ft., . 16·54
Total head pumped against, in ft., . . . 102·39

Revolutions—

Total number per counter, ... 20,375
Per minute, ... 33·958

Steam used—

Air-pump discharge, in lbs., ... 19,684·5
,, ,, per hour, ... 1,968·45
Water discharge from jacketing traps, . . . 3,281·6
,, ,, ,, per hour, . . 328·16
Total steam used, ... 22,966·1
,, ,, per hour, ... 2,296·6
Water discharged from separator trap, . . . 350·5
Drains from steam pipes, ... 69
Water discharged from safety valves and other leaks, . 1,646·4
Percentage of steam unaccounted for to total steam generated, 8
Percentage of jacket to total steam generated, . . 14·3

Average Indicated Horse-power—

High-pressure cylinder, top, ... 21·4
,, ,, bottom, ... 20·5
Intermediate-pressure cylinder, top, . . . 27·9
,, ,, ,, bottom, . . 29·1
Low-pressure cylinder, top, ... 38·9
,, ,, bottom, ... 41·8
                    Total, ... 179·6

Pump Horse-power—

High pressure, ...	55·5
Intermediate pressure, ...	53·7
Low pressure, ...	55·5
Total, ...	164·7

Friction horse-power, per cent., . . .	6
Delivered horse-power in water lifted, . . .	156
Mechanical efficiency, ...	86·8
Actual steam used per I.H.P. per hour, *i.e.*, air pump, *plus* jacket discharge, ...	12·76
Actual steam used per delivered horse-power per hour, .	14·72

Relative Volumes of Cylinders—

High-pressure cylinder, ...	1
Intermediate, ...	2·43
Low pressure.	6·75

Expansions, etc.—

Mean cut-off in high-pressure cylinder, . . .	·36
,, ,, intermediate ,, . . .	·45
,, ,, low-pressure ,, . . .	·33
Mean total expansions, ...	18·3
Mean terminal pressure absolute, . . .	7·1

Pump Test—

Total gallons lifted, ...	3,015,500
,, ,, per hour, . . .	301,550
,, ,, per minute, . . .	5,025·9
Average height of lift in ft., . . .	102·39
Total ft. lbs., ...	3,087,570,450
Duty per cwt. of coal used, . . .	131,730,000
Duty on 1000 lbs. of steam, . . .	134,333,000
Pounds of coal per I.H.P. . . .	1·46

Pumping-Engine at Hampton.—Fig. 231 represents the outline of a compound engine by Mr. Restler at the Hampton station of the Southwark and Vauxhall Water Co. Professor T. Hudson Beare's trial of it is reported in the *Proceedings of the Institution of Mechanical Engineers,* 1894, to which report we are indebted for the following particulars.

The cylinders are 32 in. and 52⅝ in. in diameter, each having a stroke of 84 in. The piston rods are each 6 in. in diameter.

The engines are provided with ordinary D-slide valves, one at each end of each cylinder, each D-valve having a Meyer expansion valve. The valve chests are prolonged to the ends of the cylinders, thereby securing short ports and very small clearance volumes.

The main pumps are of the vertical piston type placed one under each cylinder, and are worked direct from the engine cross-heads by means of vertical rods arranged so as to pass the crank shaft. Each pump is 19 in. in diameter, and has a 7-ft. stroke and works against a head of about 180 ft. of water.

The air pump and feed pump are actuated by a beam from the cross-head of the high-pressure cylinder, and the circulating pump by a similar beam from the low-pressure cross-head. Each engine has a surface condenser containing 553 tubes 7 ft. long and 1 in. in outside diameter, with a total cooling surface of 1040 sq. ft. The diameter of the feed

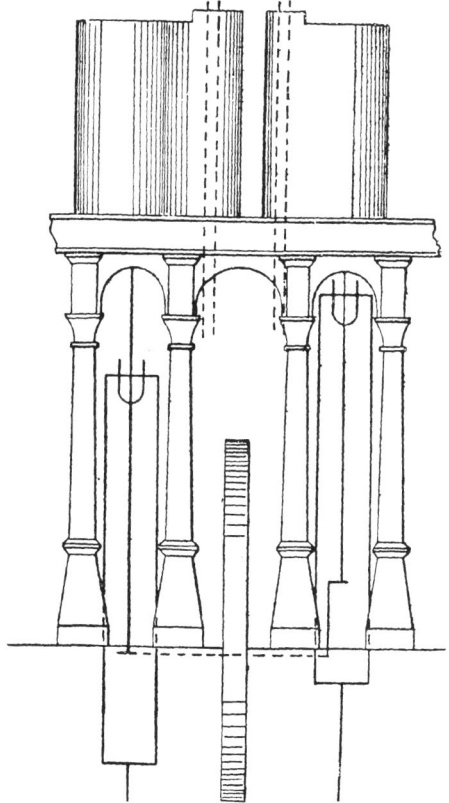

FIG. 231.—Compound Pumping-Engine, Southwark and Vauxhall
Water Co., Hampton.

pump is 4¾ in., of the air pump 24 in., and of the circulating pump 13 in., the first has a 27 in., and each of the latter has a 39 in. stroke.

Before the feed water enters the boiler it passes through an economiser, and is raised some 70° or 80° in temperature.

The bodies and ends of both cylinders are jacketed; also the receiver, steam passages, valve chests, and covers. The cylinders form liners within the body jackets. The jackets are supplied with steam by separate pipes leading from the main steam pipes. The cylinders and steam pipes are coated externally with 2 in. of Leroy's non-conducting composition and lagged outside with wood.

Duration of trial,	5 hours.
Number of expansions,	8·75

Steam Pressures—

In boilers—above atmosphere,	94·7	lbs. per sq. in.
,, high-pressure steam jacket above atmosphere,	92·9	,,
,, low-pressure steam jacket above atmosphere,	91·1	,,
,, intermediate receiver above atmosphere,	13·5	,,
,, condenser, absolute,	2·3	,,
Barometric pressure, absolute,	14·8	,,
Mean effective pressure H.P. cylinder,	32·16	,,
,, ,, L.P. ,,	10·31	,,
,, ,, total, reduced to L.P. cylinder,	22·07	,,
Revolutions per minute,	21·23	
Piston speed, ft. per minute,	297	
I.H.P. high-pressure cylinder,	228·9	
,, low- ,, ,,	200·6	
,, total,	429·5	

Dryness Fraction of Steam—

In high-pressure cylinder after cut-off,	79·4	per cent.
,, ,, before release,	92·1	,,
,, low-pressure ,, ,,	85·9	,,

Air-pump Discharge Water—

Per hour,	6284	lbs.
,, I.H.P. per hour,	14·63	,,

Jacket Water—

From high-pressure jackets, per hour,	240	,,
,, low- ,, ,, ,,	238	,,
,, receiver ,, ,,	67	,,
Total jacket water per hour,	545	,,
,, per I.H.P. ,,	1·27	,,
,, in per cent. of total feed water,	8·0	

Feed Water, including Jacket Water—

Total per hour,	6829	lbs.
,, per I.H.P. per hour,	15·90	,,
Saved per lb. of jacket water,	1·39	
Percentage *less* feed water, with steam in all the jackets,	10·0	

Absolute Pressures, lbs. per sq. in. measured from Indicator Diagrams—

Maximum initial in high-pressure cylinder,	99·2	lbs. sq. in.
At cut-off in high-pressure cylinder,	79·4	,,
,, release ,, ,,	26·1	,,
Minimum exhaust in high-pressure cylinder,	19·1	,,
Maximum initial in low-pressure cylinder,	23·3	,,
At release in low-pressure cylinder,	8·0	,,
Minimum exhaust in low-pressure cylinder,	3·5	,,

Temperature range in high-pressure cylinder,	.	.	102° Fahr.	
,, ,, low- ,, ,,		.	88° ,,	
,, of air-pump discharge, .	.	.	98° ,,	
Boiler steam, total heat per lb.,	.	.	1184 Th. U.	
Air-pump discharge, total heat per lb.,		.	66 ,,	
Jacket steam, latent heat per lb.,	.	.	879 ,,	

Heat passing through Engine—

Through cylinders per stroke,	.	.	2758·0 ,,	
,, jackets ,,	.	.	239·5 ,,	
Total through engine per stroke,	.	.	2997·5 ,,	
Equivalent of I.H.P. ,,	.	.	432·4 ,,	
Thermal efficiency, .	.	.	14·4 per cent.	

Pumping-Engines at the Copenhagen Waterworks.—At the Copenhagen Waterworks there are a considerable number of compound condensing pumping-engines of the beam type. The following is an account of the experiments made by Mr. F. Ollgaard on a pair of engines of this type, coupled together on the same crank shaft, and working four water pumps, two from each beam. Each of them consists of high- and low-pressure cylinders, the walls of which are jacketed with steam, but not the top covers. The dimensions of the engines are as follows :—

	Diameter.	Stroke.
High-pressure cylinder, ...	19 in.	3 ft. 8·6 in.
Low- ,, ,, . . .	31 ,,	5 ,, 6 ,,
Main water pumps, ...	20 ,,	4 ,, 7·3 ,,

Two Cornish boilers furnished the steam used during the trials, which was supposed to contain 5 per cent. of priming water; the feed-water was measured into these boilers by a Kennedy water-meter.

The water pumps gave an efficiency of 98 per cent. of their contents.

The subjoined table gives details of two sets of experiments, one with steam in the jackets, and one without. In the latter set it was not possible to run the engines faster.

The jacket water was carefully measured. When the jackets were on, the pressure on the jacket of the high-pressure cylinder was 40 lbs., in that of the low-pressure cylinder 37 lbs. per sq. in. For each pound of steam condensed in the jackets about 5 lbs. of steam were economised in the cylinders. It will be seen that there is a considerable gain in economy due to the jackets. The quantity of steam condensed in the high-pressure cylinder during admission was calculated at 28 per cent. of the total. The total expansion in both cylinders was about 10. It was reckoned that about 5 per cent. of the total heat was withdrawn from the walls of the cylinder, and carried off to waste in the condenser. The table gives the chief points of interest in these experiments.

TABLE OF EXPERIMENTAL RESULTS WITH AND WITHOUT STEAM
IN JACKETS.

	With Steam in Jackets.	Without Steam in Jackets.
Number of revolutions per minute, ...	24	19
Mean boiler pressure, ... lbs.	53	54·3
Mean temperature of feed water, ...	72° Fahr.	74° Fahr.
Consumption of dry steam per H.P. per hour of water raised (jackets included), ...	22·7 lbs.	28·5 lbs.
Steam (with 5 per cent. water) per lb. of Newcastle coal, ... lbs.	7·57	7·48
Consumption of coal per I.H.P. per hour of water raised, ... lbs.	3	3·8
I.H.P. for each engine, ...	81·1	65·7
Efficiency of engine (proportion between work done in water raised and indicated work), . . .	85·22 per cent.	83·45 per cent.
Consumption of dry steam per I.H.P. per hour (jackets included), ... lbs.	19·3	23·8
Work due to the high-pressure cylinder in percentage of total, ...	52½	60
Work due to the low-pressure cylinder in percentage of total, ...	47½	40
Proportion of consumption of steam in the jackets to total, ... per cent.	6·25	...
Condensation of steam in the cylinder during admission, ... per cent.	28	40·4*

Losses due to Working Conditions. — If we take the case of a mining or waterworks engine running continuously, the loss in economy due to the irregularities of stoking, temporary imperfections in working, and management, will not in well-organised works amount to more than from 10 per cent. to 15 per cent. of the total coal used, compared with that computed from a twenty-four hours' well-conducted trial, but in most cases this loss is much larger.

In the London Hydraulic Power Supply Works, where the demand for power is chiefly during twelve hours out of the twenty-four, but of a varying quantity, and where the steam must always be kept up, Mr. Ellington calculates the losses as follows:—

	Per cent. of total coal.
Coal utilised at efficiency of trials, calculated on total output, .	60
Coal wasted through intermittent running, based on an experiment at Falcon Wharf, ...	20
Coal used in keeping steam up in boilers and engine-jackets when stopped during nights and Sundays, and changing over boilers, .	12
Steam used for other purposes, variation in quality of coal, defective stoking, etc., ...	8
	100

In pumping establishments, where the engines are only worked during

* *Proc. Inst. Civil Engineers*, vol. civ. p. 389.

the daytime, and not on Sundays, 30 per cent. more coal is often consumed than that calculated from the results of a twelve hours' trial; but much depends on the stoking and the boiler losses during the time the engines are stopped. The steam pipes may be of great length, and both pipes and boilers may be imperfectly clothed.

No exact rules can be laid down to meet the variations which occur.

Load Factor.—Various definitions have been given of the term 'load factor,' but for certain purposes the following will be found convenient.

The load factor is the ratio between the average output per hour and the maximum output in any one hour during the year.

Let *a, b, c, d,* fig. 232, represent to scale an area equal to the total work

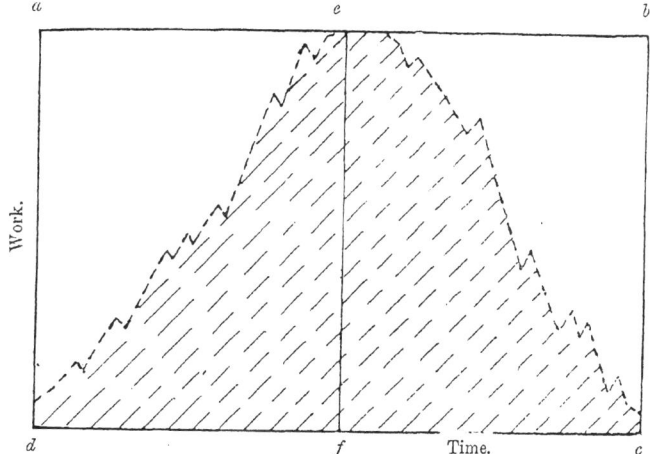

FIG. 232.—Diagram illustrating the term 'load factor.'

The shaded area represents work done, and the rectangle *a, b, c, d* the work which would have been done if the engines had worked fully loaded the whole of the time. The load factor may be applied to a day's or a year's work.

which the engines would have done if the mean power had been equal to the maximum during the period, the vertical line representing work and the horizontal line time; then if the engines work fully loaded half the time, the work done would be *a, e, f, d,* or 50 per cent. of the whole. The load factor on this basis might then be expressed by ·5 (the unit being 1), or by '50 per cent.'

It would not, however, have a definite meaning if the rectangle *a, b, c, d* represented one year's possible work, and *a, e, f, d* half a year's work when no work was done during the other half of the year. The load factor is only applicable for the period during which the machinery is working.

If the speed of the engines varies, then the work done would be represented by the area enclosed within the dotted line, and the load factor by its percentage value relatively to the total area of the rectangle.

The diagram may be made for a single day's work, but to be of commercial and practical value it must represent the average day's work taken over a long period. Load factors may have different constructions, according to the purpose to which they are applied.

It will be seen that unless the load factor is simply applied to questions relating to the saving in fuel or steam, the reserve power must be taken into account, as it has a direct bearing on the capital charges and cost of upkeep of the whole power, part of which is represented by reserve power.

It is obvious that the term 'load factor' may be made use of in several ways. It is exceedingly useful in showing at a glance what is not so readily understood when expressed in figures.

The Birmingham hydraulic power supply is produced by means of gas engines, and the following table shows to what extent the economy of gas has been secured by increase in the output.

Year ending	Total Water Pumped into Hydraulic System, Million Gallons.	Consumption of Gas in Cubic Feet per 1000 Gallons Pumped.
31st March 1894, . .	2·443	555
,, 1895, . .	3·182	439
,, 1896, . .	3·585	386
,, 1897, . .	4·726	308

Large engines of the same type and quality are more economical than smaller ones. In small works the general establishment charges and wages are together often far more than the coal bill; all these commercial considerations must be taken into account in designing the works.

Influence of Load and Speed of an Engine on Economy.—In collieries it is seldom that the pumping shaft has to be deepened after the colliery has been put to work, therefore the pumping-engine generally starts with its full load; but in metalliferous mining it is quite different, as in these the shafts are frequently deepened and the load on the engine is consequently increased. The friction of the engine is not largely affected by the increase or decrease of load. Assuming that 3 lbs. per sq. in. is all that is necessary for friction, then, if the total indicated load on the piston is 25 lbs., the mechanical efficiency of the engine would be 88 per cent.; but assuming that the friction remained the same, and the total indicated load was reduced to 12 lbs., then the mechanical efficiency would be reduced to 75 per cent. The real economy of a pumping-engine should be estimated not on the indicated, but on the pump H.P., and, taking the engine and pump together, in H.P. of water actually pumped.

The load then may remain constant or be an increasing one, according to the conditions under which the engine is being worked.

Variations in speed also affect the economy, but when the variation does not exceed 20 to 25 per cent., it is not very important, except in the case of centrifugal pumps, where the best results are obtained at a critical speed suited to the lift. For this reason the efficiency of reciprocating pumps under variations of speed and lift is much more uniform than that of centrifugal pumps.

In ordinary pumping-engines the importance of a high average pressure consistent with steam economy per I.H.P. is very great, for not only does it increase the mechanical efficiency, but it also reduces the percentage loss from back pressure. An additional back pressure in the condenser of 1 lb. would be 3·3 per cent. only if the mean indicated pressure was 30 lbs.; but it would be 10 per cent. if the mean pressure was only 10 lbs. The mean pressure usually employed in Cornish and old-fashioned beam engines was 13 lbs. per sq. in., and the economy fell off very rapidly with increase in the back pressure of the condenser.

It has been shown in the chapter on Cornish Engines that the modern double cylinder or compound Cornish engine is worked with a water load of 30 lbs. or, say, 33 lbs. indicated. This engine starting with only half its load would have an indicated load of 16·5 lbs. more than the full load of the old type of engine. Its mechanical efficiency is therefore not so much reduced by light loads and back pressure.

CHAPTER XIV.

Low-Lift Pumps.—Low-lift pumps are required for the drainage of lands and for many other purposes. The quantity of water to be lifted is large compared with the power. The problem presented is that of the best application of power under the circumstances. The general impression has been that a piston pump largely decreases in efficiency with low lifts, and that for very small lifts its efficiency is very low. If, however, the lift exceeds 12 or 15 ft., the mechanical efficiency of a properly designed piston pump is as high as that of any other kind of pump for the same lift, and generally it is higher. The valid objections, however, to piston pumps for very low lifts are their larger size and the number of valves necessary.

The Scoop Wheel is a rotary pump much used for low lifts in the drainage and irrigation of lands, etc. Its application has been much restricted since the introduction of the centrifugal pump. The scoop wheel is a kind of reversed undershot water-wheel, and is illustrated in fig. 233.

Signor Cuppari, in a paper in the *Proceedings of the Institution of Civil Engineers*, vol. lxxv., gives the following particulars of scoop wheels in use in Holland.

The Halfweg pumping station has six float wheels with a combined width of breast of 38·70 ft. The external diameter of one of the wheels is 23 ft., that of the other five, 21·33. The internal radius is 5·92 ft. The floats, twenty-four in number, are inclined to the radius, so as to be tangential to a circle concentric with the wheel, and having a radius of 2·85 ft. The centre of the wheel is at 5·6 ft. above datum. The steam-engine has four separate valves, with expansion regulated by hand (of the old double-acting Cornish type). The steam cylinder is 3·33 ft. diameter by 8·0 ft. stroke. There are three boilers always in steam, each having 538 sq. ft. of heating surface, and carrying a maximum steam pressure of three atmospheres. The driving axle is connected on each side by toothed gearing to a shaft carrying three wheels. The speed is reduced in the ratio of 13·5 to 6. The wheels are all upon the same shaft, built up of several pieces, which can be coupled up as required.

Three systems are adopted in the construction of water-wheels, which differ in the method of transmitting the force. In the first system the force is transmitted by a driving-axle and spokes acting as struts; in the second by a wheel with teeth upon the circumference, the axle and spokes acting as struts; and in the third by a similarly toothed wheel, but with a double set of rods in tension instead of spokes. The Dutch still adhere to the first system, although, according to Redtenbacher, it is not suitable when the power is 10 or 12 H.P., as large wheels of this kind are relatively very heavy. Thus the Italian wheels at Bresega, near Adria, which are on

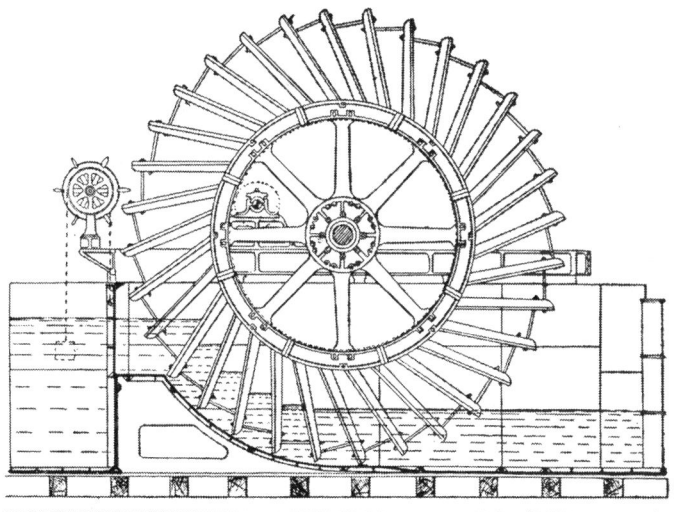

Fig. 233.—Scoop Wheel.

the third system, have an external diameter of 39·4 ft., an internal diameter of 26·25 ft., and a breadth of 6·56 ft., and are the largest known in Holland or Italy. The displacement of water is about 5300 cubic ft. per minute per wheel. The axle of these wheels has a diameter at the thickest part of 1·41 ft. and 1·25 ft. at the bearings. The wheels at Katwijk, which are on the first system, are only 29·5 ft. in diameter, yet the axles of the furthest wheels are of the same diameter as above.

At Zeeburg, near Amsterdam, there are eight wheels of the most recent construction, 26 ft. 3 in. in diameter and 10 ft. 8 in. across the breast. The driving-axles of the wheels furthest from the motor are 1 ft. 6 in. in diameter, and weigh over 6 tons. Each wheel has four sets of spokes, each set weighing 4 tons. The driving-axles nearest the motor are nearly twice as heavy.

There are 282 cubic ft. of oak and 222 cubic ft. of pine timber in each wheel. Compared to the mass and weight of material, the volume of water raised, amounting to 7063 cubic ft. per minute per wheel, seems small. The velocity of a point on the periphery is about 208 ft. per minute, which in Holland is considered moderate, but in Italy high. In this system of construction the axle is subject both to bending and torsion; in the suspension system to bending only.

Mr. Zangirolami of Adria constructs wheels with curved buckets capable of raising water to a height equal to $\frac{4}{5}$ of the radius; they give excellent results.

Instead of two toothed wheels, one at each side of the wheel, he puts one toothed wheel in the middle, thus avoiding the practical difficulty and expense of two cogged wheels of precisely similar pattern on the one hand, and the twisting effect produced by a single cogged wheel when placed at one end of the wheel on the other.

At Halfweg the driving-axle is of cast iron, solid, 1 ft. 2 in. in diameter near the motor and 10 in. at the further end, with enlargements at the joints. The framework of each wheel is formed of three sets of spokes, which are cast in one piece with the nave and ring, or rather each set is cast in two parts and bolted together. The whole weight of one wheel with its axle is probably about 15 tons.

For the proportioning of scoop wheels Mr. Forster gives the following formula :—

$$D = 9.82 \sqrt{H},$$

in which D is the diameter of the wheel and H the height from the bottom of the wheel to the highest level to which the water has to be raised, but Mr. Wheeler gives $8.75 \sqrt{H}$ as more nearly representing English practice.

Sometimes the scoops are curved. The wheels at Zuidplas in Holland have curved scoops. Diameter of wheel, 32·8 ft.; dip of scoops, 3·28 ft.; head of water, 11·8 ft. H in the above formula is then—

$$11.8 + 3.20 = 15.08 \text{ ft.}$$

The largest scoop wheel in England is 50 ft. in diameter.

The maximum speed of a point in the periphery is usually 8 ft. per second; 30 ft. wheels usually make 4 to $4\frac{1}{2}$ revs. per minute. The displacement efficiency is found to be about 80 per cent., and when driven by steam engines, the total efficiency, that is, water H.P. divided by indicated H.P., has been found to be about 60 per cent. $\dfrac{\text{W.H.P.}}{\text{I.H.P.}} = \cdot 60$.

Higher efficiencies have been obtained, but much depends on the conditions of application and working.

A wheel proportioned for a high lift will give a low efficiency at a low lift, and naturally so. An engine working a wheel with 2 ft. lift has been found to use 5·5 lbs. of coal per W.H.P. ; but when the lift was reduced to 1 ft., it used 14 lbs. of coal, and at 6 in. lift 50 lbs. per W.H.P.

Of course the increased consumption was principally due to the reduced efficiency of the steam and the reduced mechanical efficiency of the whole plant ; very little was probably due to the reduced efficiency of the wheel itself. The same thing happens with centrifugal pumps, though not to the same degree. Mr. Barker gives the following results of trials with a centrifugal pump where the lift was varied :—

Lift	= 1	2·40	3·30	7·20
Coal per W.H.P.	= 9·37	8·22	7·71	6·60

The angle for the scoops of scoop-wheels, or floats as they are sometimes called, is determined by the following consideration.

The scoops usually dip from the radial line about 40 degrees, and from tangents to a circle concentric with the wheel. The centre of the wheel is fixed so as to divide the head and dip equally, except when the lift is great. The dip is seldom more than 6 ft.

The efficiency or ratio of discharge to displacement of bucket and piston pumps has been found for lifts of about 13 ft. to be 89 per. cent. ; but much will depend on the design and mode of application of the pump. Signor Cuppari gives the following as his experience in Holland.

Total useful effect, .	$= \dfrac{\text{W.H.P.}}{\text{I.H.P.}}$
Float or scoop wheels,	67 per cent.
Lift and force pumps,	70 ,,
Centrifugal ,,	45 ,,

Signor Cuppari gives the relative cost per W.H.P. of pumping stations in Holland in the following table.

	Buildings.	Machinery.	Total.
	£	£	£
Scoop wheels, ...	46·14	46·28	92·42
Screw pumps, ...	...	...	94
Centrifugal pumps, ...	34·20	36·8	71
Piston pumps, ...	...	...	72

Piston and Bucket Pumps.—The chief disadvantage (apart from the complication of the machinery itself) in the use of piston and bucket pumps, as compared with the centrifugal pump, is that the delivery is constant for a given speed at all lifts, whilst with the centrifugal pump a reduction in the height of the lift increases the rate of delivery. There is thus a saving in the average time occupied in pumping.

Centrifugal Pumps.—There are two forms of centrifugal pumps, one having a vertical and the other a horizontal spindle ; both forms are illustrated in figs. 239 and 242.

The horizontal spindle pump is usually termed a 'cased' pump, is self-contained, and is driven from the engine by means of a belt or gearing, or is actuated direct from the engine shaft. The pumps having vertical spindles are so arranged as to be always charged or submerged in the well or shaft below the engine. To keep the horizontal spindle pump always charged it is necessary to place it in a dry well below the lowest water level, the water being supplied by a suction pipe through the side of the well.

When the pump is above the water level a charging apparatus is required, and that involves a foot valve in the suction pipe. Before the pump can deliver water it must be fully charged and all air excluded. The charging of the pump is sometimes effected by means of a steam ejector which ejects the air, and thus draws water into the pump through the suction pipe.

The chief application of the vertical spindle pump is in draining lands, emptying and filling docks, etc.

Examples of this class of pump are given further on.

The action of centrifugal pumps may be thus described.

If the pump and pipes are full of water, and the fan is worked at a sufficient velocity, the centrifugal force imparted to the water in the fan will cause the water to be discharged from the fan casing into the discharge pipe, and to be drawn in from the suction pipe, thus giving rise to a continuous stream of water.

To ensure a discharge, the centrifugal force must be more than is sufficient to support the full column of water. To support the column without any discharge, the velocity of the periphery of the fan must be at least equal to the velocity which a body would acquire in falling through the height of the column.

Calling the velocity in feet per second V, and h the height, then $V = 8 \sqrt{h}$, but in practice the shape of the blades and the size of the fan influence the periphery speed. The velocity of the periphery of the fan necessary to support the column without discharge varies from $V = 8 \sqrt{h}$ to $V = 9 \cdot 5 \sqrt{h}$. As the velocity increases so does the discharge, but the discharge for a maximum efficiency is a matter of experiment. For results of experiments with centrifugal pumps, see the papers by Thompson and Parsons in the *Proceedings of the Institution of Civil Engineers.* The former author says: " The best duty is obtained when the speed of the periphery of the fan exceeds the velocity of a falling body due to' the height of the lift by from 6 to 8 ft. per second, but the duty is not very much reduced when the excess of speed falls to $4\frac{1}{2}$ ft. per second or rises to 14 ft. per second, as determined by experiment with lifts varying from $5\frac{1}{2}$ to 20 ft."

The same author says : " When a centrifugal pump, properly proportioned, is worked by a steam engine, the maximum duty that may be expected is about 55 per cent. of the indicated power in the smaller pumps, rising in the larger ones to 70 per cent."

Centrifugal pumps of modern construction vary very much in form of

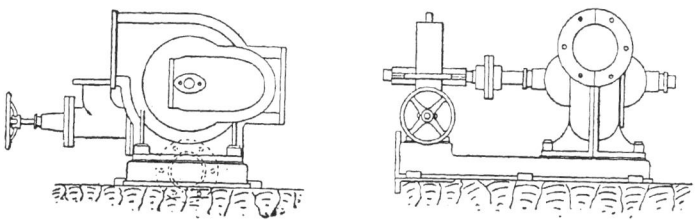

FIG. 234.—Centrifugal Pump driven direct by means of Pelton Wheel.

blades; some are made to curve forward at the tips. It does not, however, appear that higher efficiencies are obtained than were secured by the best pumps of former times.

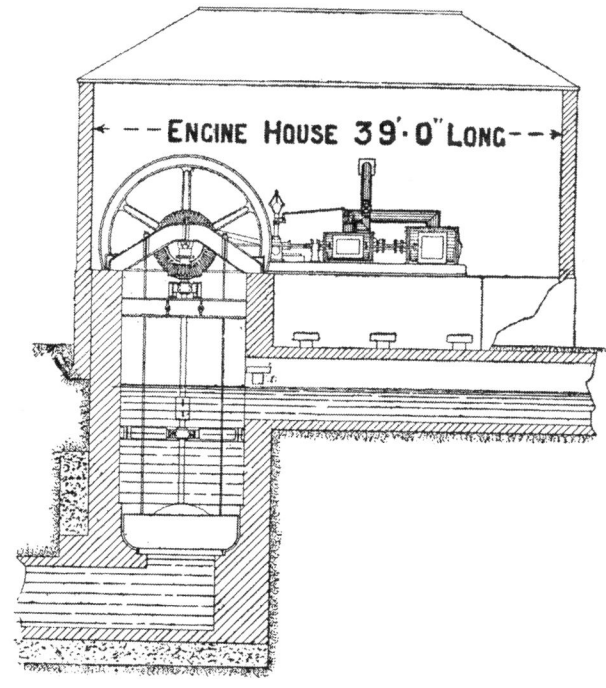

FIG. 235.—Centrifugal Pump at Burnt Fen.

For ordinary pumps of medium size working under the usual conditions, 50 per cent. efficiency is as much as is realized, but 60 per cent.

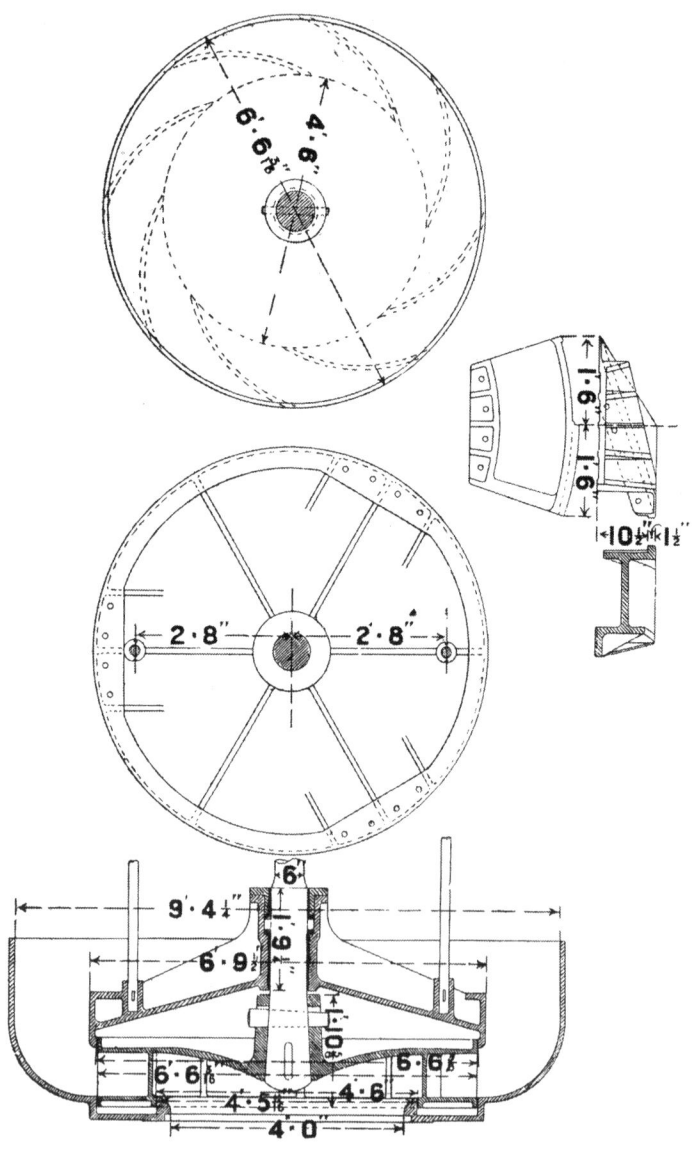

F𝐢ɢ. 236.—Centrifugal Pump at Burnt Fen.

is obtained with properly proportioned large pumps working under favourable conditions. An allowance must always be made from the result of special trials. Although a higher percentage may be obtained under special conditions of lift and speed, such conditions cannot be maintained in ordinary work. When using the term efficiency, we must distinguish between the efficiency of the pump itself and the efficiency of the pump and engine combined. The efficiencies just named are meant to be the combined efficiency of engine and pump, that is

$$\frac{\text{Water H.P.}}{\text{Indicated H.P.}}$$

Let the efficiency of the pump itself be 70 per cent., and that of the engine itself 80 per cent., then the combined efficiency is 56 per cent.

The centrifugal pump is not generally used for lifts of more than 20

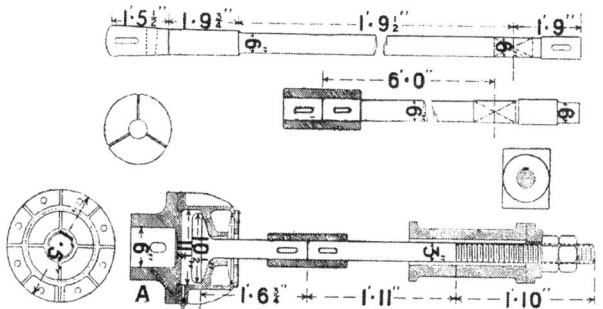

FIG. 237.—Details of Pump Spindle and Onion Bearing of Centrifugal Pump at Burnt Fen.

to 30 ft., and experiments are wanting as to the efficiency with higher lifts. Experiments are being made with combined pumps in which several fans are mounted on one shaft with a casing forming several pumps, one discharging into the other, the total lift being the combined effect. Thus three pumps, each equal to a head of 30 ft., would give a combined effect of 90 ft.

We are without any reliable data as to the efficiency of such pumps. It is only for special purposes, where the efficiency is not of great importance that they are likely to be used.

In special cases centrifugal pumps may be driven by electric motors or a Pelton wheel directly fixed to the spindle of the pump. Fig. 234 illustrates a pump driven by means of a Pelton wheel used in one or two instances by the author in waterworks, for lifting the water from the storage reservoir to the filter beds, a height of from 6 to 10 ft., the power for working the Pelton wheel being obtained from the town main.

Fig. 235 is an illustration of a large centrifugal pump designed by the author for the Cambridgeshire fens. Figs. 236 and 237 give details of the pump and spindle. The fan is 6 ft. 6 in. in diameter, and 10 in. deep. It lifts 120 tons of water per minute to a height of 12 ft., and is driven by means of a compound condensing engine. The gearing between engine and pump is in the proportion of 1·54 to 1. The pump has a free vortex. Above the fan is a shield to relieve the pump from pressure.

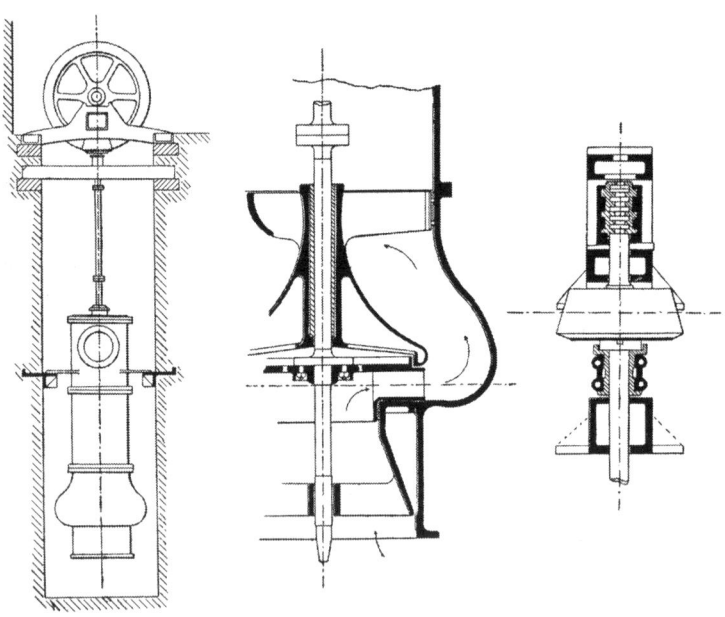

Fig. 238.

The shield is suspended by means of tie rods from a girder thrown across the well, and is held in position by means of three brackets bolted to the outer casing, the brackets having inclined surfaces to minimize the resistance offered to the water in the well during its upward spiral motion. So largely does the shield take off the pressure from the top of the fan, that in the normal condition of running the weight of the fan, shaft, and wheel, is almost entirely supported by the water passing through the pump. There is no thrust block on the fan spindle. The spindle is supported by the 'onion' bearing A, fig. 237, suspended from a girder. The onion is of steel. The bearing is of gun metal fastened on the top of the pump spindle. The bearing forms an oil cup. As the pump is so well balanced, there is little weight on the 'onion' after the pump begins to deliver its water.

The following table of the results of trials made with pumps having different forms of arms or blades is given in Thompson's paper :—

APPOLD'S CENTRIFUGAL PUMP.

	Height of lift in ft.	Discharge in Gallons per minute.	Revolutions per minute.	Velocity of Circumference in ft. per minute.	Ratio Percentage of power to effect.*
With radial arms, . .	18·0	474	720	2262	24
With straight inclined arms, ...	18·0	736	690	2168	43
With curved arms, . .	8·2	2100	828	2601	59
,, ,, . .	9·0	1664	620	1948	65
,, ,, . .	18·8	1164	792	2988	65
,, ,, . .	19·4	1236	788	2476	68
,, ,, . .	27·6	681	876	2751	46

The above experiments were made with the object of ascertaining the best form of fan blade.

Experiments were also made by Thompson with the pump illustrated in fig. 238 at the Leith Docks, with the following results :—

R.	I.H.P.	S.	S'.	E.S.	H.	E.*
169·0	116	42·03	34·08	7·99	12·0	49
172·2	120	42·84	39·30	3·54	16·0	59
182·0	102	45·26	43·38	1·89	19·6	69
195·0	81	48·50	43·94	4·56	20·0	70

R. = Revolutions of fan per minute.
I.H.P. = Indicated H.P. of the engine.
S. = Speed of periphery of fan in ft. per second.
S'. = Speed of periphery of fan due to lift, as per formula $9·825 \sqrt{h}$.
E.S. = Excess speed over that given by the formula.
H. = Height of lift in ft.
E. = Efficiency per cent.

Parsons, in a paper (*Proceedings of the Institution of Civil Engineers*, vol. xlvii.), gives the following results of experiments with the pump illustrated in fig. 242.

* The ratio of power to effect was measured by means of a dynamometer.

Experimental efficiencies obtained by R. C. Parsons on a 14-in. revolving fan, 10-in. suction, and 10-in. discharge, made on the Appold principle.

Revolutions per minute.	Gallons per minute.	Lift in ft.	Ft.-lbs.		Efficiencies per cent.
			Water raised.	Indicated Power.	
392	1012	14·67	148·461	298·438	49·74
394	1108	14·70	162·875	317·158	51·35
395	1197	14·65	175·364	332·136	52·80
400	1431	14·75	211·073	374·954	56·20
405	1695	14·75	251·987	419·790	60·17
425	1108	17·20	190·576	388·316	48·97
431	1431	17·40	248·994	447·552	53·63
435	1695	17·60	298·310	486·050	61·37

The casing had a spiral form and the arms of the fan were curved as shown.

Other experiments were made with the different forms of fan-blades or arms shown in the other figures and with a concentric casing, but with inferior results.

Experiments were also made with the vertical spindle pump (fig. 239).

One series was made with the Rankine form of fan (fig 240), and another with the Appold fan (fig. 241). The latter gave the better result, but the efficiencies were not equal to the best results now usually obtained with vertical spindle pumps. The best result with the Appold fan was 50 per cent. of the dynamometer power, and the best with the Rankine fan 39·6 per cent. Vertical spindle pumps and engines on a large scale give efficiencies from 50 to 60 of the indicated H.P.

Experimental Tests of Centrifugal Pumps by Mr. W. O. Webber.—Mr. Webber explains that the use of the term efficiency indicating the value of

$$\frac{\text{Water H.P.}}{\text{Indicated H.P.}}$$

for such pumps as are driven by an engine direct, does not therefore show the full efficiency of the pump, but that of the combined pump and engine. It is, however, a very simple way of showing the relative values of different kinds of pumping-engines, the motive power of which forms a part of the plant.

In calculating the efficiency of the pump, the cubic feet of water passing over the weir, as measured by the hook-gauge, is converted into lbs. by multiplying by 62·3 ; this multiplied by the height from the level of water in the tank when the pump is running to the centre of the discharge pipe gives the number of lbs., which number, when divided by 33000, indicates the quantity of water horse-power being developed.

The power utilized is measured by the dynamometer in terms of dynamometer horse-power ; the water horse-power divided by the dynamometer horse-power shows the efficiency of the pump being tested ; or, to formulate,

$$\frac{\text{Water H.P.}}{\text{Dynamometer H.P.}} = \text{the efficiency of such pumps as are driven by a belt.}$$

Fig. 243 shows the efficiency curves for different velocities, plotted from tests made of two pumps with 5-in. apertures.

These tests were made under an average elevation of 17 ft., the pumps in both cases draughting about 9 ft. and discharging 8 ft. higher. The upper curve a b was the result of tests made with a pump that was very clean and smooth inside. The lower curve c d was made with a pump in

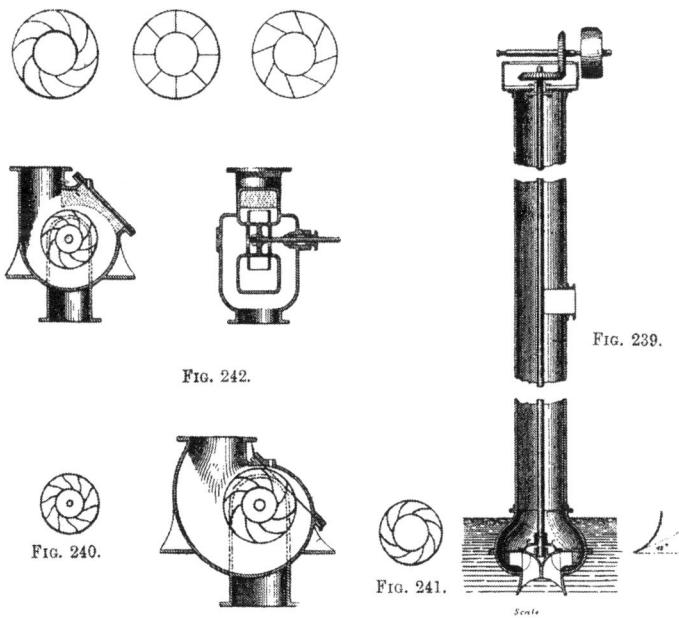

FIG. 242.

FIG. 239.

FIG. 240.

FIG. 241.

FIGS. 239 to 242.—Illustrations of Centrifugal Pumps used in Parson's Experiments.

which, through carelessness in the foundry, the core sand had been allowed to burn into the inside face of the volute or casing and water passages.

The difference between these two curves (which, by the way, are remarkably uniform) shows the absolute necessity of having the inside of all such pump castings very smooth and free from the slightest roughness. These tests seem to show that the efficiency rises very gradually and uniformly until the water reaches a velocity equal to $11\frac{1}{2}$ ft. per second. The highest efficiency with a pump of this size is apparently obtained with a velocity of 12 ft. per second, after which point the efficiency falls very rapidly.

Comparative Trials of Piston Pumps and Centrifugal Pumps.*—The establishment of Messrs. Schaeffer, Lalance & Co., at Pfastatt, uses daily from 8000 to 10,000 cubic metres of water, drawn (until 1885) from the river Doller by means of ten centrifugal pumps. To obtain a supply of water with a more constant temperature, a well was sunk in the water-bearing strata near the river, and 20 to 45 gallons per second drawn by means of two 'conjugated' centrifugal pumps supplied by Messrs. Nent & Dumont. Subsequently the means of supply were extended, and piston pumping-engines were constructed capable of drawing and delivering 90 gallons per second into the tank already constructed for the centrifugal

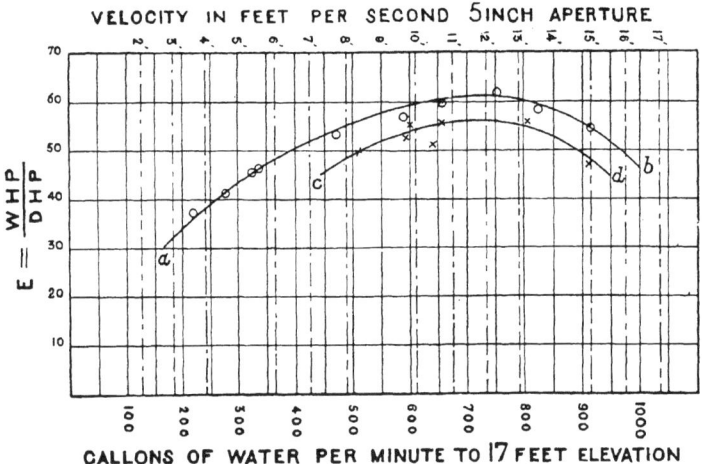

FIG. 243.—Webber's Experiments with Centrifugal Pump.

pumps. The opportunity was embraced of making test-trials of the two systems of pumping—by centrifugal pumps and piston pumps—under identical conditions, from the same well with the same lift, by the same discharge pipe into the same tank of large capacity. The discharge pipe delivered the water over the edge of the tank, and the head of pressure was therefore constant, whatever might be the level in the tank.

The centrifugal pumps were erected in 1872, and had worked till the time of the trials in May 1888 without needing repair. The suction pipe was about 12 ins. in diameter, and the delivery pipe 10 in. But to deliver the required quantity, 37 gallons per second, through an average height of 39·4 ft., the pumps were 'conjugated,' so that one followed the other in succession, and could perform the required duty at a speed of 520 turns per minute. The steam engine employed to drive the pumps had a cylinder 21·65 in. in diameter, with a stroke of 39⅝ in. The revolutions of the pumps were recorded by a counter.

* *Proc. Inst. Civil Engineers*, vol. xcvi. p. 423.

Each trial lasted from $4\frac{1}{2}$ minutes to $7\frac{1}{2}$ minutes. The leading results were as follow :—

DUTY OF CONJUGATED CENTRIFUGAL PUMPS.

Trial.	L.	M.	N.	O.
Speed of engine in turns per min., . . .	45·9	49·6	52·6	60·0
,, pumps ,, ,, . . .	470	506	538	611
Mean lift, ... ft.	35·51	35·59	35·64	35·71
Duty in water lifted per turn of engine, . .	15,549	16,448	17,741	19,589
Indicated work in cylinder for one revolution, ft. lbs., ...	32,125	37,105	43,000	56,419
Duty, percentage of engine power, . . .	48·4 per cent.	44·3 per cent.	41·25 per cent.	34·72 per cent.

The bucket pumps and steam engine consisted of a horizontal compound receiver steam engine, connected by bell cranks to four pumps in the well. The first cylinder was 19, and the second 33 in. in diameter; the stroke common to both cylinders 36 in. The buckets of the pumps were 22 in. in diameter, with the same stroke as the engine, 36 in.

The pumps worked well and without shocks at speeds up to twenty revolutions per minute. The engine and pumps, together with the carriage, duty, and erection, cost £2160.

The total weight was 69 tons, costing £31, 6s. per ton.

The trials, nine in number, lasted from $3\frac{1}{2}$ minutes to 7 minutes. The leading results are given in the annexed table.

DUTY OF PISTON PUMPS AND ENGINES.

Trial.	A.	C.	D.	E.	F.	G.	H.	I.	K.
Speed in turns per minute, . . .	13·44	20·3	18·4	22·7	24·5	26·1	18·7	16·3	13·3
Mean lift, ft., . . .	35·45	36·03	36·23	36·41	36·52	36·60	36·21	36·31	36·19
Duty in water lifted per turn, ft. lbs., .	69,586	71,079	71,434	71,789	71,753	72,195	70,716	70,919	71,405
Indicated work in cylinder for one turn, ft. lbs.,	99,035 per cent.	103,689 per cent.	101,311 per cent.	106,220 per cent.	106,680 per cent.	111,686 per cent.	102,326 per cent.	99,202 per cent.	96,715 per cent.
Duty, percentage of engine power,	70·26	68·55	70·5	67·9	67·4	64·6	70·0	71·46	73·8
Ratio of water actually lifted to calculated quantity, .	0·992	1·0	0·998	0·998	0·995	0·998	0·998	0·990	1·0

It is notable that the percentage of duty decreases as the speed of the engine and pumps increases, thus—

In the trials—

    K   A   I   D   H   C   E   F   G

the speeds in turns per minute were—

    13·3 13·4 16·3 18·4 18·7 20·3 22·7 24·5 26·1

and the percentages of duty were—

    73·8 70·2 71·4 70·5 70·0 68·5 67·9 67·4 64·6.

CHAPTER XV

Hydraulic Rams are hydraulic machines by means of which water is raised by utilizing the energy of a moving column of water. If the motion of a column of water in a pipe be suddenly arrested, the energy of the column will be spent in increasing the pressure in the pipe, and if the pipe be provided with a branch leading to a higher elevation, some of the water will rise in the branch, due to the increase in pressure. The energy of the column is equal to $\dfrac{W\ V^2}{64\cdot4}$, in which W = the weight of the water and V the velocity in ft. per second. Let the column be 40 ft. long with a capacity of 4 lbs. of water per ft., and the velocity 5 ft. per second, then $\dfrac{160 \times 25}{64\cdot4}$ equals an energy of 60 ft. lbs. or that required to raise 1 lb. of water a height of 60 ft.; if the utilized energy = 70 per cent., then ·7 of a lb. of water would be raised 60 ft. each time the water column acquired the velocity of 5 ft. per second, and was suddenly arrested. The construction of the hydraulic ram is illustrated in fig. 244.

A is a pipe from a stream or reservoir, B a valve opening downwards (which we will call the *momentum valve*) into the pipe C. A branch from the pipe C has an air vessel fixed on to it, and containing a non-return valve D. A pipe E conveys the water raised to the cistern. Let the valve B be opened, then the water in the pipe A will flow through until its velocity is sufficient to lift the valve B. The said valve will then suddenly close and the energy of the moving column in the pipe A will be spent in raising some of the water through the air vessel to the cistern.

When the energy has been thus spent, a reaction is set up, the valve B falls by its own weight, and the operation repeated. During the reaction a little air is drawn in through the snifting valve S, thus keeping the air vessel charged.

Hydraulic rams have been said to have given an efficiency as high as 60 to 70 per cent., but in ordinary practice the efficiency is low. The length of the pipe A is important in relation to the height to which the

water has to be lifted.   For a given fall its length should be proportional
to the height of the delivery.   The lift of the valve B should be adjustable.

Rams have been made to lift water by suction, by means of an

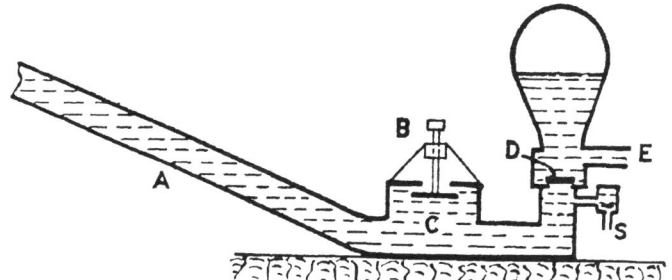

FIG. 244.—Simple Hydraulic Ram.

independent water supply.   Leblanc's ram is of that kind, and is illustrated in fig. 245.

G is the independent source of supply of power water.   It is required

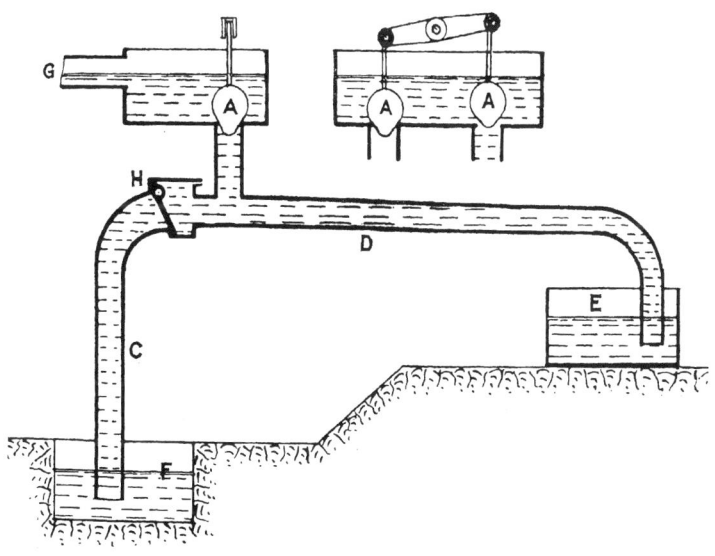

FIG. 245.—Leblanc's Syphon Hydraulic Rams.

to lift from F to E.   A is a floating momentum valve, and H a non-
return valve.   The rush of water through the pipe D from G causes the
valve A to close.   The energy of the water in D carries the water forward,

S

and tending to form a vacuum in front of the valve H, causes the water in C to follow the water in D; some of the water in F passes into the pipe D, and from there to the cistern E.

When the energy of the water is spent, it commences a retrograde motion because of the syphon action between F and E. The valve H then closes, and the valve A is forced open, when the action just described is repeated. This ram is made double-acting by putting two rams side by

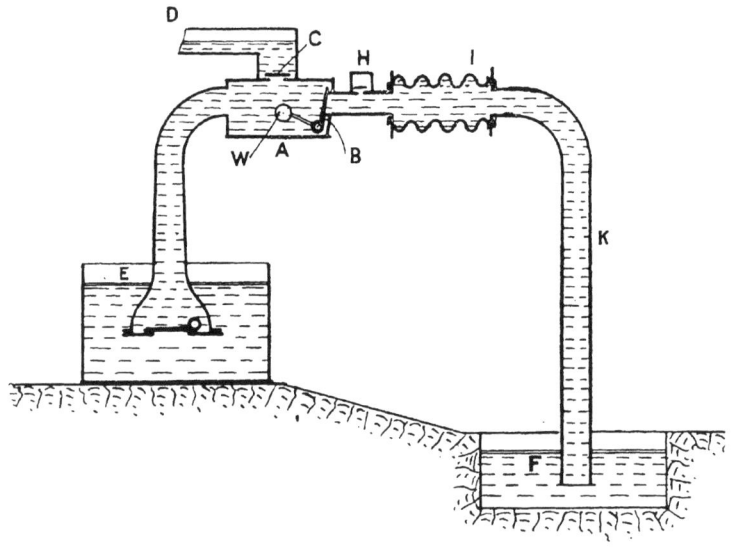

Fig. 246.—Lemichel's Hydraulic Syphon Ram.

side, and connecting the valves A by a beam, so that one goes down when the other goes up.

Syphon rams have been constructed in which the ram is above the source of supply, as in fig. 246.

The ram is at A, the momentum valve at B, the non-return valve at C, and the discharge at D. Water is syphoned from E to F, and part is delivered by the action of the ram at D. To ensure the continued action of the ram, it is necessary that the syphon shall be kept free from an accumulation of air, and the inventor, Lemichel, accomplishes this by providing an outlet air-valve at H, and a flexible diaphragm in the pipe at I. The result is that the reaction in the pipe K, after the closing of the momentum valve, expels the air through the valve H.

Pearsall's Hydraulic Ram.—Large rams have been made with momentum valves operated, not by the flow of the water in the supply pipe, but by independent means. Pearsall's ram is of that class. It is made to pump a larger quantity of air than that required to keep the air-vessel charged, and the surplus air is used to work a small air engine

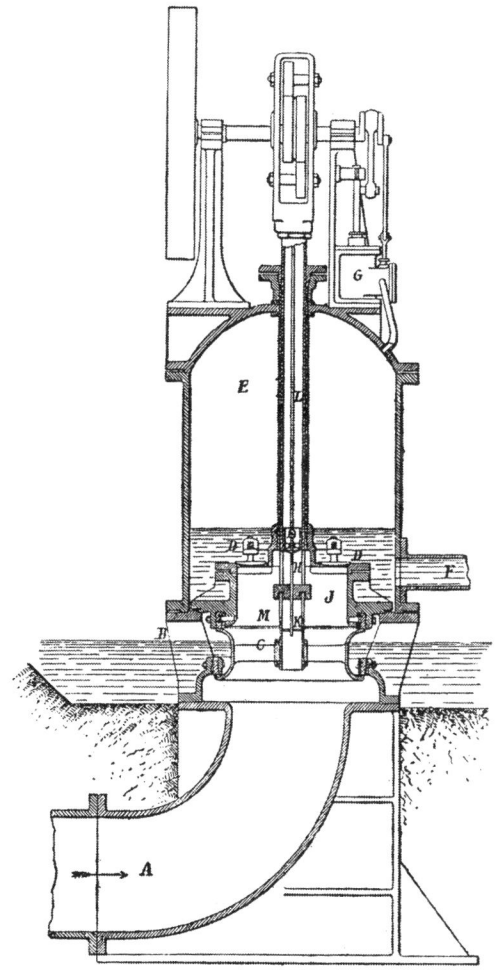

Fig. 247.—Section of Pearsall's Hydraulic Ram for forcing water.

which operates a large equilibrium momentum valve, quietly without the hammering action of the ordinary momentum valve.

This ram is shown in fig. 247. A is the flow-pipe (of a certain length depending on the fall and other circumstances of the case) conducting water from the source to the tail-race B. C is the main valve, opening and closing communication with the tail-race. D D are delivery valves opening into the air-vessel E. F is the delivery pipe. The main valve is opened and shut by means of a small motor G, worked by the compressed air from the air-vessel. H is an air-valve carrying a float J, the distance of which from H is adjustable by means of a screw K and rod L.

The action is as follows :—

The flow-pipe being full of water, the main valve is opened by the motor, and water flows into the tail-race, thus putting into motion all the water in the flow-pipe, the chamber M also emptying itself into the tail-race and being filled with air through the valve H.

After the flow has continued for a certain time—say, for example,

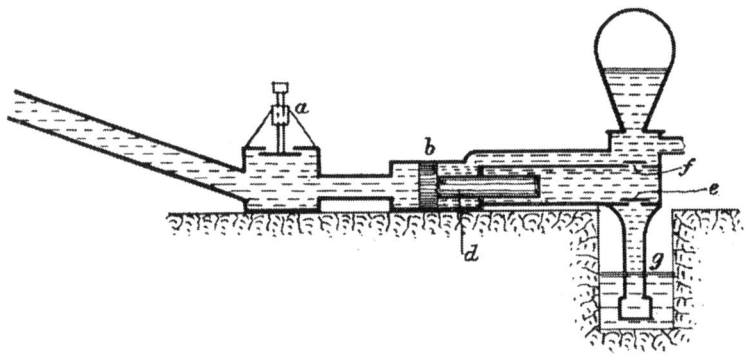

FIG. 248.—Hydraulic Ram in which the water used as motive power is made to pump other water.

a, is the momentum valve ; *b*, a piston in a cylinder ; *d*, a pump plunger ; *e*, the suction, and *f*, the delivery valve of the pump ; *g*, the suction pipe for clean water. There are many other ways, sufficiently obvious, of applying hydraulic rams for this purpose.

two seconds—the main valve is closed by the motor. During the closing of the valve the flow of the water is not checked, as it can rise without resistance in the chamber M, the air freely escaping by the valve H. The motion of the valve C need not, therefore, be rapid. When the main valve is closed and the water has reached a certain height in the chamber, it raises the float J, and closes the air-valve H. If the float be adjusted so as to close the air-valve, while there is still some air in the chamber, this air is then compressed (by the energy of the column of water) till its pressure equals that in the air-vessel, when it and some of the water is forced into the air-vessel. Water thus continues to flow into the air-vessel until the energy of the column of water in motion is exhausted by the resistance in the air-vessel, when the column of water comes to rest, and the delivery valves gently close. The action of the motor is timed, so that after this has taken place the main valve is again opened, and the cycle of operations is repeated.

Of course water flows out of the air-vessel continuously under pressure through the delivery pipe F.

It is not necessary, ordinarily, to confine and compress more than a very small quantity of air in the chamber.

If the small quantity necessary to supply the motor be confined, it will suffice for all the cushioning that is desirable. In starting the machine, however, it is an advantage to adjust the float so as to force in a considerable quantity of air, to *fill* the air-vessel with compressed air, and reduce the water level to a height of a few inches only above the valves, at which height it can easily be kept, thus avoiding the inconvenience arising from a mass of water above the valves, and thus utilizing the full capacity of the air-vessel.

Hydraulic rams supplied with impure water are used for raising clean water, the source of each kind of water being distinct. The power may be applied to a piston actuating the piston of a pump. A very obvious arrangement is that shown in fig. 248.

The Taylor Hydraulic Air Compressor is described by the inventor in the following terms.

Fig. 249 shows a complete compressor ; its details are as follows :—

A. Penstock or water supply pipe.
B. Receiving tank for water.
C. Compressing pipe.
D. Air chamber and separating tank.
E. Shaft or well for return water. (The required pressure is proportional to the depth of the water in this shaft.)
F. Tail-race for discharge water.
G. Timbering to support earth.
H. Blow-off pipe.
I. Compressed air main.
J. Headpiece, consisting of—
    a. Telescoping pipe with
    b. Bell-mouth casting opening upwards.
    c. Cylindrical and conoidal casting.
    d. Vertical air supply pipes. (Each pipe has at its lower end a number of smaller air inlet pipes branching from it towards the centre of the compressing pipe.)
    e. Adjusting screws for varying the area of water inlet.
    f. Hand-wheel and screw for raising the whole headpiece.
K. Disperser.
L. Apron.
M. Pipes to allow of the escape of air from beneath apron and disperser.
N. Legs by which the separating tank is raised above the bottom of the shaft to allow of egress of water.
P. Automatic regulating valve.

The water is conveyed to the tank B through the penstock A,

where it rises to the same level as the source of supply. In order to start the compressor, the headpiece J must be lowered by means of the hand-wheel f so that the water may be admitted between the two castings b and c. The supply of water to the compressor, and consequently

FIG. 249.   FIG. 250.

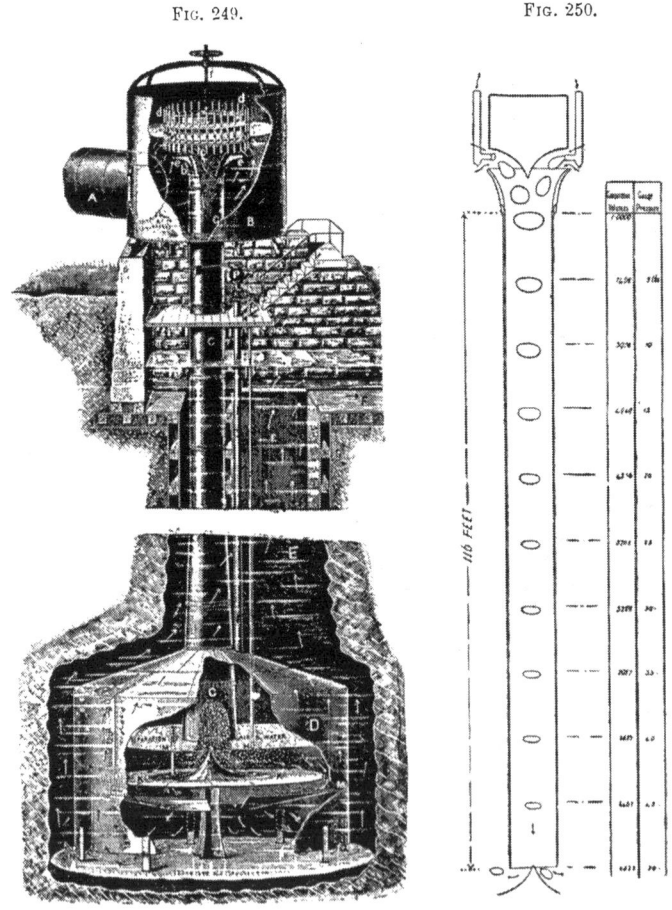

FIGS. 249 and 250.—Apparatus for compressing air by means of a fall of water—Taylor's system.

the quantity of compressed air obtained, is governed by the depth to which the headpiece is lowered into the water. The water enters the compressing pipe between the two castings b and c, passing among, and in the same direction as, the small air-inlet pipes. A partial vacuum is created by the water at the ends of these small pipes, and hence atmospheric pressure drives the air into the water in innumerable small bubbles, which are carried by the water down the compressing pipe C. During their downward course with the water the bubbles are compressed, the final pressure being proportional to the column of return water sustained in the shaft E and tail-race F. Fig. 250 shows the relative size of the bubbles as they descend in a compressing pipe, 116 ft. in length. When they reach the disperser K their direction of motion is changed, along with that of the water, from the vertical to the horizontal. The disperser directs the mixed water and air towards the circumference of the separating tank D. Its direction is again changed towards the centre by the apron L. From thence the water flows outward, and, free of air, passes under the lower edge of the separating tank. During this process of travel in the separating tank, which is slow compared with the motion in the compressing pipe C, the air, by its buoyancy, has been rising through the water and pipes M M from under the apron and disperser to the top of the air chamber D, where it displaces the water. The air in the chamber is kept under a nearly uniform pressure by the weight of the return water in the shaft and tail-race.

The air is conveyed through the main I up the shaft to the automatic regulating valve (of which a diagram is given), and from thence to the engines, etc. The air pressure in the main and air chamber increases 1 lb. per sq. in. for each 2 ft. 3½ in. that the water is displaced downwards in the air chamber by the accumulating air. The variation in pressure from this source will not be more than 3 lbs. per sq. in. in a working plant. As the automatic valve requires a change of only 1 lb. per sq. in. pressure to close it completely, it will be evident that by properly adjusting the valve some air can always be retained in the air chamber, and that the water can be prevented from ever reaching the inlet to the air main.

If a large quantity of air has accumulated in the chamber, the valve allows of its free passage along the main, but when the air is being used more quickly than it is accumulating, and the pressure decreases below a certain point, because the chamber is nearly emptied of air, the valve shuts partially or completely, adjusting itself to the supply from the compressor.

When the air has displaced the water almost to the lower end of the compressing pipe, it escapes through the blow-off pipe H.

Syphons.—A syphon-pipe is a pipe used to convey water from one level to a lower level over an intervening higher level automatically, thus avoiding tunnelling or excavating.

Let A, fig. 251, be a hill over which water is to be taken from reservoir *b* to reservoir *c*, and let the height *d* be less than that of a column of water at atmospheric pressure; then, if the ends of the pipe be submerged, and the air be exhausted from the pipe at *e*, the pipe will become filled with water. As the pressure of the water column *e c* is greater than that of the water-column *e b*, water will descend from *e* to *c*, and ascend from *b* to *e*, under atmospheric pressure, and a stream will be set up from *b* to *c*. To maintain this stream it is necessary that no air shall be allowed to collect in the pipe. Air collecting at *e* would destroy the action by reducing the effective atmospheric pressure sustaining the column of water in the shorter

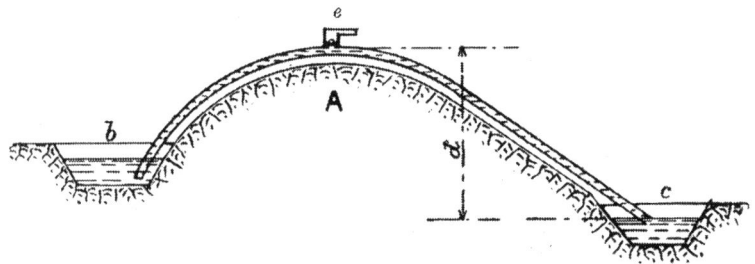

FIG. 251.—Ordinary Syphon Pipe.

leg of the syphon *e b*. To exhaust the air from the syphon, and to prevent its subsequent accumulation therein, an air pump is fixed at *e* and worked as occasion requires.

The accumulation of air is guarded against in various ways.

Let the longer leg of the syphon *c*, fig. 252, be at a small angle with the horizon, then a moderate velocity of water through the pipe will be

FIG. 252.—Syphon in which air will not collect, but is entrapped with the water and discharged.

sufficient to catch up and carry with it the air which otherwise would collect at *e*.

It constantly occurs in engineering practice that special cases have to be dealt with. The following is an interesting case met with in the author's experience.

A pumping main was to connect a reservoir on one side of a hill with a reservoir on the other without tunnelling, and the water was to be pumped through during the day only. A syphon arrangement would meet the requirements, provided air could be prevented from accumulating at its apex during the cessation of pumping. This apex was too far away from the pumping station to allow of an apparatus requiring personal attention being placed there, therefore an automatic control of the syphon action had to be devised. In fig. 253 *d* is the pumping station, *e* a stand

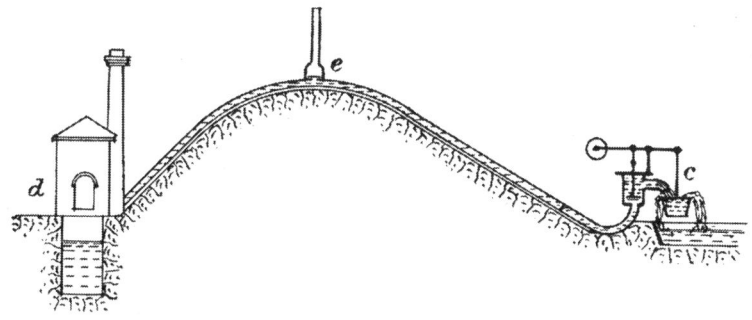

FIG. 253.—Pumping Main over a hill acting as a syphon in reducing the pumping head.

pipe on the apex of the syphon, *c* the discharge end of the pipe. An air valve is placed in the stand pipe *e*, and a weighted valve at *c*. When the main was charged, the air escaped from *e*, and water rose in the stand pipe. The pressure thus produced lifted the valve at *c*, and the syphon action was started. The water flowed into a bucket, which, when full, further raised

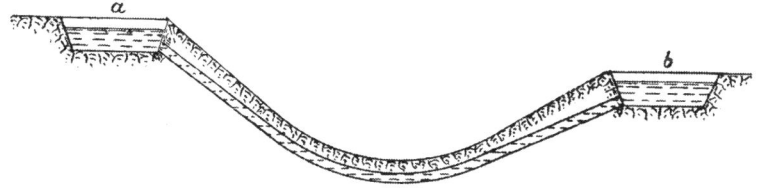

FIG. 254.—Hydraulic Main crossing a valley, usually termed an inverted syphon, the flow being from *a* to *b*.

the valve at *c* by means of the lever, and kept it fully open. When pumping ceased the water in the bucket escaped by a small hole in the bottom, and the weight on the valve caused it to close again while the syphon was still filled with water.

Inverted Syphons so-called are pipes carried across valleys, as in fig. 254, and are merely depressions in the pipe. They present no special features or difficulties in themselves, but special applications often require special provisions. If the syphon is a long one of such a capacity as to involve a risk of flooding from its bursting, a valve should be placed at *a*, which will close the pipe automatically directly the velocity of the water is increased by the bursting of the pipe. This may involve an overflow into a water-course at a to take away the water when the pipe is closed.

Syphons have been used to drain fen lands. Mr. Wheeler, in his book on the *Drainage of Fens*, thus describes the Middle Level syphons.

"When an accident occurred to the Middle Level sluice on the river Ouse, in 1862, it became necessary to place a solid dam of a very substantial character across the drain ; and in order to afford means of discharging the water from the drain into the river, syphons were erected under the direction of Sir John Hawkshaw.

"The syphons erected at the Middle Level were sixteen in number, laid across the dam at an inclination of 2 to 1 on either side, each end terminated by a horizontal length containing the upper and lower valves.

"The upper surface of the lower pipes was laid 1 ft. 6 in. below low water of spring tides, and the top of the syphon was 20 ft. above the same level. The syphons were of cast iron, $1\frac{1}{8}$ in. thick, 150 ft. in total length, and 3 ft. 6 in. in diameter. They were put into action by exhausting the air from the inside by an air pump worked by a 10 horse-power steam engine. These syphons continued in use fifteen years. Owing to their capacity not being sufficient to cope with heavy floods, and to discharge the water with sufficient rapidity, there was frequently a difference of more than 4 ft. between the level of the water in the drain and that in the river, the average varying from 2 to 3 ft., a very serious loss in such a flat district. It being found that the cost of adding a sufficient number of syphons to drain the fens effectually would be greater than that of building a new sluice, Sir J. Hawkshaw reluctantly advised the latter course, although contending that the syphons were right in principle."

The discharge of a syphon running full and free from air is that of a pipe of the size and length of the syphon with a head of water equal to the difference in level between the inlet and the outlet. Valves, bends and other obstructions must, of course, be taken into account and allowed for.

Hydraulic Mains.—When water is at rest in a main or pipe the pressure on the sides of the pipe is that due to the head of water above that particular part regardless of the cross section of the pipe, but when the water is in motion that is not the case, and the lateral pressure on the pipe will be influenced by the altered condition.

Assume that a main is under pressure from a service reservoir.

With the water at rest, the pressure is that due to the head of water above the pipe, but when the water is in motion part of the pressure head is changed into velocity head, and the lateral pressure on the side of the pipe becomes less ; but if the velocity be suddenly checked the pressure becomes more, because the *vis viva* of the water is spent in creating an additional head or pressure. The flow of water in a pipe of varying capacity is illustrated in the following figure (255).

Let the water be flowing from the reservoir a to the reservoir b through a pipe of varying capacity. The difference in the levels of the water in a and b, together with the friction through the pipe, will determine the rate of delivery; but as the pipe has a varying cross section the water will move in different parts of the pipe with different velocities. In the narrow parts it will move quickly and in the wide parts slowly, but in all parts of the pipe (neglecting friction) the sum of the pressure and the velocity heads will be the same. If vertical pipes be inserted in the sides of the pipe at points of different cross section, as shown in fig. 255, then the pressure head at those points will be indicated by the height of the water column in the vertical pipes.

Neglecting friction, the total head is h and the velocity head is h^1.

Where the pipe is very much enlarged then the pressure head is nearly the full head, and the velocity head is very small. (It is on this principle that the Venturé water meter is constructed.)

If a stop valve be placed in the pipe close to b, when closed the water

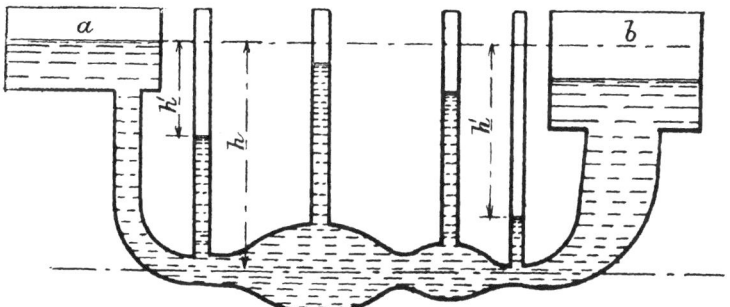

Fig. 255.—Diagram illustrating the flow of water in a pipe of varying cross section, and showing the head of water due to pressure and that due to velocity.

would stand in the vertical tubes on a level with the reservoir a. On gradually opening the valve the columns would fall because some of the pressure head h would become changed into velocity head h^1. On closing the valve again the columns would rise, and if closed suddenly the water would rise to a greater height than that of the reservoir b, because the *vis viva* of the water would be spent in creating additional head; this leads to a consideration of the effect of velocity in long mains.

In Chap. IV., fig. 69, we have considered the effect of momentum of water in mains in connection with pumping-engines from one point of view; we will now consider the question of safety of mains, the water in which is subject to sudden changes of velocity. Let us assume that a long pumping main is controlled by an air-vessel and that there is a uniform flow through the air-vessel to the reservoir.

If the supply to the air-vessel be suddenly stopped, the momentum of the water will carry it forward, reducing the pressure in the air-vessel; a reaction will then take place, and the water will then flow backward into the air-vessel, and the reciprocating action will continue till the *vis viva* of the water has been absorbed in friction.

With a long length of main and a high velocity of water the reaction on the air-vessel may be dangerous.

Assume the main to have a capacity of 5 gallons, or 50 lbs. per ft., with a length of 5000 ft., the velocity of the water being 3 ft. per second, then the energy of the water will be

$$\frac{5000 \times 50 \times 9}{64} = 35,000 \text{ ft. lbs.}$$

The area of the pipe in section would be 450 sq. ins.

$$\frac{35000}{450} = 77.$$

Therefore if a piston of the same area as the pipe were made to act as

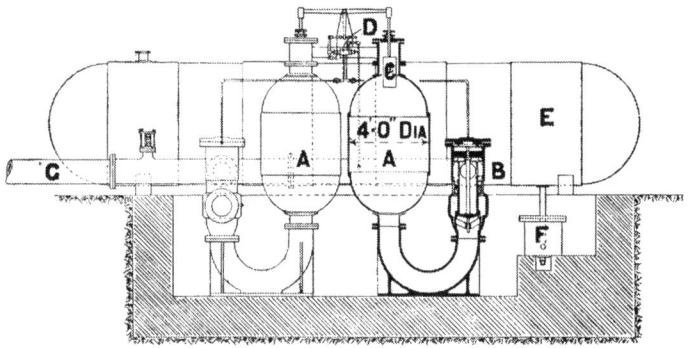

FIG. 256.—Hydraulic Air Compressor.

a buffer to absorb all the energy, it would require a mean resistance of nearly 78 lbs. per sq. in. and a movement of 1 ft.

In these calculations the friction of the water in the pipe has been neglected; but it is readily seen that in pumping into long lengths of main, care should be taken to start and stop the engines slowly.

Relief valves are used to avoid accidents.

Assume that there is no air vessel, but a retaining valve in its place, and that the supply through the retaining valve is suddenly stopped; then the reaction would produce a great shock, the water reacting on the valve with the effect of the 'water hammer.'

Hydraulic Air Compressor.—Fig. 256 represents a hydraulic air compressor designed by the author, in which a column of water of considerable height is used to compress the air, and in which the principle of the hydraulic ram is more or less brought into action.

Referring to the figure, G is the pipe bringing the water from an elevated reservoir, B is a valve box having valves of the type described and illustrated in detail in Chap. XI., fig. 192.

C is a float for the purpose of actuating the subsidiary valve D, and admitting water under pressure from the column G for the purpose of working the main valve in B. Air is compressed in the vessel A and delivered into the receiver E through a non-return valve not shown in the illustration. The mode of operation is as follows:—

Water from the pipe G is admitted into the vessel A through the valve in B. The water rushing into the vessel A meets with a resistance in the compression of the air, but as the resistance is small to begin with, the water in the pipe G acquires a velocity, and therefore momentum which is utilized in the final compression of the air and its delivery into the receiver E. The water level in A having risen to the float C, the float is pushed up,

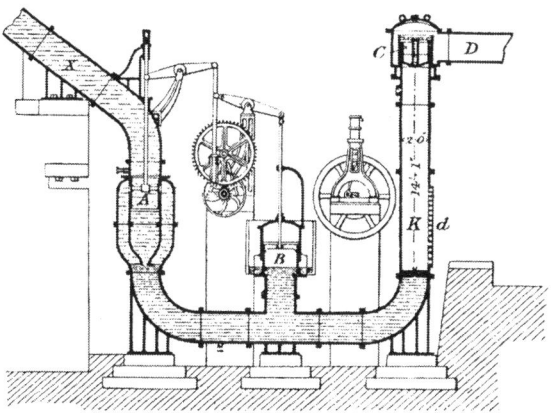

Fig. 257.—Hydraulic Air Compressor used during the construction of the Mont Cenis Tunnel.

thereby reversing the position of the subsidiary valve D. The inlet valve in B then closes, and the exhaust opens. The water in A then runs out by gravity, drawing in air from the atmosphere through suction valves not shown in the illustration, and the action is repeated. It will be seen that the momentum of the water for the final compression will depend on the length of the pipe G and the velocity the water in it acquires during the compression.

F is an automatic drain valve for letting off any water which enters the receiver E.

Sommeiller Hydraulic Air Compressor.—The early compressors used at the Mont Cenis tunnel were constructed by the Cockerill Company at Seraing, to the designs of M. Sommeiller. The chamber K, fig. 257, having been filled with air at atmospheric pressure, the valve A is opened and high-pressure water admitted at the bottom, thus compressing the air above it as it fills the chamber. As soon as the pressure of air in the chamber exceeds that in the mains, the outlet valves C are forced open, and the air is delivered to the receiver. The exhaust valve B is then opened, and as the water level falls in the chamber K, air is sucked in through the inlet valves at d. The pressure at which the air is delivered depends on the head of water available. At the Mont Cenis installation, the head of water was 85 ft. and the air pressure 75 lbs. per sq. in. It made about three strokes per minute. In many respects this compressor gave good

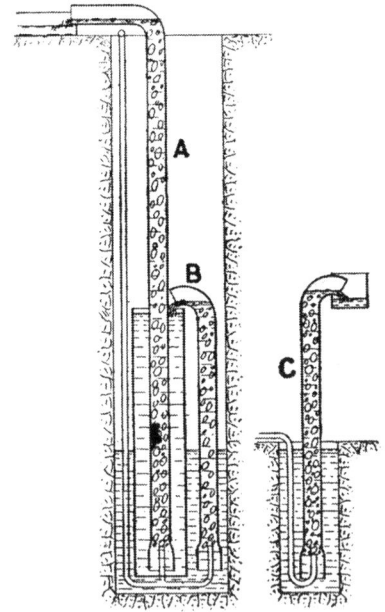

Fig. 258.—Pohle's Air-Lift Pump.

results, but it was very bulky for the quantity of air delivered and the efficiency was low.

Pohle's Air Lift Pump.—In fig. 258, C is a water column extending some distance below the water level in a well or cistern. A pipe conveying compressed air is taken down to the bottom of the column and air is injected into the water in the column in an upward direction. The air mixing with the water forms air bubbles, and thus reduces the specific gravity of the water column, and the water in consequence rises higher in the pipe than the level of the water outside. It will readily be seen that the lift must not be very great compared with the submersion.

It therefore becomes necessary to use the arrangement of double lifts, as shown at A and B, when the height to which the water has to be raised is considerable.

Mr. H. C. Behr and Mr. Rendall in America made some experiments with this pump to ascertain its efficiency. The general results are given in the following table :—

Submersion of the column in ft., . .	52	53	26	18	15
Height to which the water was raised, ft., .	75	35	62	69	36
Percentage of efficiency, . . .	34	37	18	8	9

The best efficiency was obtained when the pressure of the air did not greatly exceed that due to the submerged column and the height to which the water was lifted did not much exceed the submersion. The authors say, "That the efficiency in their trials did not take into account the efficiency of the air compressor. Therefore if that is taken as 70 per cent., then the total efficiencies in the above tables would be only 70 per cent. of the amounts given." It will readily be seen that the system is one of low efficiency.

INDEX.

T

PRINTED BY
NEILL AND COMPANY, LIMITED,
EDINBURGH

Printed in Great Britain
by Amazon.co.uk, Ltd.,
Marston Gate.